PARTICLES IN GASES AND LIQUIDS 1

Detection, Characterization, and Control

PARTICLES IN GASES AND LIQUIDS 1

Detection, Characterization, and Control

Edited by

K. L. Mittal

IBM US Technical Education
Thornwood, New York

PLENUM PRESS • NEW YORK AND LONDON

ISBN-13:978-1-4612-8085-9   e-ISBN-13:978-1-4613-0793-8
DOI: 10.1007/978-1-4613-0793-8

Proceedings of the symposium on Particles in Fluids: Detection, Characterization, and Control, held in conjunction with the Eighteenth Annual Meeting of the Fine Particle Society, held August 3–7, 1987, in Boston, Massachusetts

A Division of Plenum Publishing Corporation
233 Spring Street, New York, N.Y. 10013

PREFACE

This book documents the proceedings of the Symposium on Particles in Fluids: Detection, Characterization and Control held as a part of the 18th Fine Particle Society meeting in Boston, August 3-7, 1987. This was the Premier symposium on this topic and the response was so good that we have decided to organize it on a biennial basis and the second symposium will be held under the rubric Particles in Gases and Liquids: Detection, Characterization and Control at the 20th Fine Particle Society meeting in Boston, August 22-26, 1989.

In the modern manufacture of sophisticated and sensitive microelectronic components and other precision parts, there has been a great deal of concern about yield losses due to micrometer- and submicrometer-sized particles. These particles can originate from a number of sources including fluids, i.e., gases and liquids used in the manufacturing process. So the detection, characterization and control or removal of these undesirable particles is of cardinal importance and this symposium was conceived and organized with this in mind. The purposes of this symposium were to bring together those actively involved in all aspects of particles in fluids, to provide a forum for discussion of the latest techniques for the detection, characterization and control of particles, and to highlight areas which needed intensified R&D efforts. The printed program contained a total of 46 papers and a variety of topics dealing with various ramifications of particles in fluids were presented. In addition to the didactic presentations, these were brisk and enlightening (both formally and informally) discussions throughout the duration of the Symposium. If the comments from the attendees are a barometer of the success of a symposium, then this symposium was a grand success and the putative objectives were amply fulfilled.

As for this Volume, the readers will notice that the title is slightly different from the title of the Symposium, i.e., the word "fluids" has been replaced, after due consideration, by "gases and liquids." Also this volume is christened Volume 1 as we expect follow-up volumes on this topic in the future.

It should be recorded that all manuscripts were peer reviewed before inclusion in this volume, as the comments from the reviewers are a necessary condition to maintain high standard of publications. The present volume contains 19 papers and the topics covered include: monitoring of particles in gases and liquids; liquid particle counter comparison; particle contamination control in gases; in-situ monitoring of particles; measurement and control of particle-bearing air currents in clean room; particle deposition behavior, collection efficiency of filters, and particle retention of point-of-use filters; particle contributions from cleanroom garments; and particle generation in clean manufacturing.

I certainly hope this book will serve as a repository of latest developments and knowledge anent particles in gases and liquids. Furthermore, this book should be of interest to both the experienced practitioners in this field as well as to those who wish to make their maiden voyage in the world of particles in gases and liquids.

Acknowledgements. First I would like to record that this symposium was jointly organized by Dr. J.V. Martinez de Pinillos and yours truly, and my sincere thanks are extended to him. Thanks are due to the Fine Particle Society for sponsoring this event. I am thankful to the appropriate management of IBM Corporation for permitting me to organize this symposium and to edit this volume. The valuable comments provided by the unsung heroes (reviewers) are gratefully acknowledged. Thanks are extended to Pat Vann of Plenum Publishing Corporation for her continued interest in this project. Also I would like to express my thanks to my wife, Usha, for helping in many ways during the tenure of editing this book. Last, but not least, the interest, enthusiasm and contribution of authors is thankfully acknowledged without which this book would not have been born.

K.L. Mittal
IBM US Technical Education
500 Columbus Ave.
Thornwood, NY 10594

CONTENTS

MONITORING CONTAMINANT PARTICLES IN GASES AND LIQUIDS: A REVIEW

Douglas W. Cooper

IBM Research Division
T.J. Watson Research Center
Yorktown Heights, NY 10598

In the microelectronics and pharmaceutical industries especially, particles microns in size and even smaller diminish the quality and quantity of the products produced. To reduce losses due to contamination, we need to detect particles in air, in other gases, and in liquids, as well as on surfaces. Various methods are discussed here. The methods using the scattering of light predominate currently.

INTRODUCTION

Contamination control is a mixture of high technology and low technology. Solving today's problems often means making a painstaking survey of current procedures and contamination levels and resorting to increased care in dealing with people, equipment and materials. Solving tomorrow's problems will require all that, along with greatly improved methods of detecting and quantifying contaminant particles.

Most yield loss in the microelectronics industry today is due to contamination. More than half the defects on wafers are due to particulate matter.[1]. The products being planned for the future will be even more sensitive than current products, so that by the next decade this industry will be concerned with the impact of particles approaching 0.01μm in size. An introduction to the particulate contamination in the microelectronics industry is available in Ref. 2.

Figure 1 shows the kind of problems that particles pose: they can cause short circuits between lines or layers that are conductive and they can cause disruption of current flow in conductive lines. They may do this by their continuing physical presence or by their having been present during the various lithography steps involved in the processing of the micro-circuits. Currently, such circuits have features microns in dimension and even smaller.

Figure 2 shows two widely-used relationships to predict yield: a simple exponential distribution and Murphy's Law.[3] For more advanced yield models, see Ref. 4. In Figure 2, yield is plotted versus the mean number of particles, AD, of the critical size (or larger) expected in the critical area A. D is defect density, in units such as particles per cm^2.

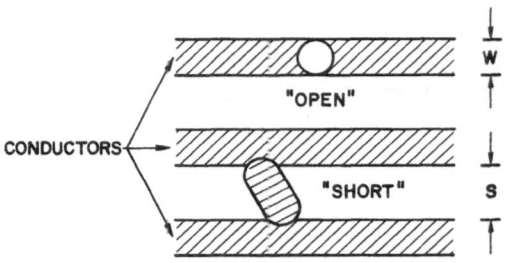

Figure 1. Schematic of "opens" and "shorts" in insulators and conductors in a microelectronic circuit element.

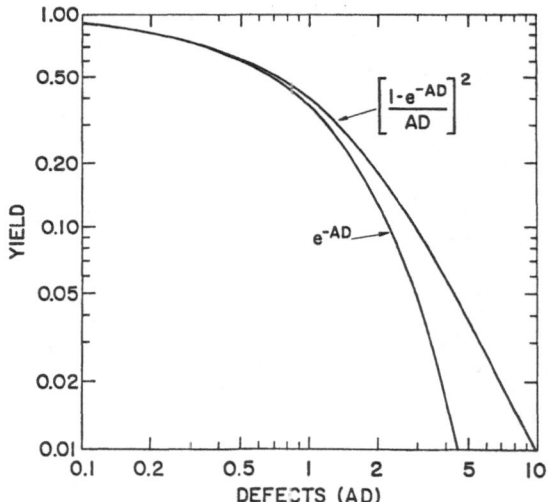

Figure 2. Exponential and Murphy's Law distribution models of yield versus expected number of defects, AD.

As AD becomes larger than 1, the yield losses are so great that the cost is strongly affected, thus affecting economic viability. Concise introductions to the importance of contamination to the microelectronics industry were written by Burnett[5] and by Faure and Thebault.[6]

Contamination control in the pharmaceutical industry centers on prevention of biologically active particles from entering the drugs. Because biologically active particle concentrations are not closely related to total particle concentrations, the most effective monitoring is that which centers on biologically active particles. Particle count data is not of particular value, although it may be useful in discriminating normal conditions from abnormal conditions.

SYSTEMS APPROACH

A variety of approaches could warrant the term "systems approach."[7] Here we refer to the following:[117]

- Determining there is a problem.
- Discovering the source of that problem.
- Establishing the criteria for what would be a successful solution.
- Generating various alternatives as possible solutions.
- Evaluating these alternatives according to the criteria.
- Implementing the alternative chosen through evaluation.
- Monitoring the alternative as it is applied.

The steps of problem determination (detection), analysis (determining source), and monitoring all require the ability to measure contamination levels as well as yield and quality.

Figure 3 shows the feed-back and feed-forward loops associated with adaptive control.[117] Adaptive control is really a continuing form of problem identification and solution. The inputs and environment are monitored, as are the conditions of the process, and the outputs are monitored, with monitoring assisting in adjusting the transformation process. Contamination monitoring is important for many such situations.

Integrated circuit fabrication facilities generally employ extensive monitoring not only of contamination levels but also of other environmental factors, such as temperature and humidity, and the results of the various fabrication steps. A good example was presented by Ligtenberg of Hewlett-Packard.[8] As another example, monitoring of liquids, gases, and surfaces was carried out to aid in prevention and removal of contamination of optical surfaces.[9]

A systematic approach to the detection and control of microbial contamination was presented by Rechen: sources, transport, detection, removal are crucial elements.[116]

GOALS OF MONITORING

What do we want to know? Usually, we are measuring the number concentration as a function of time and place, but we would also like information on particle size distribution, chemical composition, and sometimes particle charge levels or viable / non-viable particle dis-

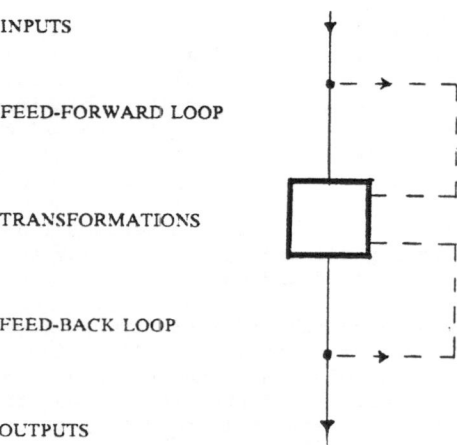

INPUTS

FEED-FORWARD LOOP

TRANSFORMATIONS

FEED-BACK LOOP

OUTPUTS

Figure 3. Adaptive control, with feed-foward and feed-back loops.

tinctions. We want this information to understand and adapt to particle impact on product and productivity.

WHEN AND WHERE TO SAMPLE

Figure 4 shows some options schematically, to emphasize that the sampling may be done in one or more locations and may be continuous, continual, or --- worse yet and not shown --- sporadic. Clearly, continuous sampling at many locations gives the best coverage of time and space.

When we cannot be comprehensive in sampling, we try to obtain representative samples. Choices have to be made: do we want typical conditions or only the extremes? This will depend on the use intended for the data. Usually, we want enough locations to characterize a region of interest and we want to take the samples at the times of interest and the conditions of interest, generally during and near production.

An important choice is whether to use multiple sensors or a single sensor with multiple sampling lines when one needs information simultaneously or nearly so from many locations. Some considerations are:
 - a sensor malfunction where there is only one sensor means loss of all the data;
 - a disruption of flow at the source(s) of vacuum will have different impacts;
 - multiple sensors require reliable means of data transmission;
 - multiple sensors require more calibration and generally one obtains less sophisticated sensors for the same cost;
 - multiple sensors allow testing for malfunctions by simply exchanging sensors;
 - multiple lines can lead to different transport losses to sensors, requiring separate corrections, although these may be designed to be negligibly small. Some of these considerations and others were discussed at length by Hope.[10]

UNBIASED SAMPLING

Representative sampling seeks also to get samples that are not biased in terms of particle concentration, particle size distribution, and particle composition. To avoid losses in inlets to probes and in the flow through sampling tubes, one prefers to use in-situ methods, where no sampling takes place. An example is directing a beam of light at the fluid to be assessed and then focussing on a portion of the beam, analyzing the light scattered by the particles in a small focal region. Unfortunately, light scattering is a complex phenomenon and rarely gives unambiguous information about particle size and concentration, much less composition. (See, for example, Ref.11.)

Many instruments cannot be used in situ. Rather, one must pull a sample from the fluid to the sensing region of the instrument. In doing so, one risks introducing bias in the concentrations of particles of different sizes due to sampling, either by selective entrance to the probe inlet or by selective losses during transit from the sampling point to the sensing volume. An especially informative review on this topic was presented by Fuchs.[12]

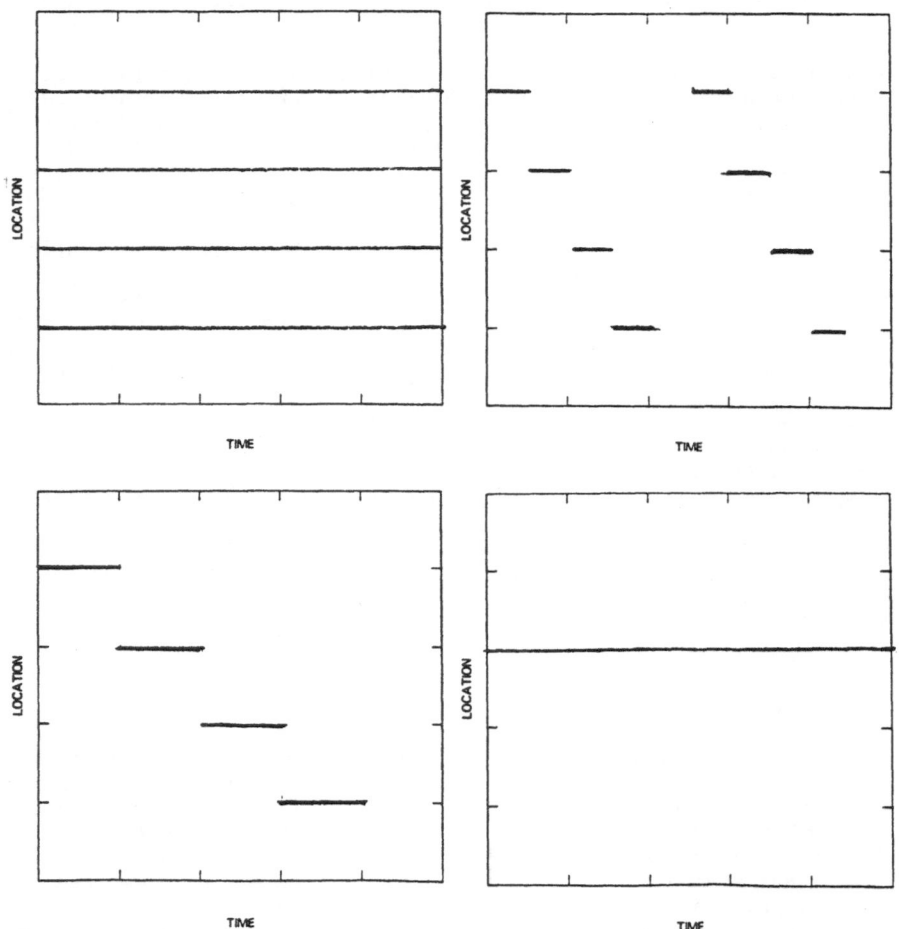

Figure 4. Some options in sampling at different locations and times.

Figure 5 shows one of several methods for obtaining nearly isokinetic sampling from a gas or liquid.[13] Isokinetic sampling is where the probe is aligned with the flow and has a flow velocity at its entrance that matches the flow velocity in the main stream. In Figure 5 there is a match between the velocity within the curved pipe and the pipe drawing the sample to the instrument. In Figure 5 there is a mismatch between the inlet to the curved pipe and the main flow. Inertial bias in particle size distribution determination, which often predominates, is related to the ratio of the velocity mismatch and to the diameter of the sampling pipe. It is possible to choose a large value for the curved pipe's diameter and greatly lessen the inertial bias. The "centrifugal force" is lessened by using the larger pipe and gravitational loss in the sample is prevented by aligning the sampling tube vertically. It is not always necessary or convenient to use a set-up like that in Figure 5, however. More details on sampling and losses are available in the aerosol literature (e.g. Ref. 14) and in a tutorial article by the author.[13] Good practice is to use smooth, conductive tubing to transport particles to sampling instruments;[13] data on transport efficiencies under relevant conditions have been published.[15] Recent reviews and research on aspiration efficiencies give results for thin-walled and blunt sampling probes that should be of use.[16] [17] [18]

MONITORING TO MEET FEDERAL STANDARD 209 (CLEANROOMS)

Federal Standard 209, dealing with the certification and monitoring of air cleanliness in cleanrooms, was recently amended and was promulgated in 1988 by the General Services Administration. The major changes have been outlined elsewhere[19]. They are primarily: definition of some even-cleaner air cleanliness classes, along with the particle sizes at which the cleanliness is to be measured; specification of sample volume size, number of locations; specification of statistical criteria to be met in order to satisfy the standard. The primary measurement tool is the optical particle counter, which may be used for particles as small as 0.1μm. An example of the certification of a Class 10 facility according to the revised standard was presented by Hellander.[20]

OTHER MONITORING ATTRIBUTES

Tolliver presented a list of desired attributes for a particle sampling and monitoring (sensing) system:[21]

- fully automated sampling, sensing, data analysis, data retrieval
- many parameters besides concentration (e.g., temperature, RH)
- immediate ("real-time") information continuously (24 hours per day)
- multiple locations
- computer compatibility

Accuracy, precision, and reliability are crucial.

COMPARISON: COLLOIDS VS. AEROSOLS

Although much that is written here holds for both colloids (particles in liquids) and aerosols (particles in gases), there are instances where we will have to distinguish between the two.

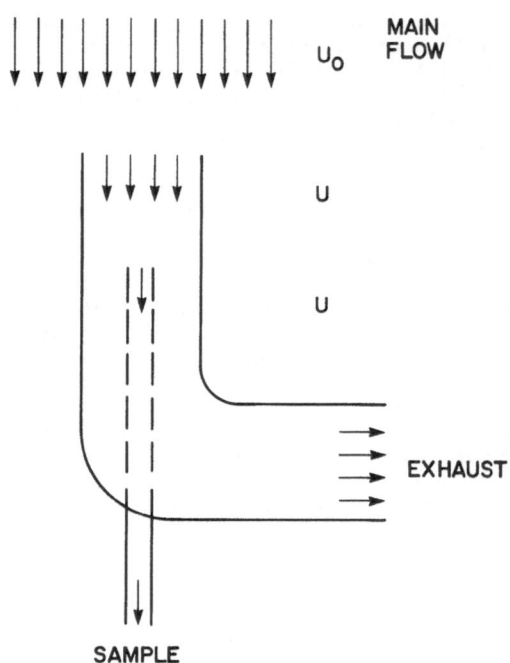

MAIN FLOW

U_0

U

U

EXHAUST

SAMPLE

Figure 5. A configuration for obtaining an isokinetic sample.

Some general remarks on the differences between them are:

1. The higher viscosity of the liquid medium reduces transport of colloidal particles in comparison to aerosol particles of the same size, and this difference is made greater by "slip" for aerosol particles comparable in size to the mean free path of the gas molecules, 0.06µm in air at NTP.

2. The buoyancy effects of the liquid help reduce colloidal migration velocity in gravity and acceleration fields, in comparison with aerosols.

3. The higher refractive index of the liquid reduces the optical contrast between the colloidal particle and its medium, in comparison with aerosols.

4. The presence of bubbles in liquids produces background noise that makes optical detection of particles more difficult.

5. The electrical conductivity of some liquids can be used in the electro-resistive particle detection and sizing technique, but it also makes it more difficult to subject the particles to an electric field.

6. The generally higher concentration of ions in liquids create an electrical double-layer around charged particles that complicates their interactions with each other and with surfaces.

7. The greater scattering by liquid molecules and the tendency for liquids to be dirtier than gases makes optical detection of particles in liquids more difficult.

8. Aqueous solutions can have biological activity that produces particles, including organisms and their fragments. A high filtration efficiency may be insufficient if conditions are appropriate for organism reproduction downstream from the filter.

9. Evaporation and condensation can create, destroy, or simply change aerosol particles; solvation and precipitation are similar processes that occur in liquids.

In practice, the major differences are that we do not have convenient methods of absolute filtration of liquids that compare with the high efficiency particulate air filters (HEPA filters) for air, that we cannot conveniently detect particles in liquids much smaller than 0.5µm, that we have more difficulty in obtaining a sample from a colloid without filtration, and that the colloids we deal with have much higher particle concentrations than the aerosols.

METHODS OF PARTICLE SORTING AND DETECTION: GENERAL

Figure 6 gives a general schematic for particle detection instruments. The volume of interest is isolated, sometimes in situ, more often by sampling and transporting. The particles may then be sorted by size or charge or some other characteristic or not sorted, then sensed. The signals from the particles may again be sorted or not, then quantified.

To detect the particles, one needs some difference between the particles and the medium in which they occur. Particle size and shape and refractive index, for example, provide contrast for optical methods. Other information is available through special techniques using dark-

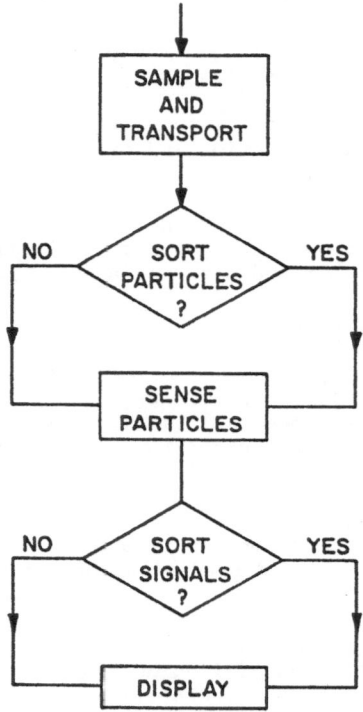

Figure 6. General schematic for possible particle detection instruments.

field rather than bright-field, polarized light, fluorescence detection. Charged particles have a different mobility from gas or liquid ions, so this electrical mobility may be used to segregate them, in mobility analyzers for aerosol particles 0.01μm in size to less than 1μm, and in electrophoresis equipment for colloids. Particles generally will have a different electrical resistance from that of the medium, exploited in the electro-resistive technique more familiarly known as the Coulter counter. Physical and elemental differences can be used in electron microscopy, useful on particles microns in size to not much smaller than 0.01μm. Occasionally, just the geometric size difference between particles and molecules is used and the particles are sieved from the suspending medium. These differences can be used to sort the particles, to isolate them from the "background", and to detect them.

In general, the technology for sorting electrical signals is more advanced than that for sorting particles themselves. Methods that sort particles include impaction, electrical mobility analysis, and diffusion "batteries". Methods that sort signals include optical particle counters, electro-resistive counters, microscopy in many forms (parallel processing of signals received simultaneously).

METHODS OF PARTICLE COLLECTION FOR ANALYSIS

In-situ monitors do not need to collect particles to analyze them, and various other devices can operate by drawing a flowing sample past an analysis region. Some methods do rely on particle collection, however, so we discuss it here.

Filtration is the capture of particles on a filter. For microscopy, one wants the particles to be on the surface of the filter, although this is not always necessary. Getting almost all the particles to be on the front surface requires highly efficient filters. Various matted filters can provide very high filtration efficiencies, greater than 99%, but they may rely on capture deep in the filter matrix to achieve this. To obtain a sample for microscopy, one generally needs --- whether for gas or liquid --- filter pores that are small compared with the particle size of interest. This can be achieved with certain membrane filters or with Nuclepore (TM) filters.

One can deposit aerosol particles onto surfaces using electrostatic or thermal forces or inertially, by directing a jet at a surface, and analyze these deposits with microscopy (Ref. 11). A detailed review by the late Prof. N.A. Fuchs of inertial, thermal, and electrostatic deposition was published recently.[22]

Leaving a surface unprotected in a volume of interest will result in a deposit, but calculating the surface concentration of the deposit from the appropriate volume concentration of the material in the liquid medium is difficult. However, for microelectronics industry contamination control often the surface deposit density is of more interest, being a better surrogate measure of contamination on the product than is aerosol or liquid concentration. Mechanisms in gases and liquids that tend to deposit particles are: gravity, diffusion, electrostatics, thermal gradients, and turbulence. Deposition velocities can be estimated by combining these effects.[11] [23]

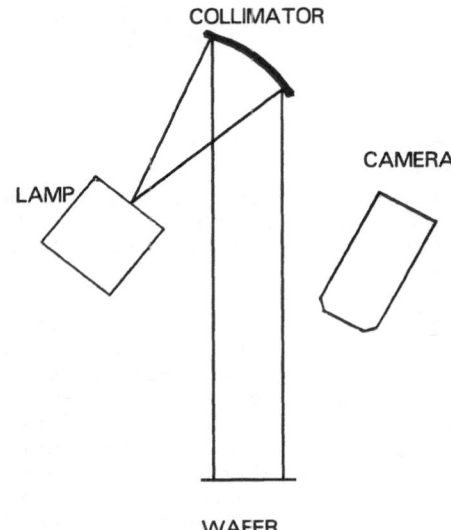

Figure 7. Schematic drawing of one kind of surface monitor.

METHODS OF PARTICLE DETECTION: MICROSCOPY

Having captured the particles on a suitable surface, one can bring them to a microscope for analysis. The topic of microscopy is immense and cannot be done justice here. Transmitted or reflected light may be used. Transmitted electrons or scattered electrons may similarly be used. Most work in contamination control is being done with optical microscopy or electron microscopy. New techniques, including scanning tunneling microscopy, scanning thermal microscopy and scanning acoustic microscopy, are being transformed from experimental set-ups to commercial devices.

Identification of the particle elemental composition can often be achieved by choosing the proper surface analysis technique from among: scanning electron microscopy coupled with energy-dispersive spectrometry (SEM/EDX), relatively fast and inexpensive; secondary ion mass spectrometry (SIMS), especially sensitive but destructive and hard to interpret; Auger electron spectrometry (AES), which is very sensitive (films ca. 10 A thick) but relatively expensive; further comparisons were provided by Koellen and Saxon.[24] The particular problem of counting and identifying particles in high-purity water was addressed by Balazs and Walker.[25]

METHODS OF PARTICLE DETECTION: OPTICAL SURFACE SCANNERS

Particles on surfaces are not central to this article, and the topic has been recently surveyed by Batchelder.[26] A brief summary here is all that will be attempted.

Figure 7 is a schematic of a commercial surface monitor (similar to one sold by Inspex, Inc., Waltham, MA). The surface is illuminated; light scattered away from the specular reflection direction is detected. In this instrument, the light is incident perpendicularly and the scattering is detected on a vidicon, indicating the location of the scatterers and the approximate magnitude of light they scatter, thus the approximate size.[27][28][29] Other surface monitors (available from Tencor, Inc., and PMS, Inc., for example) use a small bright beam of light to scan the surface; the scattered light is detected and matched with beam position, with information about the particle size inferred from the magnitude of the light pulse.

There is a growing literature on such devices.[26][28][31] Particle size, shape, and index of refraction all affect scattering, as do the reflectivity, roughness, and refractive index of the surface. The angles of light incidence and detection are important. A few generalizations can be made:
- the smoother the surface, the smaller the particle that can be detected;
- the devices have difficulty detecting single particles much smaller than the wavelength of light currently;
- light scattering intensity is an approximate surrogate for particle size;
- devices that work with scanning beams rather than total surface illumination need data processing to avoid miscounting due to beam overlap.[31]

As this is written (early 1988), surface scanners for optically smooth surfaces can detect particles a few tenths of microns in size and larger. For rough surfaces, the particles have to be larger than the characteristic roughness length.

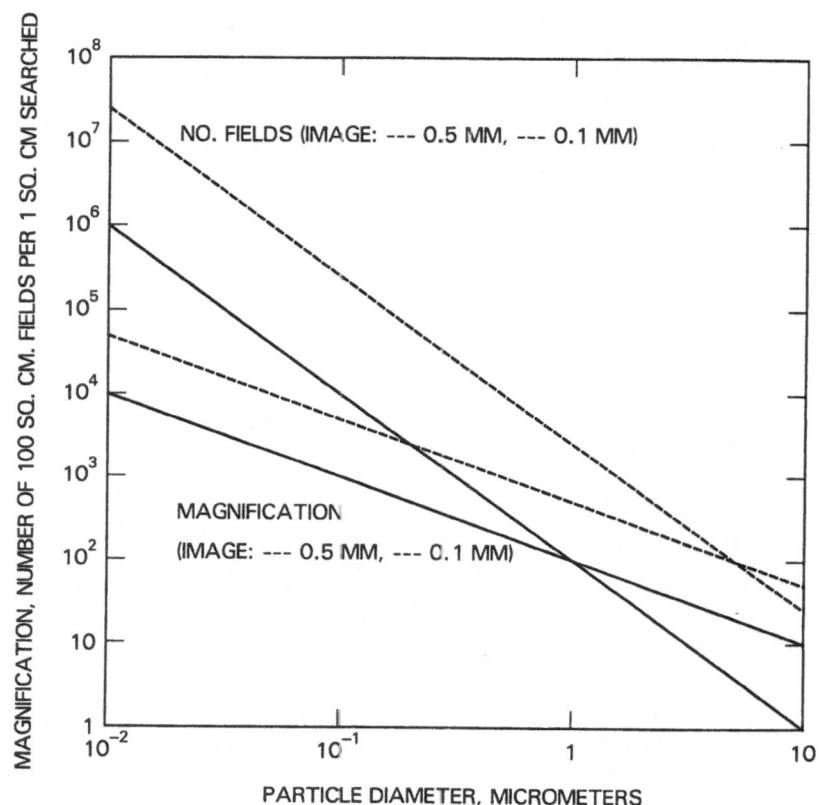

Figure 8. Number of viewing fields (100 cm²) needed per cm² to detect (image > 0.1mm) or determine size (image > 0.5 mm) of particles.

The problem of light scattered by a sphere on a plane is much more difficult than the already complex problem of light scattering by a sphere suspended in an infinite medium. The former problem has only been recently solved, just for light that is perpendicular to the surface,[32] whereas the latter problem was solved by Mie just after the turn of the century.[33]

Figure 8 suggests the limits of traditional microscopy.[2] The magnification and number of optical fields needed to see or size particles grows by a power law relationship with the reciprocal of the particle size of interest. Analysis by microscopy with current technology of a microelectronic component of 1 cm^2 becomes extremely time consuming as the particle size of interest approaches 0.1μm. Sensitive, accurate, robust alternatives to microscopy are needed.

METHODS OF PARTICLE DETECTION: OPTICAL PARTICLE COUNTERS

Currently, optical particle counters are the instruments used most often to measure particle concentrations and size distributions in liquids and gases.

Figure 9 is a schematic of an optical particle counter. The gas or liquid is drawn through an illuminated region, and the light that is scattered from the particle is sensed with a detector. Pulses from the detector are processed to give a count of particles and a size estimation from the magnitude of each pulse. Light's interaction with a particle that has a different index of refraction than that of the medium is a complex phenomenon when the particle is neither very much larger nor very much smaller than the light wavelength. Light undergoes diffraction and scattering and may undergo absorption as well. (Many texts are available, Ref. 11 gives a good introduction.)

The light received by the detector will depend on particle size, shape, refractive index difference from the medium, the light wavelength and polarization, and the angles of incidence and detection. Particles much smaller than the wavelength of light scatter light in proportion to their volumes squared, dipole scattering. Particles much larger than the wavelength of light scatter light in proportion to their cross-sectional areas, geometric optics. In between these limits is the Mie scattering regime, where the scattering dependence on size goes from the Rayleigh (dipole) to the geometric, with many resonances, thus oscillations, between. For Mie scattering it becomes difficult to associate a unique particle size with a certain magnitude of light scattering, although choices of instrument angles and illumination wavelength(s) can help.

Although most optical particle counters operate by drawing a sample through an inlet to the sensing region, the Particle Flux Monitor (High Yield Technology, Mountain View, CA) measures particle flux rather than concentration, allowing the ambient gas flow to determine the flow past the sensing volume, an open region. The first such device, the PM-100, was found to sense a few percent of the particles of roughly 1 micron diameter to nearly 100% of the particles nearly 10 microns in diameter that passed through the illuminated zone.[34] Its operation and use in sampling process equipment and a vacuum environment have been described, demonstrating the value of instantaneous flux measurements.[35] Fujii et al.[36] and Hayakawa et al.[37] have used a laser beam and an optical system to allow detection of particles in the ambient.[37] Several in-situ optical particle counters for liquids were described by Montgomery, who indicated the lower limit of sensitivity was about 0.3 micron particle size in 1987.[38]

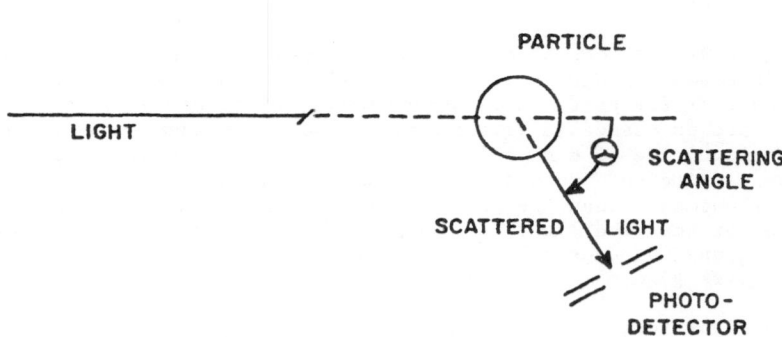

Figure 9. Schematic of an optical particle counter.

Articles comparing the capabilities of various optical particle counters have appeared from time to time in the technical literature, which should be consulted.[39][40] Knollenberg[41] explained the effects of particle and medium refractive indices on light scattering, described optical particle counters for liquids and gases and monitoring surfaces, and provided approximate solutions and added insight on the problem of predicting the scattering of light from a sphere on a surface. A technical discussion of the resolution of such instruments and its impact on particle sizing and counting was presented by van der Meulen et al.[42]

Current optical particle counter technology for aerosols allows detection of particles larger than 0.1μm in diameter at flow rates of 1 cubic foot per minute (cfm), 472 cm^3/s. By incorporating a virtual impactor (impaction from a fast-flowing stream to a slower one) upstream from the optical particle counter, the concentration of particles larger than a critical aerodynamic diameter can be markedly increased in the slower stream, which can then be monitored, and an instrument of this type has been reported.[43] Design criteria for minimal losses in virtual impactors were developed and tested by Chen and Yeh,[44] among others. Within a few years, the optical particle counter technology will likely allow higher flow rates and even smaller particle sizes, although the strong dependence (approaching volume squared) of scattering on particle size will make detecting much smaller particles very difficult. By condensing vapor onto the particles, various condensation nucleus counters can detect particles as small as 0.01μm, at flow rates up to 0.1 cfm.[45][49] In his article, "Measurement of Nanometer Aerosols," Sinclair gave a comprehensive review of the design and calibration of various diffusion battery and condensation nucleus counter types and the analysis of data from them.[50] A screen-type diffusion battery[119] and a continuous-flow condensation nucleus counter are available from TSI, Inc., St. Paul, MN. An assessment of the use of the diffusion battery in clean rooms was presented by Locke et al. [51] A fine review of the use of the condensation nucleus counter in contamination control applications was presented by Fisher.[52] A method of calibration was presented by Liu and Pui.[53] The performances of several models were tested and compared in depth by Bartz et al.[54] The condensation nucleus counter has been used to study condensation nuclei in semiconductor cleanrooms.[55] Using a conventional optical particle counter along with a condensation nucleus counter, for example, Cheung and Roberge[56] explored particle generation in equipment due to shedding during changes of temperature in a diffusion furnace.

Current optical particle counter technology for liquids cannot detect particles much smaller than one-half micron and uses much lower flow rates than do aerosol optical particle counters. Khilnani stressed the importance of contaminated liquids, listed the chemicals commonly used in microelectronics manufacturing, and discussed typical contamination levels as well as methods of measuring them and lowering them.[57] Grant and Schmidt demonstrated an improved methodology for determining submicron particle concentrations in liquids. This methodology compensated for refractive index of the liquid and the particles, coincidence, the response versus particle size, as well as prevented the formation of bubbles by operating at high pressure.[58] The roles of pressure drop during flow and vapor pressure in the formation of bubbles were outlined by Fisher et al.[59], who demonstrated bubble production that would cause errors in an optical particle counter. Equipment modifications for pressurizing liquids before counting in an optical particle counter were also presented by Csikai and Schware.[60] Methods and results of monitoring the

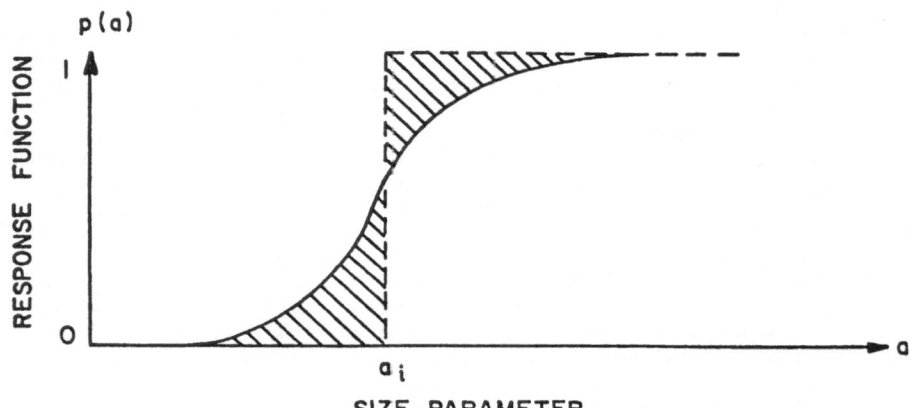

Figure 10. Ideal and non-ideal instrument responses.

particle contamination in liquids were given in a series of articles by Dillenbeck.[61] [62] [63] A particularly complete description of the methods and the results obtained in measuring particle size distributions and concentrations by optical particle counter for over a dozen processing liquids were described by Csikai and Barnard. [64]

The non-volatile residues in liquids can be studied by atomizing the liquids and counting and sizing the particles formed as the drops dry. This was suggested by Wen et al.[65] and developed further by Kousaka et al.[66] who used polydisperse atomization followed by diffusive size classification and a condensation nucleus counter, and by Blackford et al.[67], who used monodisperse atomization, followed by analysis of the particle size distribution either using the Differential Mobility Analyzer or the Aerodynamic Particle Sizer (both manufactured by TSI, Inc.).

Optical particle counters for gases and liquids are under development, in some cases commercially available, for hostile environments: high pressures, corrosive media.

Manufacturers of optical particle counters for gases and liquids include the following:

- Particle Measuring Systems, Inc., Boulder, CO;
- ppm, Inc., Knoxville, TN;
- Met One, Grants Pass, OR;
- Faley International Corp., Laguna Hills, CA;
- Atcor, Mountain View, CA;
- Hiac/Royco, Menlo Park, CA;
- TSI, Inc., St. Paul, MN;
- Climet, Redlands, CA;

This list is not an implied endorsement of these manufacturers, is in no particular order, and is not claimed to be complete.

METHODS OF PARTICLE ANALYSIS: OTHER AEROSOL INSTRUMENTS

We discuss next, briefly, several aerosol measurement instruments that are not yet widely used in the contamination control field, but are becoming so: diffusion batteries, electrical mobility analyzers, aerodynamic sizing devices.

Diffusion batteries are structures, such as parallel channels, pores, screens, that are designed to capture particles primarily due to diffusion, which is caused by the Brownian motion of particles in response to fluctuations in their bombardment by the molecules of the medium in which they are suspended.[11] [68] Diffusion batteries help us infer diffusivity, which is a function of temperature, particle size, and medium viscosity and mean free path. The smaller the particle, the greater the diffusivity. Unfortunately, the collection of these particles depends not only on these factors but also on their initial positions and on chance. These factors combine to give a response to particle size that is not sharp, as shown in Figure 10.

Figure 10 illustrates the response (penetration) of a diffusion channel as a function of particle size. For size discrimination, one wants a curve which is nearly a step function (dashed lines). As the response becomes less and less sharp, size resolution decreases. There are a variety of computer programs available to try to use mathematical

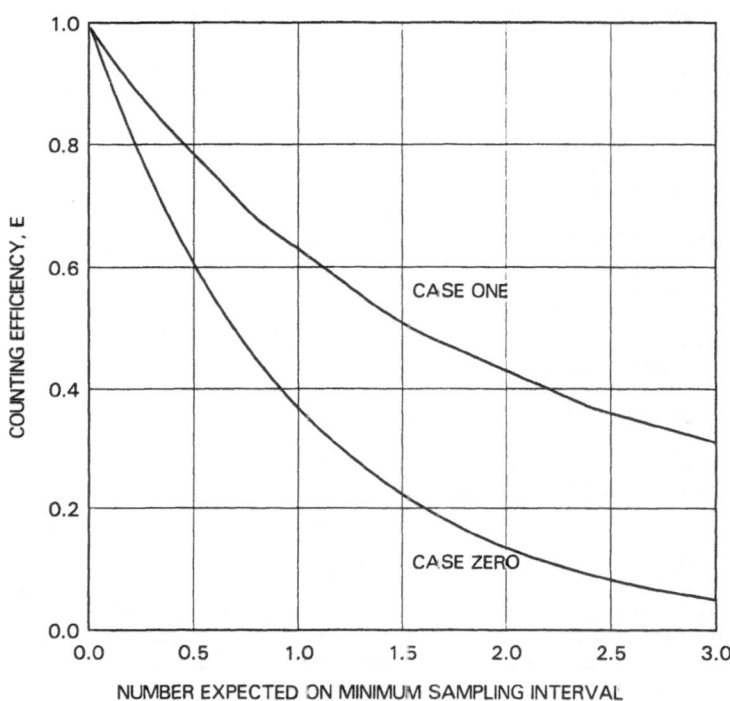

Figure 11. Counting efficiency versus expected count in the minimum
sensing interval. Poisson statistics assumed (uniform, constant
concentration) and coincidence either produces a count of one
(Case One) or zero (Case Zero).

processing to overcome this lack of resolution (e.g., Twomey's iterative algorithm[69], Nelder and Mead's search procedure[70], the iterative EM algorithm[71]). However, it is easy to fool oneself with such analyses: they tend to work well for data that does not have errors, synthetic data, but they have various problems that become important when they are applied to real data.[73] [75] It is hard to obtain a sufficient particle count in a clean room environment to use a diffusion battery and a data inversion algorithm but it has been done.[72] Although there are examples where the processed data agreed well with data from optical particle counters in the size range of overlap, there is anecdotal evidence that this is rarely the case.

Electrical mobility analyzers separate particles on the basis of their electrical mobilities, the product of their charges and their mechanical mobilities. Mechanical mobility is the terminal velocity per unit external force applied to the particle, a function of particle size and medium viscosity and molecular mean free path. (See Ref. 11 and 76.) Early electrical mobility analyzers were designed for an integral mode of operation, where they gave discrimination of particles with mobilities larger than a certain value. The latest instrument in wide use (TSI's Differential Mobility Analyzer) employs a differential mode, which can select particles in a narrow range of mobilities.[77] The integral mode may be more appropriate where particle concentrations are low, trading off some mobility precision for a greater total count, thus a greater counting precision, at a few mobility cut values. The recently developed perforated plate mobility analyzer seems to have advantageous characteristics for this use.[78]

The size range of particles these devices are best for is from hundredths of microns to tenths. For the TSI Model 3071 differential mobility analyzer, losses become appreciable as particle size decreases to 0.01 micron and smaller.[79] In use in contamination control, these instruments often do not allow sufficient particle counts to be obtained over periods of interest, however. A more generic problem is that the charging of the particles is not a unique function of particle size, and data processing methods must also be used, perhaps with somewhat fewer problems than for diffusion battery analyses.[80] An approach to the data inversion problem was presented by Hagen and Alofs, who reviewed several previous methods.[81] Data inversion is improved when the charge distribution developed by Fuchs is used to model the aerosol rather than the Boltzmann distribution, as the Fuchs distribution is monotonic and more nearly correct, especially for particles much smaller than 0.1μm. in electrical equilibrium.[82]

Aerodynamic particle sizers discriminate among particles on the basis of aerodynamic equivalent diameter, the diameter of a sphere having the density of water that would fall at the same rate under gravity as the particle in question if both were present in air at NTP. This diameter governs particle deposition due to gravity and various acceleration fields (such as impaction and bends in tubing). The most common instrument of this type is the impactor, which directs a jet of air or a set of jets of air at a surface. The smallest particles avoid capture by the surface, but the largest have too much inertia to escape. Many jets in parallel can be used in an impaction stage. Series of stages can be used, with different characteristic dimensions, to produce selected particle sizes ranges on each stage. The collected particles are then analyzed by various methods, including microscopy, chemical analysis, even piezoelectric detection on a vibrating collection crystal. Examples of particle collection by a multi-stage impactor followed by scanning

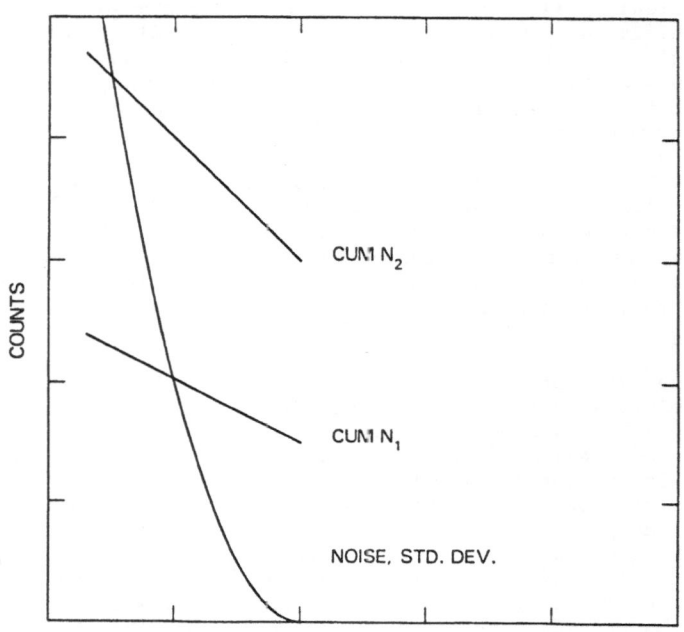

Figure 12. Counts versus the particle size threshold of a hypothetical instrument. Shown are the standard deviation of the noise and two different cumulative count size distributions.

electron microscopy (SEM) and energy-dispersive x-ray (EDX) analysis were presented by Barnard.[115] There is also a commercially available collection device that uses a jet directed at the surface of a silicon wafer, which is then analyzed with a surface counter; the device (Clean Room Sentry, by VLSI, Inc. of Mountain View, CA) is described as relying on turbulent diffusion for capture, but impaction may also be significant.

Two commercial devices use particle inertia to distinguish particle size, followed by the use of light scattering to detect the particles. An even earlier use of this approach was a variable slit impactor in series with an optical particle counter[74], but this was never made into a commercial instrument. The TSI Aerodynamic Particle Sizer uses a nozzle to accelerate particles past a pair of laser light beams.[83] The time of flight is used to determine acceleration, which is primarily a function of aerodynamic diameter. Its calibration and use were discussed by Baron.[84] The device begins to lose size resolution as the particles decrease to a half-micron in aerodynamic diameter, roughly its lower limit, according to the evaluation by Chen et al.[85] A second device, SPART, determines aerodynamic size by particle response to gas motion, sensed by light scattering.[86]

A recent review of instruments for measuring single submicron particles in aerosols in "real time" gives additional information on optical particle counters, the aerodynamic particle sizer, particle electrical mobility analyzers, diffusion batteries, and condensation nucleus counters.[77]

METHODS OF PARTICLE ANALYSIS: OTHER COLLOID INSTRUMENTS

The optical particle counter is often used for monitoring liquids, as is filtration followed by microscopy. These techniques have been described above. Microscopy is often the definitive measuring technique, used for calibration of other techniques and for the identification of particles. Optical particle counters are much more convenient, but their data are less reliable. More details on how these two approaches compare are available in Ref. 87.

High-purity water is crucial to microelectronics and pharmaceuticals. Particles are not the only important contaminants, however. Examples of production disruption, discussion of past and current contamination levels in final filter water, and some techniques for measuring contamination were described by Balazs.[88] Techniques have been developed and are in use for measuring the following: dissolved inorganics (conductivity, atomic absorption) dissolved organics (infra-red spectroscopy, high pressure liquid chromatography), and microorganisms or their fragments.[89] Craven et al.[90] recommend monitoring pure-water supplies for total oxidizable carbon, silica, dissolved ions, bacteria and particulate material, and they described the techniques they used to survey pure-water systems for these. They presented data showing correlations between some of these measures and yield loss or product contamination.

Biological particles can be assessed several different ways besides through microscopy: various viable-count methods such as the use of nutrient surfaces and the counting of colonies; chemical tests for pyrogens, adenosine triphosphate (ATP), fatty acids, or total organic carbon.[91] Bacteria were found at high concentrations in water used in microelec-

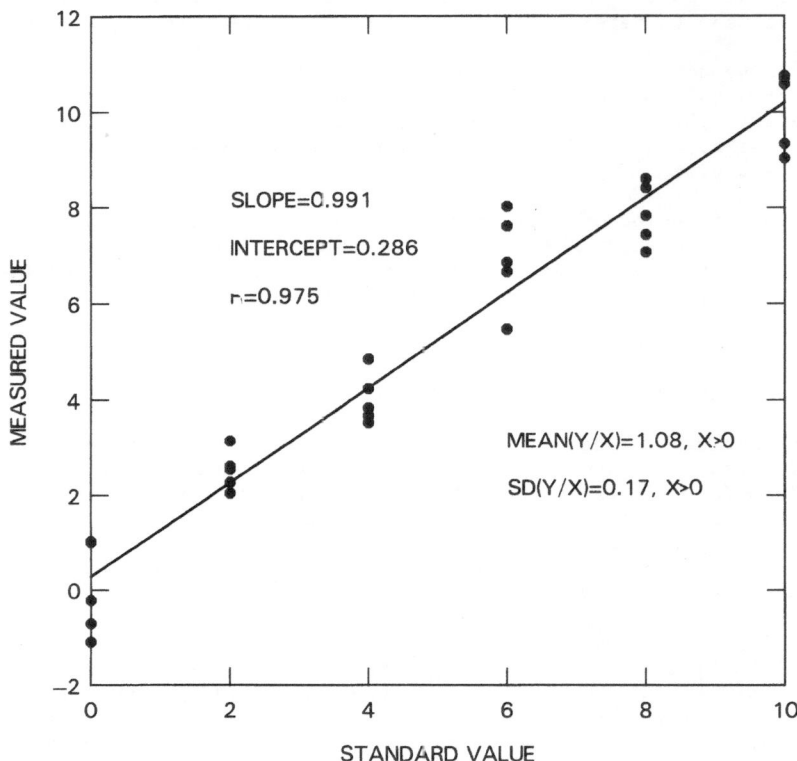

Figure 13. Hypothetical example of calibration data taken for an instrument with a range of 0 to 10. Data have Gaussian (normal) noise added with mean of 0 and standard deviation of 1.

tronics manufacturing (tested by a plate count method adapted from ASTM F60-68) and suggestions were offered by Harned to aid in reducing bacteriological contamination.[92] Although total oxidizable carbon (TOC) assays are widely used to assess biological contamination, they are not highly correlated with bacterial counts, at least in some instances.[93] However, correlation is a measure of the variability of the dependent variable in comparison to the variability of the independent variable, and will depend on the range of the measurements as well as the strength of the relationship and the random variations.

Besides the optical particle counter and microscopy, another technique gives information on the physical size of a colloid particle: the electro-resistive particle counter (or "Coulter counter") passes a stream of liquid through an insulating orifice. The liquid itself is conductive (it may have to be made to be conductive) and the particles that can be sensed are those that are non-conductive. These particles partially block a current path that is created through the liquid and through the orifice. The current is generated from electrodes placed on each side of the orifice. Blockage of the current is proportional to the volume of each insulator particle, and a volume distribution is obtained. Total blockage of flow is a problem, so the orifice size has to be chosen carefully. The larger the orifice, however, the larger the minimum detectable particle size. The minimum particle size is somewhat smaller than the minimum for optical particle counters, at present. As with optical particle counters, bubbles may cause false counts. These volumetric sensors are reported to operate best at clean conditions, < 100 particles per liter [87] although this restriction --- as with many other instrument limits --- can be expected to change appreciably within even a few years' time.

SPECIAL SITUATIONS: VERY HIGH, VERY LOW CONCENTRATIONS

We define very high concentrations as those where one has a non-negligible probability of getting more than one particle in a sensing interval, causing the instrument to count more than one particle as one or even as no particles. This error is called "coincidence."

For aerosols in contamination control situations, coincidence is not usually a problem because the airborne concentrations are low compared to the concentrations for which the instruments were designed. An exception to this could occur in sampling from manufacturing processes during periods of unusual operation.

For liquid particle counting, this situation can occur sometimes due to the presence of bubbles in addition to particles and sometimes just due to particles themselves.

For counting particles on surfaces, it is rare that they are so concentrated that one particle lies atop another, which would produce coincidence in microscopy. An exception would be samples drawn onto filter surfaces. Where the counting is done by instruments that use a scanning light beam, the concentration must be kept so that it is rare to have more than one particle in the beam at any time. Where the counting is done by instruments that analyze vidicon pixels activated by light, coincidence will depend on the fraction of pixels activated: nearly 20% coincidence losses can occur where as few as 5% of the pixels are activated.[30] Since "blooming" of the signals into adjacent pixels is used to size the particles, and one particle may have many pixels, this 5% condition is not unusual.[30]

One way to detect coincidence is to note that doubling or halving the concentration does not nearly double or halve the count. Raising the concentration increases coincidence losses and lowering it decreases the losses, so the counts will not change proportionately. Figure 11 shows the counting efficiency to be expected when sampling a uniform concentration under the assumptions that coincidence produces a single count (Case One) or that coincidence produces a zero count (Case Zero), which would happen where the combined signals were beyond the range accepted.

To reduce coincidence, steps must be taken to reduce the concentration of particles. Where the concentration is on a filter or a surface, the concentration can be lowered by reducing the volume filtered or the time the surface is exposed. Where the concentration is in a gas or liquid, one has to dilute with clean gas, a relatively easy task, or dilute with clean liquid, which can be more difficult because liquid filtration is rarely as complete as gas filtration. The dilution flow or volume must be known accurately and held fixed. If the concentration cannot be reduced, then the sensing interval must be made smaller. Where time is the critical element, the flow past the sensor can be slowed, making it less likely that two signals will overlap in time. Where space is the critical element, the sensing volume may be reduced by such means as changing the focal region size or the illuminating spot size.

By low concentration measurements, we mean those where it is difficult to obtain many particle counts. This is frequently the case in sampling filtered air, for example, with optical particle counters. Air in clean rooms can often be near Federal Standard 209 levels of Class 1, meaning fewer than 1 particle per cubic foot that is larger than 0.5 microns in optical equivalent diameter. For such rooms, optical particle counters with flows of even 1 cfm produce few counts unless many minutes are expended per sample.

One problem presented by low concentrations is that the noise rate of the instrument may become a substantial fraction of the signal rate. We have seen this in using optical particle counters on clean room atmospheres. Recent work has quantified the noise rates, or counts per volume, of various optical particle counters.[94] It may seem paradoxical, but it can be effective to raise the threshold of a particle counting instrument to improve its use in low-concentration situations. Figure 12 shows a hypothetical situation not too unlike those we have encountered. The instrument noise can be corrected for, in the mean, but there are still fluctuations in the noise around this mean value that cannot be corrected for. In our experience the noise was in the smallest sizing channels of aerosol optical particle counters. An analysis similar to that of benefit-cost analysis in economics can be used to show that the net count above the uncertainty from the noise (count minus noise standard deviation) is maximized where the marginal true count equals the marginal noise uncertainty, which is where the slopes of the noise standard deviation and the cumulative distribution are equal. In the case shown in Figure 12, this can be achieved by raising the particle size threshold well above the value where the noise uncertainty equals the cumulative count. In practice, this means that data from channels other than the smallest may be more valuable than data that includes the smallest sizing channel.

A second problem with low-concentration measurements is that data with relatively few counts is inherently more variable, on a percentage basis, than data with many counts. As the expected number of counts becomes much larger than 10 or 20, the standard deviation for a Poisson distribution, which distribution is expected when sampling constant and uniform concentrations, becomes equal to the square root of the expected count. The relative standard deviation of 25 counts is about 5/25 = 20%; the relative standard deviation of 100 counts is 10/100 = 10%; the relative standard deviation of 400 counts is 20/400 = 5%, etc. For counts not much larger than 20, the square root estimate is not as good and the relative standard deviation is actually not too different from 50%, a great deal of uncertainty. Especially where the counts are few, statistical techniques, such as chi-square analyses for count data, should be brought to bear.

Factors peculiar to particle counting in monitoring liquids at low concentrations downstream from filters were discussed further by Krygier et al.[95]

SPECIAL SITUATIONS: SAMPLING FROM HIGH-PRESSURE SOURCES

Because most aerosol particle counters are designed to operate near atmospheric pressure, one must generally direct any sample from a high-pressure gas line through a pressure-reduction device, perhaps an orifice, before bringing it to the instrument that will count it.

Gas flow across an orifice that reduces the absolute pressure downstream to half or less of the absolute pressure upstream will produce a flow in the orifice that is sonic. Such high velocities lead to a turbulent jet downstream, with deposition of the particles not only on the upstream side of the orifice but also on the downstream side and on the walls of the pipe or plenum downstream.[13][96][99] Deposition is lessened, generally, by using the largest practical diameter of pipe upstream and downstream from the orifice. Enough length must be allowed downstream for the jet to expand to the wall of the pipe and for the particles accelerated in the orifice to reach approximately the mean flow velocity in the pipe.

CALIBRATION

Calibration should not come toward the end of monitoring but toward the beginning and at intervals throughout. Instruments may arrive damaged from their manufacturers, may wear out or just drift. Although some types of degradation may lead to a random component of error, many failures lead to bias, which does not "average out" to zero.

Calibration of the instruments against standards is best. Somewhat less valuable is calibration against instruments believed to be correct. Less valuable still is calibration by measurements of situations believed to give a constant value, where that value is not known with certainty.

Calibrating an aerosol instrument against a standard aerosol is done by generating an aerosol of a known size distribution, preferably monodisperse, and determining the concentration in an independent and accurate manner, often done by sampling isokinetically and transporting the aerosol with negligible losses to a filter from which a particle count is obtained, which is used with an accurate measure of sample volume to give a concentration value. Unfortunately, although there are methods of generating an aerosol of known particle size, the generation of a predictable concentration remains to be achieved. Techniques and results of calibrations performed on several optical particle counters were very usefully summarized in a recent review by Liu et al. [100] and Liu and Szymansky presented the results of tests of counting efficiency versus particle size for eight popular optical particle counters.[94] Methods of calibration and results were also presented by Dahneke and Johnson.[101]

Polydisperse aerosols, those of a spread of particle sizes (rather than a selected group of discrete sizes) should be avoided, because it is difficult to know whether counts in a particular size range come from particles of that size or from larger or smaller particles the sizes of which were mistakenly measured.

Aerosols of known particle size can be generated by atomizing suspensions of latex spheres (polystyrene latex, polyvinyltoluene, etc.). Such particles a few microns and smaller have been available for years. Much information has been published on these spheres, including monographs by Bangs.[102] Recently, latex spheres larger than a few microns have been made and proposed for use in calibration.[103] Other methods of obtaining monodisperse aerosols involve the generation of droplets with the vibrating orifice atomizer[104] [105] or the vibrating reed atomizer or the spinning disk atomizer[106] [107], all of which may be operated to give monodisperse aerosols once any troublesome secondary aerosol component is eliminated. These secondary components include droplets that have no latex spheres and have dried down to the non-volatile impurities, secondary droplets of a size different from the primaries at certain flow rates and frequencies for the vibrating orifice aerosol generator, and secondary droplets of a size different from the primaries that are characteristic of the spinning disk aerosol generator but can be removed as a second stream of aerosol and discarded.

Another approach to generating aerosols of a single size is to produce a polydisperse aerosol by atomization or condensation, selecting from the aerosol a fraction with a narrow size distribution. This selection can be done on the basis of electrical mobility with the Differential Mobility Analyzer (TSI, Inc.). For example, monodisperse aerosols of silver particles and of NaCl particles were made this way and had diameters between 0.002 and 0.3 microns.[108] The gradual change in penetration versus particle diameter that is characteristic of diffusion batteries has thus far prevented their use in preparing a nearly monodisperse aerosol through tailored transmission of particles having a narrow range of diffusivities.

Liquid particle counters are most often calibrated by atomizing very dilute suspensions of latex spheres. Concentration is determined from counting a sample taken from a known volume deposited on a filter, or deposited completely in some other manner. A pitfall is the possible presence of bubbles to confound the measurement. A change in particle count or size due to a change in liquid temperature or pressure is a sign that bubbles are probably present. The light scattering signal from bubbles is characteristically different from particles. [61] Putting the liquid under high pressure (e.g., a few atmospheres) will diminish or eliminate the bubbles. Calibration techniques and the results of such calibration studies on optical particle counters for liquids were presented by Peacock et al.;[109] they corrected for coincidence by extrapolating counting efficiencies to near-zero concentration.

The monitors of particles that rely on evaluating light scattering from surfaces on which the particles deposit can be calibrated by generating a monodisperse aerosol and causing it to deposit, but the determination of the true count on the surface often requires tedious microscopy. There are commercially available wafers of silicon that have known numbers of particles of a given size deposited on them, and these wafers can be used for calibration until they become unacceptably contaminated, after which they are difficult or impossible to clean. Also available are Si wafers with etched pits in them that scatter light somewhat like particles do. The count is known but the connection between the response to etched pits and the response to real particles is not straightforward; such wafers seem to fall in the category of a comparison or relative standard that is useful to indicate change over time, if it occurs. Both types of wafer are available from VLSI, Inc. (Mountain View, CA). The calibration of surface monitors is discussed further in the papers by Berger and Tullis[110] and by Tullis.[111]

Figure 13 shows hypothetical data that result from a calibration of an instrument over the range (0 to 10) it is intended to measure. The error was chosen here to have a mean of 0 and a standard deviation of 1. One approach is to fit the least-squares best-fit straight line to the data and use this relationship to determine true values from measured values. This is reasonable, although the regression predicts the best-fit measured value to a known value, rather than vice-versa. Often it suffices to re-adjust the zero on the instrument based on the intercept of the best-fit line, unless the slope is statistically different from 1.00. Useful measures obtained from this procedure include the standard deviation of the data around the regression line and the mean of the ratio of the measured and true values. Analysis of the mean and standard error of the mean of this ratio can indicate whether a multiplicative calibration factor different from 1.00 is warranted. Even without sophisticated analysis, this calibration procedure will give insight into the random component of error to be expected and will help adjust the instrument to offset non-random (systematic) error.

DATA COLLECTION AND ANALYSIS

The trend in data collection and analysis clearly favors electronic transmission to memory devices that can be read by computer and analyzed by statistical and graphical analysis programs. While there is value in preserving the data, it is important to condense it into forms that are readily understood: simple statistical terms such as mean and standard deviation (and standard error of the mean), simple graphical presentations such as trend lines, histograms, and pie charts, and compound presentations such as various quality control charts.

Presentation of data is facilitated by summary statistics and charts. Analysis may require sophisticated statistical techniques, but often one of the following fairly elementary techniques will suffice:
 - linear regression to test for trends and relationships
 - Student's t test to compare means with each other or to a standard
 - analysis of variance to test for effects of several factors
 - chi-square analysis to test count data for effects of factors.

As contamination levels decrease, the background counts of the instruments used become relatively more important. Correcting for background counts can involve not merely adjusting for the mean rate, but also taking into account, through statistical analysis, the variability.[112]

Awareness of the importance of sound experimental design and statistical analysis is increasing, leading to the publication of such articles as the series by Tullis on measuring contamination in process equipment [113] and by Bzik on the estimation of contamination rates or the acceptance or rejection of hypotheses, such as "this meets the criteria," based on particle counts (assuming Poisson data).[114]

Data worth collecting are generally worth analyzing. It is often worthwhile getting a statistician involved before the data are collected. At least, a statistics text should be consulted.

CLOSING REMARKS

Monitoring contamination is valuable for determining the sources and for following the progress of efforts at cleaning up. The data can sometimes be correlated with product quality and yield, perhaps aiding in assessing the value of the clean-up activities and choices.

This review is aimed at giving the reader an overview of the topic and providing a list of references from which additional information can be obtained. The overview is from a microelectronics industry perspective. No doubt, some important work has been overlooked. The field has been greatly advanced by the work of others, and their contributions are truly appreciated, even those not specifically referenced.

REFERENCES

1. S. Gunawardena, U. Kaempf, B. Tullis, and J. Vietor, Microcontamination 3(9), 55-62 (1985).
2. D.W. Cooper, Aerosol Sci. Technol. 5, 287-299(1986).
3. B.T. Murphy, Proc. IEEE 52, 1537-1545 (Dec. 1964).
4. C.H. Stapper, IBM J. Res. Develop. 30(3), 326-338 (1986).
5. J. Burnett, Microcontamination 3(5), 32-36 (1985).
6. L.-P. Faure and H. Thebault, Microcontamination 5(3), 16-20 (1987).
7. D.W. Cooper, Microcontamination 3(8), 48-54,73 (1985).
8. A. Ligtenberg, J. Environ. Sci. 29(6), 41-45 (1986).
9. M.S. McClellan, J. Environ. Sci. 29(1),41-44 (1986).
10. D.A. Hope, Proceedings of the 1987 Institute of Environmental Sciences Annual Technical Meeting, San Jose, CA, May 1987, pp. 409-416.
11. W.C. Hinds, "Aerosol Technology," Wiley, New York (1982).
12. N.A. Fuchs, Atmos. Environ. 9, 697-707 (1975).
13. D.W. Cooper, Solid State Technol.29(1), 73-79 (1986).
14. S.P. Belyeav and L.M. Levin, J. Aerosol Sci. 5(4), 325-338 (1974).
15. J. West, Microcontamination 3(11), 87-89, 162 (1985).
16. D.C. Stevens, J. Aerosol Sci. 17(4), 729-743 (1986)
17. S.J. Dunnett and D.B. Ingham, J. Aerosol Sci. 18(5), 553-561 (1987).
18. J.H. Vincent, J. Aerosol Sci. 18(5), 487-498 (1987).
19. D.W. Cooper, J. Environ. Sci. 29(2), 25-29 (1986). Reprinted in J. Soc. Environ. Engrs. 26(1), 29-33 (1987). Reprinted (Chinese translation by X. Liu) in Clean Technol., 7-14 (December 1987).
20. R.D. Hellander, Microcontamination 5(9), 45-49, 76, 78 (1987).

21. D.L. Tolliver, Microcontamination 2(3), 12-15 (1984).
22. N.A. Fuchs, Aerosol Sci. Technol., 5, 123-143 (1986).
23. D.W. Cooper, M.H. Peters, and R.J. Miller, to be published in Aerosol Sci. Technol. (1989).
24. D.S. Koellen and D.I. Saxon, Microcontamination 3(7), 47-55 (1985).
25. M.K. Balazs and S. Walker, Semiconductor Intl. 5(4), 101-114 (1982).
26. J.S. Batchelder, SPIE Vol. 774, 8-12 (1987).
27. E.C. Douglas, IEEE Trans. Electron Devices, ED-22, 224 (1975).
28. D.W. Cooper and H.R. Rottmann, Particle sizing from disk images by counting contiguous grid squares or vidicon pixels, paper presented at the 18th Annual Meeting of the Fine Particle Society, August 1987, Boston, MA, and J. Colloid Interface Sci., in press (1988).
29. P. Lilienfeld, Aerosol Sci. Technol. 5, 145-165 (1986).
30. D.W. Cooper and R.J. Miller, J. Electrochem. Soc. 134, 2871-2875 (1987).
31. J. Pecen, A. Neukermans, G. Kren, and L. Galbraith, Solid State Technol. 30(5), 149-154 (1987).
32. G.L. Wojcik, D.K. Vaughan, and L.K. Galbraith, SPIE Vol. 774, 21-31 (1987).
33. G. Mie, Ann. Phys. 25, 377 (1908).
34. R. Caldow et al., Particle Technol. Lab. Pub. No. 645, Mech. Engg. Dept., U. Minn., Minneapolis (1987).
35. P.G. Borden, Y. Baron, and B. McGinley, Microcontamination 5(10), 30-34 (1987).
36. S. Fujii, K.Y. Kim, and I. Hayakawa, Proceedings of the 1987 Institute of Environmental Sciences Annual Technical Meeting, San Jose, CA, May 1987, pp. 422-427.
37. I. Hayakawa, S. Fujii, and K.Y. Kim, Aerosol Sci. Technol. 7, 47-56 (1987).
38. C.N. Montgomery, Proceedings of the 1987 Institute of Environmental Sciences Annual Technical Meeting, San Jose, CA, May 1987, pp. 353-359.
39. H.Y. Wen and G. Kasper, J. Aerosol Sci. 17, 947-961 (1986).
40. J. Gebhart et al., Staub-Reinhalt. Luft (German Edn.) 43, 434-438 (1983).
41. R.G. Knollenberg, J. Environ. Sci. 30(2), 50-58 (1987).
42. A. van der Meulen, B.G. van Elzakker, and A. Plomp, Aerosol Sci. Technol. 5, 313-324 (1986).
43. J. Keskinen, K. Janka, and M. Lehtimaki, Aerosol Sci. Technol. 6(1), 79-83 (1987).
44. B.T. Chen and H.C. Yeh, J. Aerosol Sci. 18(2), 203-214 (1987).
45. D.S. Ensor and R.P. Donovan, J. Environ. Sci. 28(2),34-36 (1985).
46. J. Agarwal and G. Sem, J. Aerosol Sci. 11(4), 343-357 (1980).
47. B.Y.H. Liu and D.Y.H. Pui, Atmos. Environ. 13, 563-568 (1979).
48. K.L. Rubow and B.Y.H. Liu, Microcontamination, 3(3), 39-43 (1985).
49. R.P. Donovan, B.R. Locke, D.S. Ensor, and C.M. Osburn, Microcontamination 2(6), 39-44 (1984). A. Lieberman, Microcontamination 3(3), 31, 61, 62, 64, 65 (1985).
50. D. Sinclair, Aerosol Sci. Technol. 5, 187-204 (1986).
51. R.G. Knollenberg, J. Environ. Sci. 28(1), 32, 41-47 (1985).
52. W.G. Fisher, TSI J. Particle Instrum., 2(1), 3-19 (1987).
53. B.Y.H. Liu and D.Y.H. Pui, J. Colloid Interface Sci., 47, 155-171 (1974).
54. H. Bartz, H. Fissan, C. Helsper, Y. Kousaka, K. Okuyama, N. Fukushima, P.B. Keady, S. Kerrigan, S.A. Fruin, P.H. McMurry, D.Y.H. Pui, and M.R. Stolzenburg, J. Aerosol Sci. 16(5), 443-456 (1985).
55. R.P. Donovan, B.R. Locke, C.M. Osburn, and A.L. Caviness, J. Electrochem. Soc. 132(11), 2730-2738 (1985).

56. S.D. Cheung and R.P. Roberge, Microcontamination 5(5),45-50, 94, 95 (1987).

57. A. Khilnani, Microcontamination 4(11), 24, 26-29 (1986).

58. D.C. Grant and W.R. Schmidt, J. Environ. Sci. 30(3), 28-33 (1987).

59. D.H. Fisher, S.S. Hupp, and C. Scaccia, Microcontamination 5(2), 14, 16, 18, 20 (1987).

60. N.J. Csikai and T.D. Schware, Microcontamination 5(9),40, 42 (1987).

61. K. Dillenbeck, Microcontamination 5(2), 31-38, 65 (1987).

62. K. Dillenbeck, Microcontamination 3(11), 21-30 (1985).

63. K. Dillenbeck, Microcontamination 2(6), 56-62 (1984).

64. N.J. Csikai and A.J. Barnard, Microcontamination 4(11), 44-50, 128-130 (1986).

65. H.Y. Wen, G. Kasper, and S. Chesters, Microcontamination 4(3), 32-39, 68 (1986).

66. Y. Kousaka, T. Niida, Y. Tanaka, Y. Sato, H. Kano, N. Fukushima, and H. Sato, J. Environ. Sci. 30(4), 39-42 (1987).

67. D.B. Blackford, K.J. Belling, G. Sem, J. Environ. Sci. 30(4), 43-47 (1987).

68. K.E. Brown, J. Beyer, and J.W. Gentry, J. Aerosol Sci. 15(2),133-145 (1984).

69. S. Twomey, J. Comput. Phys. 18, 188-200 (1975).

70. A. Nelder and R. Mead, Comput. J. 7,308-312 (1965).

71. E.F. Maher and N.M. Laird, J. Aerosol Sci. 16(6),557-570 (1985).

72. B.R. Locke, R.P. Donovan, D.S. Ensor, and A.L. Caviness, J. Environ. Sci., 28(6), 26-29 (1985).

73. D.W. Cooper, "The Variable-Slit Impactor and Aerosol Size Distribution Analysis," Ph.D. Dissertation, Division of Engineering and Applied Science, Harvard University, Cambridge, MA, 1974.

74. D.W. Cooper and L.A. Spielman, Atmos. Environ. 8, 221-232 (1974).

75. J.J. Wu, D.W. Cooper, and R.J. Miller, Evaluation of aerosol deconvolution algorithms for determining submicron particle size distributions, paper presented at the Annual Meeting of the American Association for Aerosol Research, September 1987, Seattle, WA.

76. B.Y.H. Liu and D.Y.H. Pui, J. Aerosol Sci. 6, 249-264 (1975).

77. R.P. Donovan, B.R. Locke, and D.S. Ensor, Solid State Technol., 28 (9), 139-148 (1985).

78. J. Schlatter, A. Schmidt-Ott and H. Burtscher, J. Aerosol Sci. 18(5), 581-583 (1987).

79. A. Reineking and J. Porstendorfer, Aerosol Sci. Technol. 5, 483-486 (1986).

80. C. Helsper, H. Fissan, A. Kapadia, and B.Y.H. Liu, Aerosol Sci. Technol. 1,135-146 (1982).

81. D.E. Hagen and D.J. Alofs, Aerosol Sci. Technol. 2, 465-475 (1983).

82. Y. Kousaka, K. Okuyama, and M. Adachi, Aerosol Sci. Technol. 4, 209-225 (1985).

83. J.K. Agarwal and R.J. Remiarz, USDHEW-NIOSH Contract Report No. 210-80-0080, NIOSH, Cincinnati, OH (1981).

84. P. Baron, Aerosol Sci. Technol. 5, 55-67 (1986).

85. B.T. Chen, Y.S. Cheng, and H.C. Yeh, Aerosol Sci. Technol. 4, 89-97 (1985).

86. M.K. Mazumder and K.J. Kirsch, Rev. Sci. Instrum. 48, 622-624 (1977).

87. S.H. Goldsmith and J.P. Barski, Microcontamination 2(3), 46-52 (1984).

88. M.K. Balazs, Microcontamination 5(1), 34-40, 62 (1987).

89. C.F. Frith, J. Environ. Sci. 27(5), 37-39 (1984).

90. R.A. Craven, A.J. Ackermann, and P.L. Tremont, Microcontamination 4(11), 14, 16-19 (1986).

91. M.W. Mittelman, Microcontamination 3(11), 42, 44, 46, 48, 50, 52, 55, 56, 58 (1984).

92. W. Harned, J. Environ. Sci. 29(3), 32-34 (1986).

93. M.W. Mittelman, Microcontamination 3(10), 51-55, 70 (1985).

94. B.Y.H. Liu and W.W. Szymanski, Proceedings of the 1987 Institute of Environmental Sciences Annual Technical Meeting, San Jose, CA, May 1987, pp. 417-421.

95. V. Krygier, M. Latham, and R. Conway, Microcontamination 3(4), 33-39 (1985).

96. D.Y.H. Pui, F. Romay-Novas, S.Z. Wang, and B.Y.H. Liu, Proceedings of the 1987 Institute of Environmental Sciences Annual Technical Meeting, San Jose, CA, May 1987, pp. 388-391.

97. W.T. McDermott and R.C Ockovic, Proceedings of the Institute of Environmental Sciences Annual Technical Meeting, San Jose, CA, May 1987, pp. 392-399.

98. H.Y. Wen and G. Kasper, Proceedings of the Institute of Environmental Sciences Annual Technical Meeting, San Jose, CA, May 1987, pp. 400-402.

99. H. Itoh, G.C. Smaldone, D.L. Swift, and H.N. Wagner, Jr., J. Aerosol Sci. 16(2), 167-174 (1985).

100. B.Y.H. Liu. W.W. Szymanski, and K.-H. Ahn, J. Environ. Sci. 28(3), 19-24 (1985).

101. B. Dahneke and B. Johnson, J. Environ. Sci. 29(5), 31-36 (1986).

102. L.B. Bangs, "Uniform Latex Particles," Seragen Diagnostics, Indianapolis, IN (1985) and L.B. Bangs, Amer. Biotech. Lab. (May/June 1987). See also, A.F. Lieberman, J. Environ. Sci. 31 (3), 34-36 (1988).

103. D.B. Blackford, Aerosol Sci. Technol. 6(1), 85-89 (1987).

104. J.P. Mitchell, J. Aerosol Sci., 18(3), 231-243 (1987).

105. R.N. Berglund and B.Y.H. Liu, Environ. Sci. Technol. 7, 147-153 (1973).

106. K.R. May, J. Sci. Instrum. 43, 841-842 (1966).

107. J.P. Mitchell, J. Aerosol Sci. 15(1), 35-45 (1984).

108. H.G. Scheibel and J. Porstendorfer, J. Aerosol Sci. 14(2), 113-126 (1983).

109. S.L. Peacock, M.A. Accomazzo, and D.C. Grant, J. Environ. Sci. 29(4),23-27 (1986).

110. J Berger and B.J. Tullis, Microcontamination 5(7), 24, 26-29 (1987).

111. B.J. Tullis, Microcontamination, 4(1),50-55, 86 (1986).

112. R.A. Van Slooten, Microcontamination 4(2), 32-38 (1986).

113. B.J. Tullis, Microcontamination, 3(11), 66-73, 160-161, 3(12) 14-21 (1985)

114. T.J. Bzik, Microcontamination 4(6), 35-41 (1986).

115. L. Barnard, Microcontamination 5(8), 34-37 (1987)

116. H.C. Rechen, Microcontamination 3(7), 22, 24, 26, 28, 29 (1985).

117. R. de Neufville and J.H. Stafford, "Systems Analysis for Engineers and Managers," McGraw-Hill, New York, 1971.

118. D.W. Cooper, in "Particulate and Multiphase Processes," T. Ariman and T.N. Veziroglu, editors, Vol. 2, pp.121-126, Hemisphere, Washington, DC, 1987.

119. D. Sinclair and G.S. Hoopes, Amer. Indus. Hyg. Assoc. J. 36, 39-41 (1975).

MEASURING AND IDENTIFYING PARTICLES IN ULTRAPURE WATER

Marjorie K. Balazs

Balazs Analytical Laboratory
2284 Old Middlefield Way
Mt. View, CA 94043

The refinements of particle counters have permitted particle counting down to a level of 0.3μm. Recent problems
in IC processing, however, have made it apparent that
those unseen particles in the sub-0.3μm sizes are causing
catastrophic "yield busts".

This paper presents state-of-the-art techniques for
evaluating <0.3μm particles in fluids. It further gives
examples of such problems caused by 0.3μm or smaller
particles and illustrates the difficulties one encounters
in identifying and resolving such problems. Included is a
specific case to illustrate how costly and difficult it
has become to identify particles smaller than 0.3μm. The
paper raises the issue as to whether companies should be
leaving quality control and system decisions to anyone
other than highly trained personnel, a policy generally
not followed today.

INTRODUCTION

During the past decade, the degree of difficulty in identifying particles
in ultrapure water has increased enormously. Ten years ago the major
difficulty with identifying particles from semiconductor industry pure
water systems or on wafers was finding the tools to obtain the analysis
that would reveal the identity of an offending particle. The particles
were large, generally greater than 5μm and often as big as 100μm in spite
of 0.2μm final filters; so lack of material was not the problem (see
Figures 1-6). Often, it was possible to pile them up on a collecting
filter so that conventional chemical analysis could be used to reveal
their composition. Success rate for identification was greater than 90%
and costs were generally below $5,000 and over 50% of the time below
$2,000. The biggest problem in those early years was to get manufactur-
ers of semiconductor devices to modify their pure water systems so that
they worked more efficiently.

Today, with the tremendous improvement in membrane technology used to
make reverse osmosis, final filters, and ultrafilters, the removal of
particles has been enormously successful. In spite of differences of
opinions on how to measure them or what the reality is of a total count

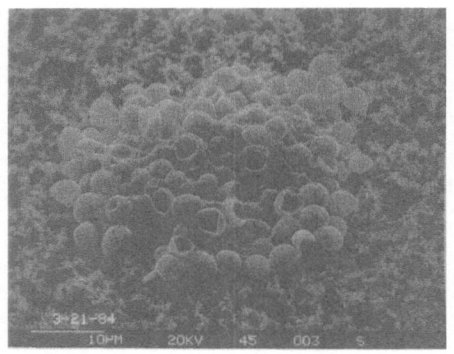

Figure 1. Bacteria colony.

Figure 2. Fiber.

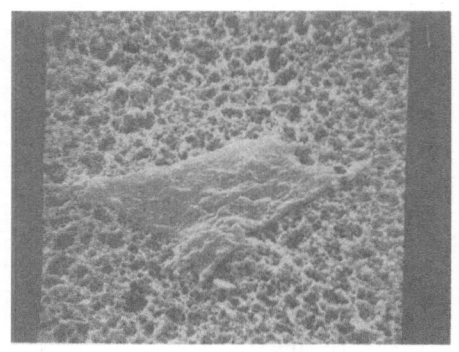

Figure 3. Sloughed polymer.

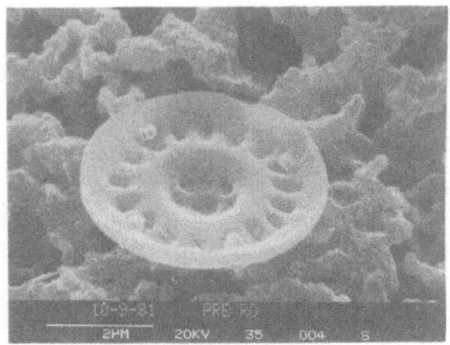

Figure 4. Diatom.

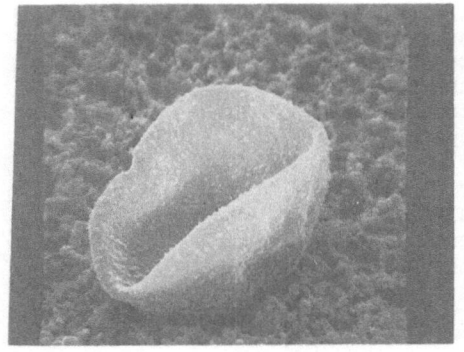

Figure 5. Spore.

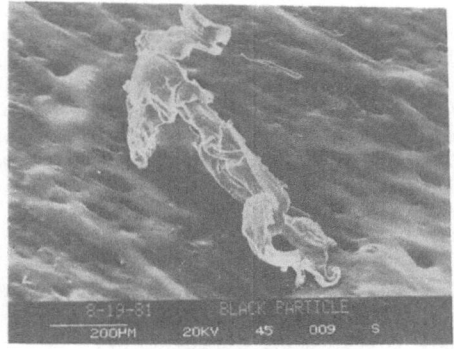

Figure 6. Rubber fragment from O-ring.

or distribution, particles over 1μm in size have virtually become non-existent. 0.5 - 1.0μm particles have also dropped to levels of <500/l by almost anyone's count and are often considered non-existent. The area of 0.3-0.5μm is the point of focus and <0.3μm is the size that creates the greatest amount of damage to the product. Particles which are <0.3μm or smaller are hard to detect because they are below the detection limits of the in-line laser particle counters (limit 0.3μm), and are difficult to identify because they are below the sensitivity level of most analytical equipment used for chemical analysis of particles.

Furthermore, the definition of a particle comes into question. For process engineers of IC's anything that leaves a detectable mark on the wafer by a laser surface scanner is a particle. This is an inappropriate definition since colloids and dissolved material in water can leave such telltale signs. Bacteria and other living organisms seemed to have also joined the ranks of being defined as a particle in water as have the fragments of these organisms. For the purpose of this paper, particles in water will be defined as any material in the water that can be filtered out of it regardless of its size, shape or composition. Dissolved material, materials of low viscosity, and colloids are left out in spite of the fact that these materials can leave blemishes on wafers.

SEM direct particle counting method (SEM-DCM) has been of value in dealing with these particles, to obtain their size distribution and composition. The technique is well described in Ref. 1.

Table I compares the number of particles found in pure water using the SEM, optical and on-line (detection limit of 0.5μm) particle counting. It can be seen that the major difference in total counts is in the number of particles that are smaller than approximately 0.7μm. Many cases have recently been published illustrating the loss of product with particles smaller than 0.5μm that went undetected by in-line particle counters (Ref. 2). This paper is concerned with the identification of particles <0.3μm that are not being detected by in-line laser particle counters.

Table I. Particle Count and Distribution Post Final Filter

Number of particles/liter

SEM		Batch (Optical)	In-line Counter	
0.2-0.5 μm	280	800	0.5 - 0.75	2
0.5-1.0 μm	52		0.75- 1.0	<1
1.0-2.0 μm	<1		1.0 - 2.0	<1
>2.0 μm	1		2.0 - 3.0	<1

New in-line counters are being developed that can detect particles down to 0.1μm and information concerning their performance is beginning to be found in journals. However, to see these small particles, information about their actual sizes and composition has been sacrificed.

It is important in particle identification to understand the usefulness and limitations of the analytical equipment available. With the drop in the size of the most abundant particles from 100μm to 0.1μm, one has to pay extreme attention to the sensitivity and capability of the analytical tools and method one is using (see Table II). Frequently, improper conclusions concerning particle identification has occurred because of a lack of total understanding of the capability of equipment used in identifying particles. Oftentimes, it is the surface the particle is sitting upon that gets identified, not the particle. Consequently, a particle is said to be siliceous because the x-ray detected only Si. The interpreter did not take into consideration the size or thickness of the particle, the detection capability of the x-ray detector, the conductive versus dielectric properties, whether there is an organic hump in the feedback pattern, whether there is a statistical number of such particles, or the composition of the surface the particle was sitting upon before drawing his/her specific conclusions. And yet, this is a popular tool that most engineers think they understand.

Table II.   Analytical Equipment Limitations
for Particle Identification

	For Physical Info.	For Chemical Data
SEM - - - - - - - - - - - - - -	60Å	--
X-ray (dispersive) - - - - - - -	--	0.2μm
X-ray (fluorescent)- - - - - - -	--	100μm
ESCA - - - - - - - - - - - - - -	--	150μm
Auger - - - - - - - - - - - - -	--	0.5μm
FTIR - - - - - - - - - - - - - -	--	20μm
Optical microscopy - - - - - - -	1μm	--
Polarized light- - - - - - - - - microscopy	1μm	--
TEM	--	0.1μm

When particles are in the sizes of <0.3μm virtually every electron beam tool becomes useless for their identification. For tools like SEM x-ray information only on physical appearance is reliable since the x-ray units cannot get sufficient feedback from the particle to determine its composition. SIMS, ESCA and Auger become totally useless as does FTIR. These tools have limited usefulness even if large numbers of these particles are accumulated on a surface. TEM has some value as does a Z number evaluation on a SEM. Consequently, to solve ultra pure water particle problems today is extraordinarily difficult, and very expensive. The success rate of experts who have been working in particle identification for solving the questions of composition and source of small particles (<0.3μm) is generally less than 20%. With greater effort and sufficient

economic backing this rate could be improved. It is hopeful that the
following case study will illustrate this point.

CASE STUDY

A large manufacturer of IC's had undergone its usual Christmas maintenance program of making changes in the pure water plumbing, moving equipment in the fab areas and sterilizing the system. Nothing was unusual
nor were any unknown materials used during the change over or steriliza-
tion. When processing resumed in January all work proceeded as normal.
During the first week of February, a worker stepped on an exposed incoming pure water line while trying to work on a unit. It broke, sending a
surge of water 60 feet across the room. The room was a foot deep in
water before the system was shut off. Absorbent materials were used to
get the water up. The area was cleaned up, the line repaired, and work
resumed. A total "yield bust" of all products in the entire building
occurred at this time. The cause was large quantities of particles that
were found on all product wafers (see Figure 7).

Two weeks of intensive effort on the parts of maintenance, the fabrication area staff and administration led to the belief that the pure water
was causing the problem. All cross checks between different buildings
revealed that the product could be made as long as the water in the
original building was not used. However, all of the on-line meters
(resistivity, particle counters, total oxidizable carbon) and laboratory
testing (see Table III) revealed that the water was of very high quality
and met the attainable specifications developed by Balazs Laboratory and
adopted by SEMI.

Table III. Results of Pure Water Tests During Shut Down
Post Final Filter

	TOC-ppb	SiO_2 ppb	Bacteria/100 ml Epifluorescence	Resistivity
Loop 1	<20	<5	14	18.2
Loop 2	25	<5	70	18.2

Particles within specification using a PMS liquid laser counter

Ion Chromatography - ppb

	F^-	Cl^-	NO_2^-	$HPO_4^=$	Br^-	NO_3^-	$SO_4^=$	Na^+	NH_4^+	K^+
Loop 1	*	<0.05	*	*	*	*	0.11	<0.05	*	<0.1
Loop 2	*	0.40	0.06	*	*	1.10	0.24	<0.05	*	<0.1

* Below detection limit

SEM-DCM

	Loop 1	Loop 2
0.2 - 0.3 μm	>100,000	>100,000
0.3 - 1.0 μm	900	750

The fabrication area investigation, however, overwhelmingly pointed to the water as the problem. Therefore, an intense effort to find particles in the water was made. SEM particle filters were placed on the system in several places, wafers exposed to a flow of water overnight were collected, and product wafers collected for review. A total review of the system and maintenance schedule was done and samples of all materials involved with the sanitization and used to clean up the water after the surge were collected.

The SEM particle study revealed over 100,000 particles/liter in the pure water post final filter (see Figure 8). They ranged in size from $0.3\mu m$ down to an estimated $0.05\mu m$ with a few particles in the $0.3\mu m$ to $0.5\mu m$ range. They were primarily carbonaceous with an occasional trace of iron, aluminum, and silicon. Their makeup was atypical to what is normally seen in pure water in that the particles were either spherical or flocculent looking. The final filter was immediately changed from a $0.45\mu m/0.2\mu m$ series to a $0.2\mu m/0.1\mu m$ series. Within two days all production lines were running with a yield even higher than before the surge and Christmas modifications. However, R&D production was not able to produce any product nor did anyone know what the contaminant was or where it came from. It was decided that positive identification of the contaminant must be made to ensure its elimination from the system and to prevent a reoccurrence in the future.

Numerous tests were run in an effort to shed light on the composition of the particles. They included:

1. SEM x-ray reviews of particles on filters, both disc and cartridge.

2. SEM x-ray reviews of wafers from production at various stages.

3. SEM x-ray and ESCA reviews of wafers rinsed with pure water overnight.

4. Auger studies of thin film interfaces on the wafers.

5. GC/MS studies on filters.

6. SEM x-ray and GC/MS studies of all materials involved in the system modifications and sanitization cleanout.

7. Evaluation of absorbent materials used to clean the fabrication area.

The SEM x-ray studies revealed that the sub $0.3\mu m$ particles found on the wafers and various disc filters from water samples were similar in physical appearance and general composition. (See SEM x-ray of particles, Figure 12). The ESCA analysis revealed only an indication of slightly higher carbon content on a 16 hour water washed wafer as on an unwashed wafer (see Table IV). High resolution ESCA (Table V) revealed the presence of hydrocarbon, ether, and ester functional groups in the organic composition on all four samples studied in Table IV. An increase in the ether functional group was found on the water washed wafer.

The presence of nitrogen and fluorine seen for the washed wafer in Table IV was interesting enough to repeat the test for a longer period of time and try to elucidate the bonding for the nitrogen. The second test by ESCA revealed trace concentrations of organic nitrogen as NR_3/NR_4^+. After ion etching of half of the organic layer, the nitrogen containing organic disappeared.

Table IV. ESCA Results: Elemental composition of the surface (approximately the top 100A) of each sample expressed in atomic percent units.

Spectrum Number	Sample Description	C	O	Si	N	F	Cl
1	Bare SiO$_2$ "as received"	10.	57.	33.	--	--	--
2	Bare SiO$_2$ after 50Å etch	0.9	58.	42.	--	--	--
3	Bare SiO$_2$ + wash "as received"	18.	51.	30.	0.2	0.6	--
4	Bare SiO$_2$ + wash after 50Å etch	1.0	59.	40.	--	--	--
5	Polysilicon "as received"	15.	33.	52.	--	--	--
6	Polysilicon after 50Å etch	--	2.7	97.	--	--	--
7	Polysilicon + back etch "as received"	14.	55.	30.	--	--	1.4
8	Polysilicon + back etch after 50Å etch	1.0	59.	40.	--	--	--

Table V. High Resolution ESCA Data: Binding energies, atom percentages and peak assignments. (Binding energies were corrected to the binding energy of the C(1s) signal at 284.6 eV. Atom percentages were calculated from the high resolution data. Peak assignments were based on the binding energies of reference compounds.)

Sample Description

	C_1	C_2	C_3	Si_1	Si_2	O_1	Cl_1
Bare SiO$_2$							
Binding Energies (eV)	284.6	286.1	289.0	---	103.3	532.6	---
Atom Percents	7.4	2.0	0.6	---	33.0	57.0	---
Bare SiO$_2$ + wash "as-received"							
Binding Energies (eV)	284.6	286.1	---	---	102.9	532.2	---
Atom Percents	13.0	5.0	---	---	30.0	51.0	---
Polysilicon "as-received"							
Binding Energies (eV)	284.6	286.1	288.9	99.1	102.8	532.2	---
Atom Percents	11.0	3.1	0.8	40.0	12.0	33.0	---
Polysilicon + back etch "as-received"							
Binding Energies (eV)	284.6	286.1	289.0	---	103.0	532.4	199.9
Atom Percents	11.0	2.8	0.7	---	30.0	55.0	1.4

C1 = C-R (R = C, H)   Si$_1$ = Si°   O$_1$ = SiO$_2$, C-O

C2 = C-OR   Si$_2$ = SiO$_2$ Cl= C-Cl

C3 = O=C-OR

Temperature programmable mass spectrometry revealed the presence of 2-butyl-N-butyrate on the final 0.2µm filter and that the sanitization solution contained $\emptyset-CH_2-$, $\emptyset-(CH_2)_3-$, $-CH_2N(CH_3)_2$, and compounds with molecular weights of 213, 241, and 269.

The next thoughts concerning the microcontaminants were directed more to the peculiarity of the abundance of 0.2µm and smaller particles with an essentially total absence of larger particles. This is a very atypical situation in pure water. Either the filters were functioning absolutely down to 0.2µm size particles or these particles were from a source where the size was controlled. These thoughts lead to two possibilities. One was that the efficiency of the final 0.2µm filter allowed only passages of 0.2µ particles. The other is the possibility that the particles were created by precipitation.

The presence of other material such as the amine/quaternary amine directed attention to the sanitizing solution (trade name "Ditrol") which is a quaternary ammonium chloride bacteriocidal solution commonly used to clean pipes. The presence of the butyl butyrate was confusing.

The final filters were examined using SEM. With water having 100-150K particles/l, these filters should have been loaded with larger particles. A review of the final filters indicated that they did not contain any large particles which are always found on the charged filter of the type used by this company nor any smaller <0.3µm particles usually found in arduous path type depth filters which these were. This finding lead to the idea that the small <0.3µm charged particles literally were rejected by the filter; and possibly, such particles could tie in with the positive charge that was accumulated on the wafers when they were exposed to water all night as was revealed by ESCA during the analysis of the surface.

Following the precipitation hypothesis, a summary was made of materials used in the water system before the surge, primarily a quaternary ammonium chloride biocide. Reverse osmosis sizing materials were also used. Lastly, at the time of the surge caused by a broken pipe, a material was spread throughout the fabrication area to absorb the water. All of these materials became possible candidates for water contamination, but the most suspicious was the Ditrol.

A study of the Ditrol was done both chemically and physically pursuing the precipitation hypothesis. Upon filtering the Ditrol and evaluating the filter by SEM, it was discovered that the Ditrol was full of agglomerated particles (millions/l) (see Figures 9 to 12). Later it was found that this sample of Ditrol was the diluted solution actually used for the cleanout indicating that the original solution had 20 times that many particles.

A small volume of Ditrol was placed in 100 ml of pure water, shaken vigorously, and filtered onto a Nuclepore disc filter for SEM evaluation. The particles looked like those seen from the water system (see Figures 9 and 10). The x-ray of these particles revealed they were organic and contained some iron (see Figure 12).

A thorough evaluation of many particles from the water system revealed that they contained iron. There is always some concern when iron is found in an x-ray unit. It may be coming from the SEM or the planchet. However, Ditrol and the water samples analyzed on a carbon planchet still revealed iron.

Knowing that the particle of concern was organic, flocculent, generally <0.2µm in size and physically resembled the particles in the Ditrol

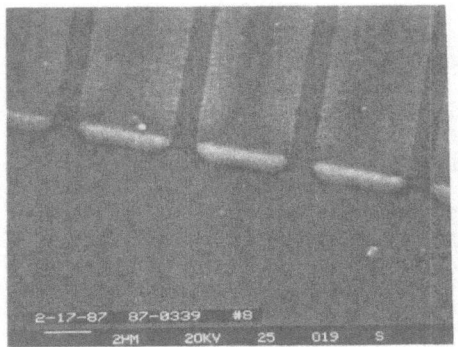

Figure 7. Particles on Product Wafer.

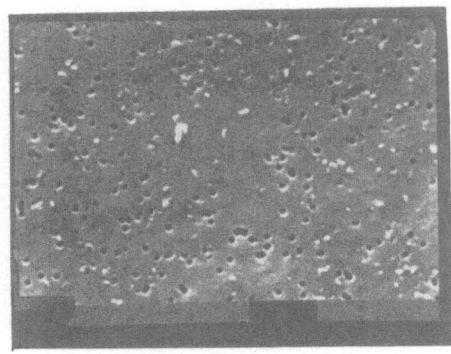

Figure 8. Particles from pure water.

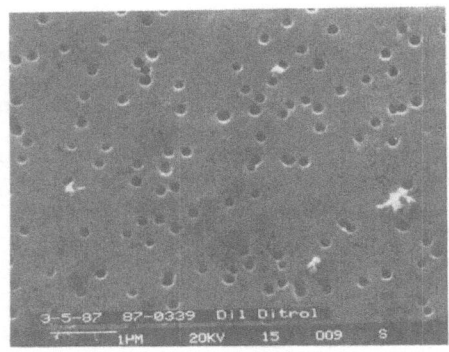

Figure 9. Particles from Ditrol
diluted 1-2000.

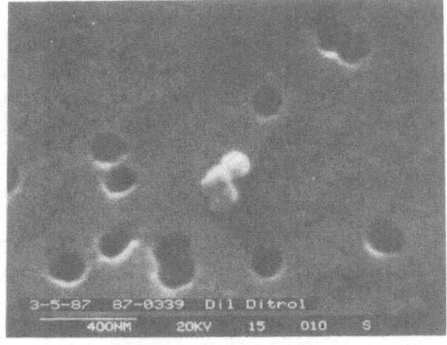

Figure 10. Enlargement of particle
from Fig. 9.

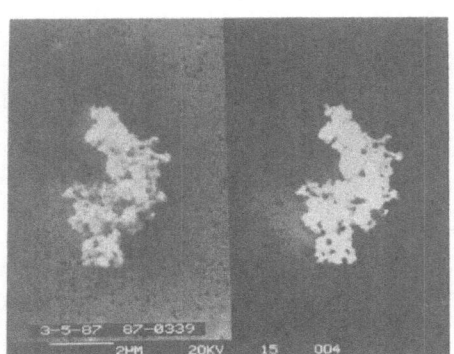

Figure 11. Large particle from Ditrol
used to sterilize the pure
water system. Left side
x-ray; right side Z number
mode.

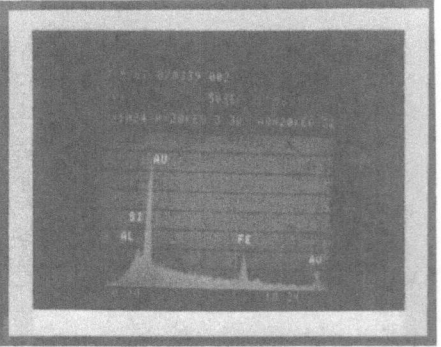

Figure 12. X-ray of Ditrol particles.
Similar spectra were also
obtained from particles
removed from pure water
and also those removed
from the wafer.

solution, it was hopeful that McCrone Associates, Inc., who are well known for their work in particle identification, could use their laser FTIR to elucidate the composition of the particles on the wafers, on the SEM filters used to collect samples from the water, and in the Ditrol. Samples of all three were sent to them.

For five weeks the evaluation went nowhere due to the following reasons:

1. During the evaluation at our laboratory and the packaging and shipping process, the samples picked up more particles. These particles mislead the McCrone staff.

2. When many particles exist on a surface, the larger particles which one sees readily draws attention from the smaller ones which often even go undetected.

3. The FTIR could not be used for identification because the particles were <1μm and this tool cannot elucidate the composition of particles smaller than 20μm.

4. Particles had to be removed from the surface and placed onto another surface to be studied by transmission electron microscopy (TEM). This task is enormously difficult and time consuming and particles after removal are frequently lost before analysis.

5. Describing particles was difficult and under optical viewing looked considerably different from what was seen on the SEM pictures which were provided.

6. The McCrone staff were having difficulty finding a statistical number of similar particles upon which conclusions could be drawn.

Recognizing that the results received from the McCrone staff (see Table VI) during this time did not agree with those found by our laboratory, directions continued to be given to focus on the <0.2μm down to 0.05μm particles. Eventually principals from both laboratories met to discuss the problem, develop a new approach, and work together until all parties agreed that they were working on the right particles.

The easier part of the problem was the evaluation of the particles removed from the Ditrol because they had agglomerated and were a much larger size. The grain size in the agglomerates was approximately 50 nm, the same as seen on the wafer. The TEM revealed iron in the carbonaceous mass.

To locate the particles on the wafer more easily, a portion of the wafer was gold coated and viewed again. The concentration of particles was found to be approximately 13/mm^2 of submicron particles mostly smaller than 0.2μm. Due to the small size a replication technique using a polycarbonate film was used to remove them. Twenty one particles were carefully marked and replicated. Of these, sixteen were successfully removed and seven analyzed (see results Table VII). Pictures of these particles and the Ditrol are shown in Figures 13 through 16. Most of the particles in the submicron range have the grainy structure found in the contaminant in the Ditrol (see Figure 17) and had an elemental composition similar to that shown in the TEM spectrum of Figure 18.

Although the evidence is not absolutely conclusive and the particle composition still unknown, it is believed by all analysts involved that the data does allow for a conclusion to be drawn. The conclusion drawn was that the contaminant came from the Ditrol. It is believed that the

Table VI. AEM* Analysis of Particles Removed from the Washed Wafer
(Particles Analyzed on a Thin Carbon Film)
(Particles Removed Individually)

Particle ID#	Size of Particle (μm)	Grain Size (nm)	Composition	Comments
1	1.2	Not determined	carbon	--
2	1.5	50	Low Z, possibly C, (60 nm Cu inclusions	Similar to Ditrol (Cu is a contaminant)
3	4.0	Not determined	Major: Si, Ca, Fe, Al, K Minor: Cl, Zn S	A non-homogeneous particle
4	0.7		Major: Fe	Fe oxide rust
5	1.5	80	Major: Ca, S	Gypsum

* AEM = Analytical Electron Microscopy

Table VII. AEM* Analysis of Submicron Particles on Washed Wafer Sample
(Particles Analyzed on a Thin Carbon Coated Polycarbonate Film)

Particle ID#	Size of Particle (μm)	Grain Size (nm)	Composition	Comments
1	0.5	60	Low Z, probably C	Similar to Ditrol
3	0.5	50	Low Z, probably C	Similar to Ditrol
6A	0.2	60	Low Z, probably C	Similar to Ditrol
6B	0.3	undetermined	Low Z, probably C	Similar to Ditrol
12	0.6	undetermined	Fe and Cl compd	Contaminant, possibly in Ditrol
19	0.5	60	Low Z, probably C	Similar to Ditrol
20	0.25	60	Low Z, probably C	Similar to Ditrol

* AEM = Analytical Electron Microscopy

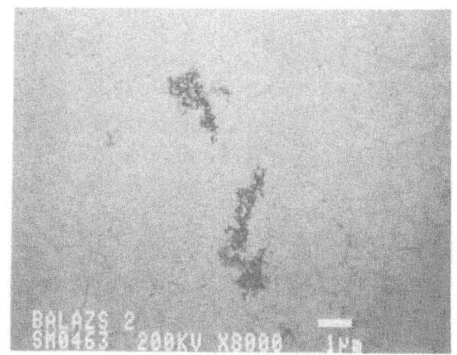

Figure 13

Figure 14

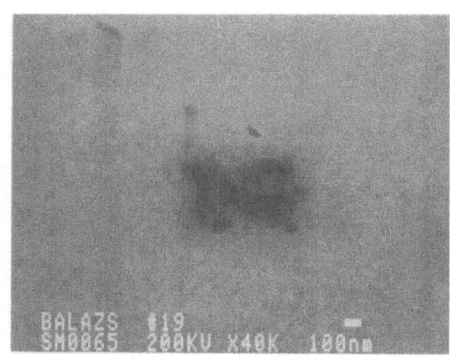

Figure 15

Figure 16

Figures 13-16. Several particles that were removed from wafer by McCrone.

Figure 17. Ditrol particle isolated and evaluated by McCrone.

surge caused a high pressure drop pulling residue Ditrol from deadly, unrinsed valves and other areas which were later located, flushed, or removed.

This problem illustrates some painful realities, however, which are:

1. Facilities operators and managers will readily flush a high purity water system with products that have not been analyzed nor understood and that do not have specifications written for them.

2. Particles undetected by in-line laser counters that are smaller than 0.2μm can completely shut down a plant.

3. Conventional particle evaluation did not indicate a particle problem and the assumption that a particle distribution curve will reveal problems is incorrect. In this case the count, which was believed to have been as high as 1 million particles/liter initially, went undetected because there were essentially no large particles.

4. The analysis and particle identification work for this problem was expensive, costing approximately $35,000.00.

5. It takes considerable experience to approach and resolve particle problems appropriately. It requires "experts" in the field of particle identification to resolve problems appropriately.

6. It takes a total cooperative team effort. This problem would not have been resolved if the total team from five different companies had not worked together and communicated with each other often, frequently daily over a 3 month period of time.

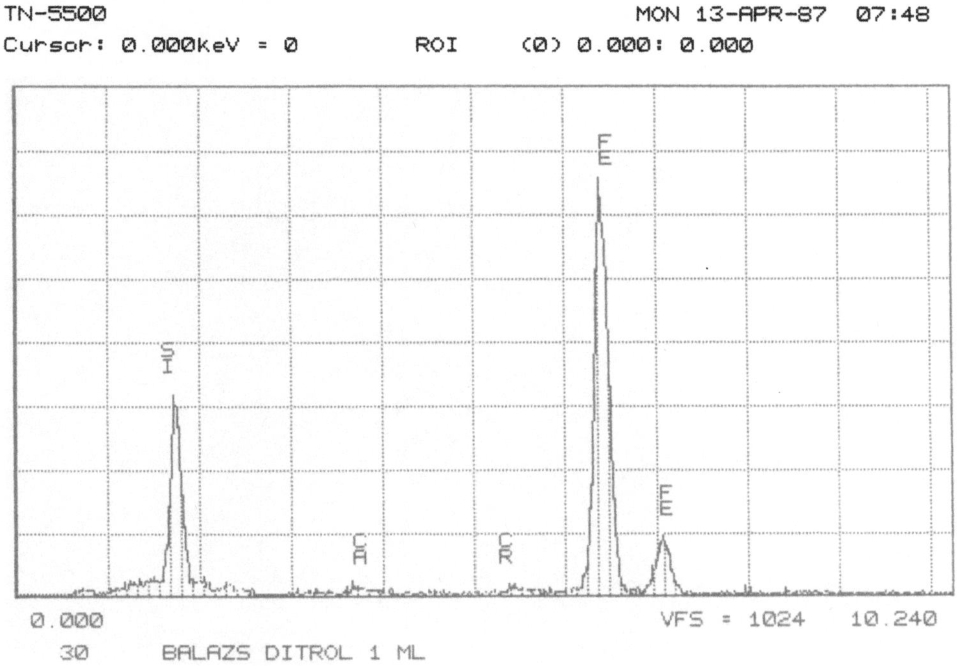

TN-5500   MON 13-APR-87 07:48
Cursor: 0.000keV = 0   ROI   (0) 0.000: 0.000

```
  0.000                                    VFS = 1024    10.240
    30      BALAZS DITROL 1 ML
```

Figure 18. TEM Results from McCrone of Ditrol particle.

CONCLUSION

Particle identification in pure water has become difficult and often impossible. Each new generation of IC products makes the understanding and identification of particles more imperative. It is time to pull in all the talent available no matter what the price. In reality, that price is the lowest fee on the crisis budget when compared to product loss, labor, and the replacement of parts that frequently do not need replacing.

New techniques for accumulating particles for identification need to be developed. It is time for industry to cooperate in R&D projects with companies and individuals who are experienced in particle identification to develop these techniques. To do less is too costly to IC manufacturers.

ACKNOWLEDGEMENT

The author wishes to acknowledge the cooperation and fine work done by Ann Teetsov and John Bradley from McCrone Associates.

REFERENCES

1. B. Stanczyk and T. Chu, The value of SEM particle counting in monitoring DI water, Proceedings of Semiconductor Pure Water Conference, Santa Clara, 1987, p. 194

2. R. Hango, B. Eldred, and B. Frith, D.I. Water: A common sense approach, Proceedings of Semiconductor Pure Water Conference, San Jose, 1984, p. 127

NON-POISSON MODELS OF PARTICLE COUNTING

R.A. Van Slooten and M.L. Malczewski

Union Carbide Corporation
Linde Division
175 East Park Drive
Tonawanda, New York 14151-0044

Repeated measurements of the number of particles in a fixed volume of high purity process gas give different values under essentially identical conditions. In this situation, it is necessary to treat the number of particle counts as a random variable, and statistical procedures are required to determine if the particle count meets specification within an acceptable confidence limit.

The counting process has been previously represented by a constant parameter Poisson model. However, occasional showers of particles observed in test data are inconsistent with this model. In situations where showers of particles persist, the procedure for computing the average number of counts will be correct but the precision of the estimate will be affected. This paper examines the application of a class of non-Poisson counting processes for modeling particle counting. Application to computer generated particle count data as well as actual test data are reported.

1. INTRODUCTION

Control of particulate contamination in process gases is a critical requirement in the manufacture of VLSI semiconductor devices. To control particulate contamination requires the ability to measure particle concentrations at the extremely low levels provided through the use of well-engineered filtration systems. At these concentrations, statistical analysis of the data is a necessity. This applies not only to data obtained from a test, but also to the background counts characteristic of the particle counters.

A constant parameter Poisson model of the counting process has been applied to low level particulate counts from a filtered process gas stream [1]. The article develops equations which allow for the estimation of the average particulate concentration λ and its variance $\sigma_\lambda{}^2$. In addition, the model allows for the correction of the test data for counter background which also is described by a Poisson process.

If particulate test data do not follow a Poisson model, then the equations for the statistics described above do not accurately characterize the gas. During a test on a filtered process gas line, the most common non-Poisson event is the occasional occurrence of a particulate shower or spike. Some of these particulate spikes are found to have an assignable cause, such as loss of sample gas flow or voltage transients appearing on the AC power line to the counters. In this case, the spikes are due to an event not related to the process gas particulate concentration and can be eliminated when treating the data. However, in most cases no assignable cause can be found for observed particulate spikes. In such cases, these spikes must be assumed to be due to actual increases in the particulate content of the process gas and must be included when analyzing the data. It is anticipated that such events will increase the uncertainty in the estimated particulate concentration.

Inclusion of the particulate spikes results in the counting process becoming non-Poisson in nature. This paper develops a non-Poisson theoretical model of the counting process which more closely represents a test in which spikes occur. Equations are developed to estimate the average particulate concentration in the process gas and its standard deviation. Comparisons to the corresponding Poisson model are also presented. Conditions are derived for which the non-Poisson distribution can be approximated by a Gaussian or normal distribution. Finally, compliance tests utilizing the non-Poisson model and the normal approximation are discussed.

Data are presented from three different sources. Data are obtained from measurement of counter backgrounds, which indicate that these measurements are best modeled by a strictly Poisson counting process. Second, data generated synthetically from the non-Poisson model by computer simulation are discussed. Finally, actual test data are presented and treated by the non-Poisson statistical analysis.

All experimental data discussed in the paper were collected using a laser particle counter Model LPC-101 (Particle Measuring Systems, Boulder, CO). Field data were collected according to the experimental methodology presented previously. [2]

2. COUNTER BACKGROUND

The background counts generated from the particle counters must be taken into account as they make a sizable contribution to the number of counts measured when observing filtered process gases. The most prevalent source of this counter background is electronic noise which occasionally causes the baseline to cross the smallest size channel in the instrument, resulting in a false count. Channels which size larger particles have higher thresholds which are rarely crossed by the baseline. The result is that these channels have little or no background associated with them.

Background measurements are carried out by drawing in air through a filter at the sampling rate of the instrument. The filter must be rated to remove all particles which can be detected by the counter. The assumption is then made that the filter provides "particulate-free gas" to the instrument and any counts observed are a result of the inherent counter background.

If the observed noise counts follow a Poisson distribution, then the interarrival times (or volumes), of the individual counts should follow an exponential probability distribution [3] given by:

$$f_X(x) = \frac{1}{k} \epsilon^{-x/k}$$

where X is the interarrival time (or volume) and the parameter k is the mean interarrival time (or volume) for the process. The cumulative distribution function (CDF) is given by:

$$F_X(x) = \int_0^x f_X(\xi)\, d\xi = 1 - \epsilon^{-x/k}$$

Normally, particulate sampling is done on a fixed sampling interval and as a result the interarrival times cannot be observed directly. However, the LPC-101 counters can be set to alarm on a single particle event, and generate a printout. When operated in this mode, the reported volume is the interarrival volume for that particle. The resolution of the counter is 6 sec. (0.01 ft^3) so that interarrival volumes of less than 0.01 ft^3 will be recorded as simultaneous events. Although theory predicts zero probability of getting exactly zero for the interarrival volume, the resolution limit permits a finite probability of getting one or more counts in a single alarm report. In fact, multiple counts may be observed whenever the interarrival volume is less than 0.01 ft^3. To facilitate data storage, these counts were handled by assigning one of the counts to the reported volume and the remaining counts a zero interarrival volume.

Two LPC-101 counters were put on a background test as described above, using the alarm sequence to directly observe the interarrival volumes. The results are shown in Figures 1 and 2. In both figures, the top plot is a histogram of the interarrival volume and the bottom plot is a sample cumulative distribution function. The abscissa in the latter plot is the interarrival time which is obtained by multiplying the interarrival volume by 600 sec/ft^3. The first counter gave a mean interarrival time of 109.0 sec (5.50 counts/ft^3). Both the experimental and theoretical CDF are shown for comparison, and the data fit the theoretical CDF very well. A six cell Chi-Square test confirmed that the data fit the exponential distribution at a 95 percent confidence level.

The data for a second LPC-101 are shown in Figure 2. Improved noise suppression circuitry and photodiode result in lower electronic noise and, as a result, a higher interarrival time (2143 sec.) and a lower background (0.280 counts/ft^3). In this case, there is a discrepency between the theoretical and experimental CDF, primarily due to a larger number of particles with zero interarrival volumes. The data came close, but failed the Chi-squared test at a 95 percent confidence level. This suggests that the attenuation of the electronic noise may have uncovered another process. Some of these multiple occurrences, which are non-Poisson, may result from actual particle counts coming from either the filter or the connecting tubing.

In spite of the foregoing discrepancy, the Poisson model closely approximates the data from most background tests. From this model and a knowledge of the counter background and sampling rate, we can predict the sampled volume necessary to achieve a given precision in the measured background count. The standard deviation of this measurement is

$$s_\lambda = \left[\frac{\hat{\lambda}}{V} \right]^{1/2}$$

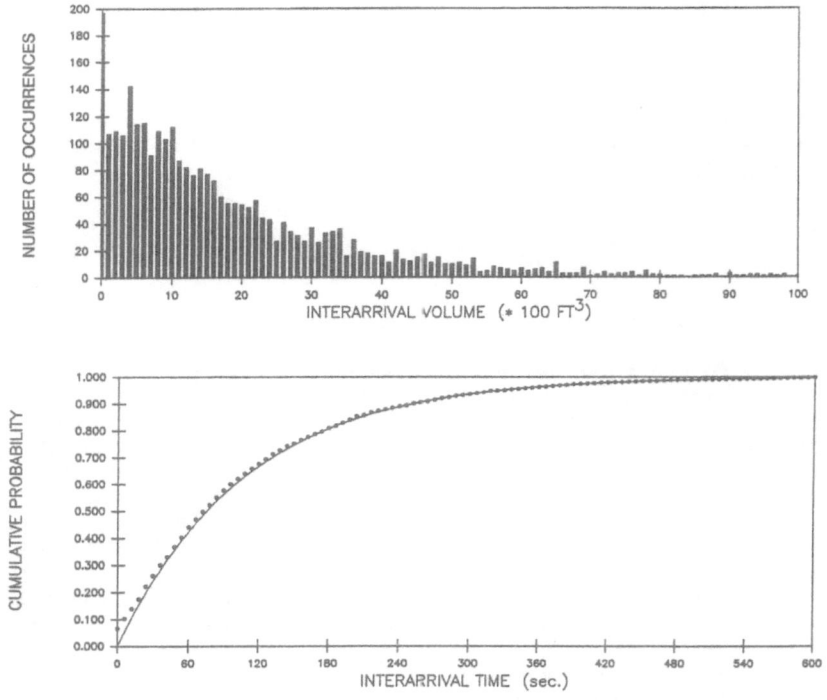

Figure 1.   Histogram and Sample Cumulative Distribution Function for
Background Interarrival Volumes from First LPC-101 Counter.

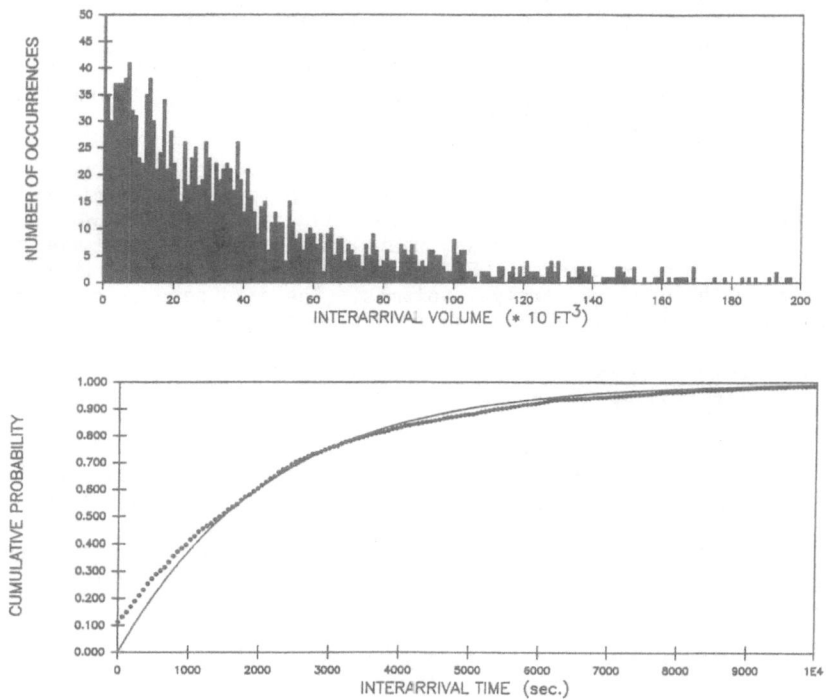

Figure 2.   Histogram and Sample Cumulative Distribution Function for
Background Interarrival Volumes from Second LPC-101 Counter.

Solved for the volume, it gives

$$V = \frac{\hat{\lambda}}{S_\lambda^2}$$

The manufacturer's stated "maximum" background count rate is $\lambda = 15$ counts per cubic foot. If the desired standard deviation is 0.1 counts per cubic foot, then one must sample 1500 ft³ of gas to achieve it, given the maximum background counting rate. The sample flow for the instrument is 6 ft³ and this volume may be sampled in a time period of 250 hours. Given the observed background count rate of 5.50 counts per cubic foot, a 0.1 count per cubic foot standard deviation can be achieved in 550 ft³ or 92 hours.

3. NON-POISSON COUNTING MODEL

Figure 3 shows a typical sample function from a particle counting test performed on a high purity nitrogen gas supply line. The gas is sampled continuously and the number of particles with diameters greater than 0.1 micron is recorded at intervals of one cubic foot of gas. Accordingly, each vertical line in the plot represents the particle count in one cubic foot of gas. Experimentally, a one cubic foot sampling rate provides the best compromise between time resolution and the amount of data which would need to be collected if using a short sampling period over a relatively long term test. Since sampling is normally carried out on fixed volume increments of one cubic foot, it is usually impossible to resolve individual counts and compute interarrival times or volumes from field data.

In earlier work, the number of particle counts in a standard volume increment was modeled as a Poisson distributed random variable. One of the basic assumptions underlying this model is that the probability of having two or more particle arrivals in a small volume increment is negligible. Figure 3 shows several occasions when 15 or more particles were observed in a one cubic foot volume increment, which is a very unlikely event for a Poisson distribution with a mean of 5 counts per cubic foot. To remove this contradiction, a new model is proposed in which the counting process is treated as the sum of a Poisson process and a batch or multiple arrival process.

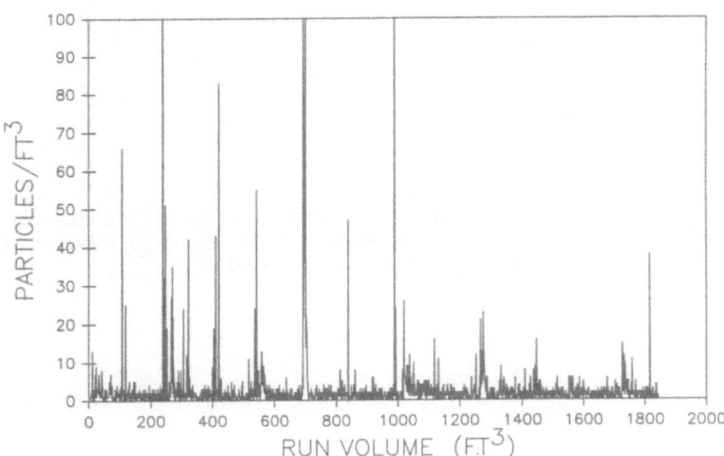

Figure 3. Sample Function from Field Data showing Total Number of Particulate Counts in One Cubic Foot Samples.

3.1 Mathematical Model

The new counting process can be expressed mathematically as follows:

$$Z(0,V) = K(0,V) + \sum_{i=1}^{N(0,V)} X_i + B(0,V) \tag{1}$$

where $Z(0,V)$ represents the total number of background corrected or real particle counts occurring in the volume interval from 0 to V, $K(0,V)$ is the Poisson component of the counting process and $B(0,V)$ is the background noise in the laser spectrometer. Since the background noise is modeled by a Poisson process, $B(0,V)$ can be combined with $K(0,V)$ to form a new Poisson process with an adjusted parameter. Consequently, $B(0,V)$ will not be included in the development of this new model. The second term in Equation (1) represents a random sum of independent random variables X_i. These variables define the number of particle counts occurring in batch arrivals; in future discussions, they will be referred to as spikes. Since the number of spikes in the interval $(0,V)$ is unpredictable, the sum extends from 1 to the random number $N(0,V)$. A typical sample function for the process $Z(0,V)$ is shown in Figure 4. Unit jumps in the function represent Poisson counts while all larger jumps represent spikes.

To develop useful statistical tools for analysis of particle counting data, further assumptions about the terms in Equation (1) are required. They are as follows:

1. $K(0,V)$ is an independent Poisson counting process with parameter $\lambda_1 V$; λ_1 defines the average number of particle counts/unit volume, excluding spikes. Its probability distribution is

$$p[K(0,V) = k] = \frac{(\lambda_1 V)^k \epsilon^{-\lambda_1 V}}{k!}$$

where $k = 0, 1, 2...$

2. $N(0,V)$ is an independent Poisson distributed random variable with parameter $\lambda_2 V$; λ_2 defines the average number of spikes/unit volume. Its probability distribution is:

$$p[N(0,V) = n] = \frac{(\lambda_2 V)^n \epsilon^{-\lambda_2 V}}{n!}$$

with $n = 0, 1, 2...$

3. The X_i are independent and identically distributed Poisson random variables with parameter θ; θ represents the average number of counts in a spike. Their common probability distribution is:

$$p[X_i = 1] = \frac{\theta^1 \epsilon^{-\theta}}{1!}$$

with $1 = 0, 1, 2...$

Other discrete distributions or even continuous distributions can be used to model X_i. This point will be examined further in Section 3.6.

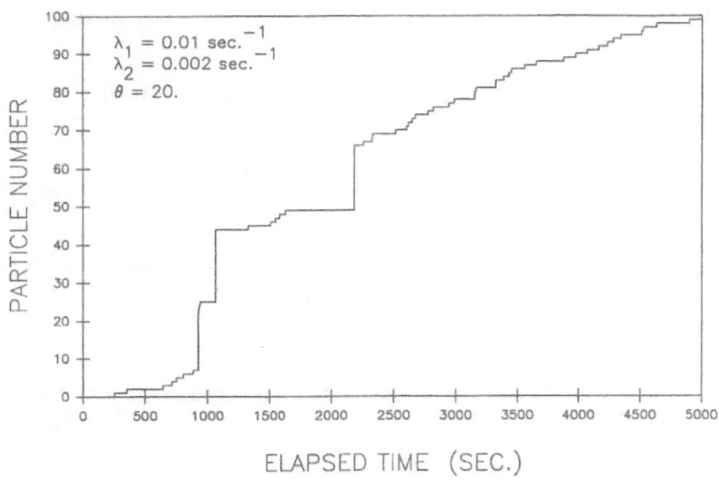

Figure 4. Sample Function from the non-Poisson Model of the Counting
Process Z (0,V).

3.2 Probability Distribution Function

The probability distribution of the new counting process Z(0,V) can be
derived by straightforward application of certain theorems from probability
theory. To start, let S(0,V) represent the random sum in Equation (1).
Then the probability distribution of S(0,V) is given by [4]

$$P\{S(0,V) = k\} = \sum_{n=1}^{\infty} P\{N(0,V) = n\}\, P\{X_1 + X_2 + \cdots + X_n = k\} \tag{2}$$

The second factor is the conditional probability of the event

$$\{X_1 + X_2 + \cdots + X_n = k\}$$

given that n terms are included in the sum; that is N(0,V)=n. The product
gives the probability of the joint event

$$P\{N(0,V) = n\}P\{X_1 + X_2 + \cdots X_n = k\}$$

and the sum over all integer values of n gives the unconditional
probability of the event $\{$ S(0,V)=k $\}$. Equation (2) is a general
relationship independent of the specific distributions for random variables
X_i and N(0,V).

Using assumption 3 and referring to a well known theorem from
probability theory [4], we note that the sum $X_1 + X_2 + \cdots X_n$ is
Poisson distributed with parameter $n\theta$. Consequently, the conditional
probability is given by

$$P\{X_1 + X_2 + \cdots X_n = k\} = \frac{\epsilon^{-n\theta}(n\theta)^k}{k!} \tag{3}$$

57

Substituting this result into Equation (2) and invoking assumption 2 gives the probability distribution for the random sum

$$P\{S(0,V)=k\} = \sum_{n=0}^{\infty} \frac{\epsilon^{-(\lambda_2 V + n\theta)}(\lambda_2 V)^n (n\theta)^k}{n!\, k!} \tag{4}$$

Note that extending the sum to n=0 adds nothing new to the right hand side because $(n\theta)^k = 0$.

The complete model, defined by Equation (1), combines a Poisson counting process $K(0,V)$ with the random sum $S(0,V)$

$$Z(0,V) = K(0,V) + S(0,V)$$

Again, referring to probability theory, the distribution of $Z(0,V)$ can be determined by convolution summation of the probability distributions for $K(0,V)$ and $S(0,V)$. To facilitate this operation, let

$$P\{Z(0,V) = l\} = c_l$$

$$P\{K(0,V) = i\} = a_i$$

$$P\{S(0,V) = k\} = b_k$$

The values of c_l are obtained from the convolution summation formula [5]

$$c_l = \sum_{i=0}^{l} a_i\, b_{l-i} \tag{5}$$

This gives the desired probability distribution

$$P\{Z(0,V) = l\} = \sum_{n=0}^{\infty} \frac{\epsilon^{-\lambda_2 V}(\lambda_2 V)^n}{n!} \frac{\epsilon^{-(\lambda_1 V + n\theta)}(\lambda_1 V + n\theta)^l}{l!} \tag{6}$$

where $l = 0, 1, 2, \ldots$ Note that this is not a Poisson distribution; consequently, the counting process is non-Poisson and, as indicated by Equation (6), its probability distribution depends on three parameters λ_1, λ_2 and θ.

Plots of this probability distribution are shown in Figures 5, 6 and 7 for three different sets of parameters. The values of the parameters are given on the plots. Figure 5 shows that the distribution can be bimodal; apparently this happens when θ is large compared to the other parameters. The log percent probability is plotted in this figure to emphasize the bimodal character of the distribution. Figure 6 shows the distribution for a moderate value of θ wherein the number of counts due to spikes is about equal to the number from the Poisson process. Figure 7 suggests that for certain values of the parameters the distribution looks like a normal curve. In fact, later it will be shown that this distribution is asymptotically normal when V is large.

58

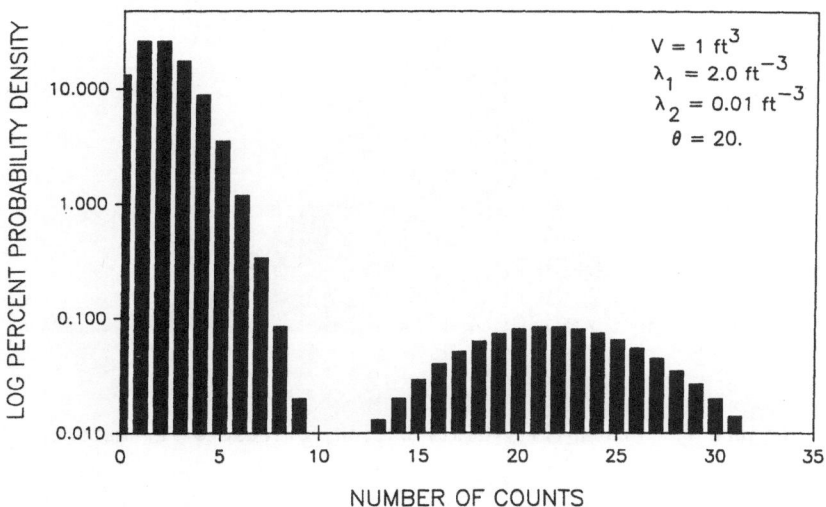

Figure 5. Probability Distribution for Case 1 Parameters.

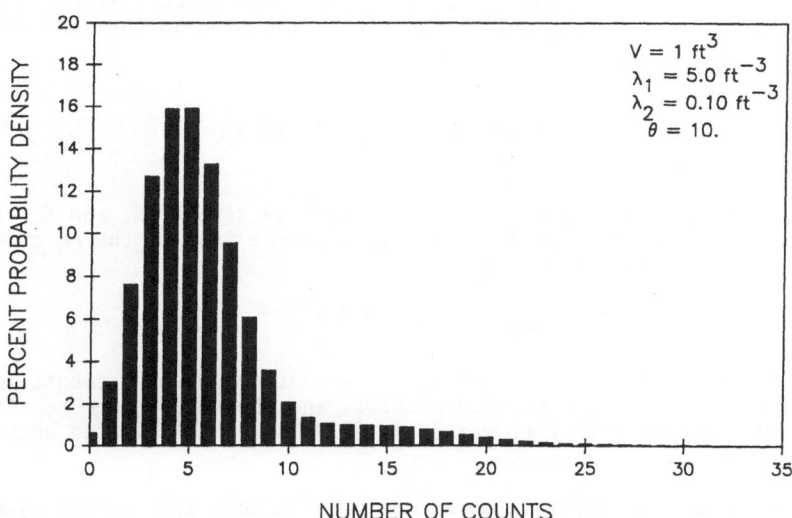

Figure 6. Probability Distribution for Case 2 Parameters.

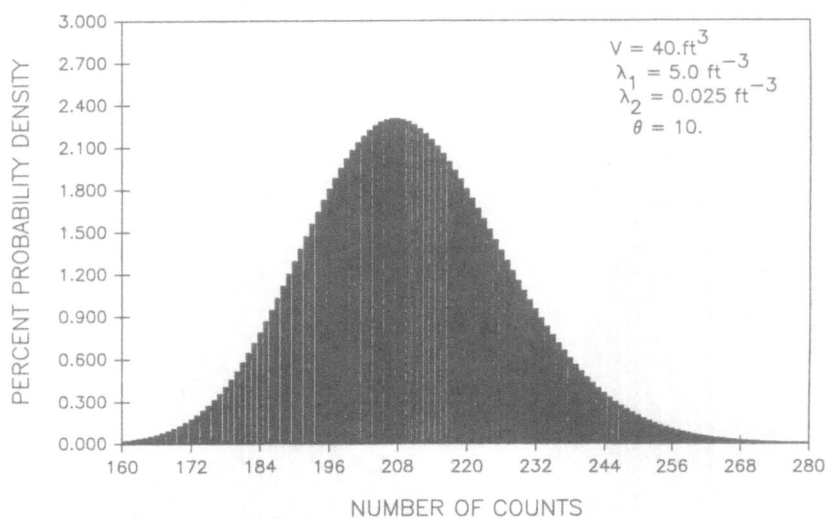

Figure 7. Probability Distribution for Case 3 Parameters.

3.3 Mean and Variance of Counting Process

The moments of the counting process $Z(0,V)$ can be calculated directly from the probability distribution function. The mean is obtained from the equation

$$\mu_z = E\{Z(0,V) = 1\} = \sum_{n=0}^{\infty} 1\,P\{Z(0,V) = 1\}$$

Substituting Equation (5) for the probability distribution, and following the steps outlined in reference 3 for finding the mean of the Poisson distribution, one can show that

$$\mu_z = \lambda_1 V + \lambda_2 V \theta \qquad (7)$$

The first term represents the average or expected number of counts in the interval $(0,V)$ due to the Poisson process, and the second term represents average or expected number of spikes in $(0,V)$ times the average number of counts per spike.

The variance of $Z(0,V)$ is obtained in a somewhat less direct manner. First, the expected value of

$$Z(0,V)\,[Z(0,V) - 1]$$

is computed from the definition

$$E\{Z(0,V)\,[Z(0,V) - 1]\} = \sum_{l=0}^{\infty} 1(1-1)\,P\{Z(0,V)=1\}$$

60

and then the variance is derived from the equation

$$\sigma_z^2 = E\{Z(0,V)\,[Z(0,V) - 1]\} + E\{Z(0,V)\} - E^2\{Z(0,V)\} \tag{8}$$

This gives

$$\sigma_z^2 = \lambda_1 V + \lambda_2 V \theta(\theta + 1) \tag{9}$$

Although the mean and variance are always equal for a Poisson process, here, we have a new term in the variance equation, $\lambda_2 V \theta^2$, which always increases the variance.

The moment generating function for the non-Poisson counting process is derived in Appendix A, and the mean and variance are obtained by differentiating this function. The same expressions for μ_z and $\sigma_z^{\;2}$ are obtained by this method.

3.4 Asymptotic Distribution of Counting Process

It was observed in discussing Figure 7 that for certain values of the parameters λ_1, λ_2, θ and V the probability distribution of the counting process looks like a normal curve. The purpose of this section is to determine conditions under which the distribution function, defined by Equation (6), converges to a normal distribution, and in the next section a quantitative index is established which tells when the normal distribution is a good approximation.

The convergence proof is based on the moment generating function, Equation (A.9), which is derived in the Appendix. To facilitate the derivation, let

$$Y = \frac{Z(0,V) - \mu_z}{\sigma_z} \tag{10}$$

This defines a new random variable with mean zero and variance one. The moment generating function for Y is

$$M_Y(t) = E\{\epsilon^{Yt}\} = \epsilon^{-\frac{\mu_z}{\sigma_z}t}\, M_z\!\left[\frac{t}{\sigma_z}\right] \tag{11}$$

Substituting Equation (A.9) for $M_z(t/\sigma_z)$ gives

$$M_Y(t) = \epsilon^{-\frac{\mu_z}{\sigma_z}t}\, \epsilon^{-\left\{\lambda_1 V\left[\epsilon^{-\frac{t}{\sigma_z}} - 1\right] + \lambda_2 V\left[\epsilon^{\theta\left(\epsilon^{-\frac{t}{\sigma_z}} - 1\right)}\right]\right\}} \tag{12}$$

Now expanding the exponentials in a power series and separating terms according to powers of t gives the desired result.

$$\begin{aligned}
M_Y(t) = \exp\Bigg[& \left[-\frac{\mu_z}{\sigma_z} + \frac{\lambda_1 V}{\sigma_z} + \frac{\lambda_2 \theta}{\sigma_z} \right] t \\
& + \frac{1}{2}\left[\frac{\lambda_1 V}{\sigma_z^{\;2}} + \frac{\lambda_2 V \theta(\theta+1)}{\sigma_z^{\;2}} \right] t^2 \\
& + \frac{1}{6}\left[\frac{\lambda_1 V}{\sigma_z^{\;3}} + \frac{\lambda_2 V \theta(\theta^2 + 3\theta + 1)}{\sigma_z^{\;3}} \right] t^3 + O(t^4) \Bigg]
\end{aligned} \tag{13}$$

But, as shown earlier,

$$\mu_z = \lambda_1 V + \lambda_2 V\theta$$

and

$$\sigma_z^2 = \lambda_1 V + \lambda_2 V\theta(\theta+1)$$

Hence,

$$M_Y(t) = \exp\left[\frac{1}{2} t^2 + \frac{1}{6} \frac{\lambda_1 V + \lambda_2 V\theta(\theta^2+3\theta+1)}{[\lambda_1 V + \lambda_2 V\theta(\theta+1)]^{3/2}} t^3 + \mathcal{O}(t^4) \right] \tag{14}$$

For fixed θ, the limit as V approaches   is

$$\lim_{V \longrightarrow \infty} M_Y(t;V) = \epsilon^{-\frac{1}{2} t^2} \tag{15}$$

This limit is the moment generating function for a normally distributed random variable with mean zero and variance one. It demonstrates that the probability distribution for the non-Poisson counting process will approach a normal distribution as the volume V approaches   for all finite values of the parameters λ_1, λ_2 and θ. This result is equivalent to the Central Limit Theorem; however, here the limit is taken for a random sum of discrete random variables.

3.5 Normal Approximation for Counting Process

Since in all physical measurements the volume V is finite, the normal distribution can serve only as an approximation to the non-Poisson distribution defined by Equation (6). For the normal distribution to be a good approximation, the parameters λ_1, λ_2 and θ and the volume V must satisfy certain functional constraints. Useful rules can be established in several different ways. Since we are ultimately interested in formulating a hypothesis test on the total number of counts per cubic foot of gas, at a high level of confidence, the normal density $f_Z(z)$ must closely approximate the discrete distribution $p_Z(k)$ in the upper tail as depicted in Figure 8. In the figure, this region is defined by $\{ z \geq 1.65\sigma_z \}$.

Here the desired functional relationship is determined by computing the probability of the event $\{ Z > z_\alpha \sigma_z \}$ for both distributions, and then requiring that the difference between these two numbers be less than some specified value ϵ. For the normal distribution, the probability is given by

$$P\{Z \geq z_\alpha\} = \int_{z_\alpha}^{\infty} f_Z(\xi)\, d\xi \tag{16}$$

This equation defines z_α; for example, if α is set to 0.05, then z_α equals 1.65. Hence, for the normal distribution

$$P\{Z \geq 1.65\sigma_z\} = 0.05$$

For the discrete distribution

$$P\{Z \geq z_\alpha \sigma_z\} = \sum_{k=k^*}^{\infty} p_Z(k) \tag{17}$$

62

where $k^* = \left[z_\alpha \sigma_Z \right] + 1$ and $\left[z_\alpha \sigma_Z \right]$ is the largest integer value less than $z_\alpha \sigma_Z$.

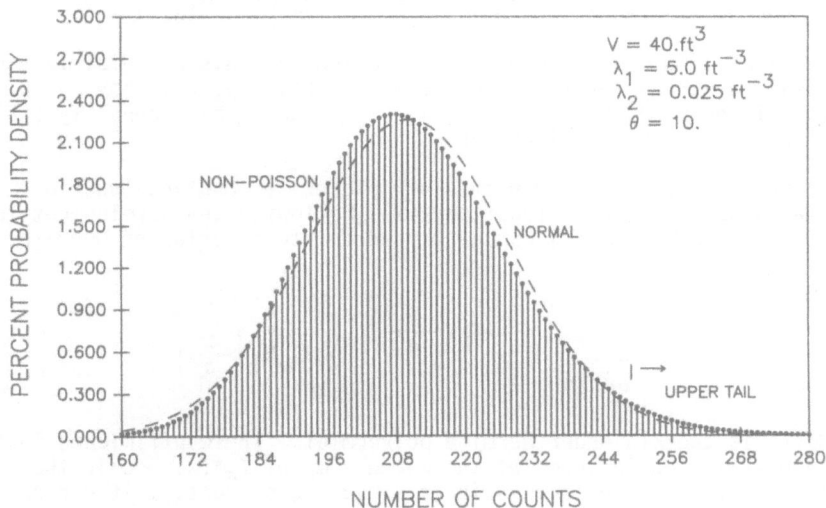

Figure 8. Comparison of the non-Poisson Distribution for Case 3 Parameters with the Normal Distribution.

The criterion for close approximation is that the difference between these two numbers be less than some specified value ε or, equivalently, some specified fraction of the normal probability α.

$$\left| \int_{z_\alpha \sigma_z}^{\infty} f_Z(\xi)\, d\xi \; - \; \sum_{k=k^*}^{\infty} p_Z(k) \right| < \rho\alpha \tag{18}$$

Although ρ can be arbitrarily small, the value used in the following development is 0.2.

The values of the parameters λ_1, λ_2, θ and V which ensure satisfaction of the above inequality provide a rule for determining when the normal approximation is acceptable. This rule, which assumes the form of a inequality among these parameters, is derived as follows.

Introduce a new random variable X defined by

$$X = Z(0,V) - \mu_z \tag{19}$$

where X has mean zero and variance $\sigma_X{}^2 = \sigma_Z{}^2$. This step simply centers the two distribution around the origin x=0. Let $f_X(x)$ denote the density function of the continuous normal distribution and $p_X(x)$ the distribution function of the new discrete counting process. To avoid mathematical difficulties in this derivation we need to replace $p_X(x)$ by a continuous distribution. Imagine a companion distribution $f_X^*(x)$ which is piecewise constant instead of discrete as illustrated in Figure 9. The probability of the event $\{ X = x_1 \}$ which for the discrete distribution is given by

$$P\{X = x_1\} = p_X(x_1)$$

now is spread over a unit interval surrounding x_1, and is replaced by

$$P\{x_1 - 0.5 < X < x_1 + 0.5\} = f_X^*(x_1) \cdot 1$$

The moments and moment generating function for this new distribution are the same as those for the discrete distribution $p_X(x)$. Any conditions imposed on the parameters λ_1, λ_2, θ and V by Inequality (18) will be true for both distributions.

Call the difference between this new piecewise constant function and the normal density an error function $\mathcal{E}(x)$ Following the development in references 4, 5 and 7, $\mathcal{E}(x)$ can be expanded into a series of Hermite polynomials.

$$\xi(x) = \frac{1}{\sigma\sqrt{2\pi}} \epsilon^{-\frac{1}{2}\left[\frac{x}{\sigma}\right]^2} \sum_{k=0}^{\infty} C_k H_k\left[\frac{x}{\sigma}\right] \tag{20}$$

where $H_k(x)$ is the kth-order Hermite polynomial. The coefficients in the series are evaluated in terms of the moments m_k of $f_X^*(x)$. Only the first nonzero term in the series is retained; consequently, it can be shown that [7]

$$f_X^*(x) \cong \frac{1}{\sigma\sqrt{2\pi}} \epsilon^{-\frac{1}{2}\left[\frac{x}{\sigma}\right]^2} \left[1 + \frac{m_3}{3!\,\sigma^3}\left(\frac{x^3}{\sigma^3} - \frac{3x}{\sigma}\right)\right] \tag{21}$$

where m_3 is the third moment of $f^*(x)$. The higher-order terms are neglected because we expect them to be very small when V is large. The third-order term, which is included, contains m_3/σ^3 and, as will be shown later, it is of the order $1/\sqrt{V}$. The next term, which is not included, contains m_4/σ^4 and it is of the order $1/V$. Consequently, for large V the third-order term is dominant. Since m_3 is the same as the third-order moment of the discrete distribution $p_X(x)$, it is easy to compute m_3 from the moment generating function of X

$$M_X(t) = E\{\epsilon^{Xt}\} = E\{\epsilon^{Y\sigma t}\} = M_Y(\sigma t) \tag{22}$$

Hence, from Equation (14), we get

$$M_X(t) = \exp\left\{\frac{1}{2}\sigma^2 t^2 + \frac{1}{6}\frac{\lambda_1 V + \lambda_2 V\theta(\theta^2 + 3\theta + 1)}{[\lambda_1 V + \lambda_2 V\theta(\theta+1)]^{3/2}}\sigma^3 t^3 + \mathcal{O}(t^4)\right\} \tag{23}$$

The moment is given by

$$m_3 = \left.\frac{d^3 M_X(t)}{d t^3}\right|_{t=0}$$

and it equals

$$m_3 = \frac{\lambda_1 V + \lambda_2 V\theta(\theta^2 + 3\theta + 1)}{[\lambda_1 V + \lambda_2 V\theta(\theta+1)]^{3/2}}\sigma^3 \tag{24}$$

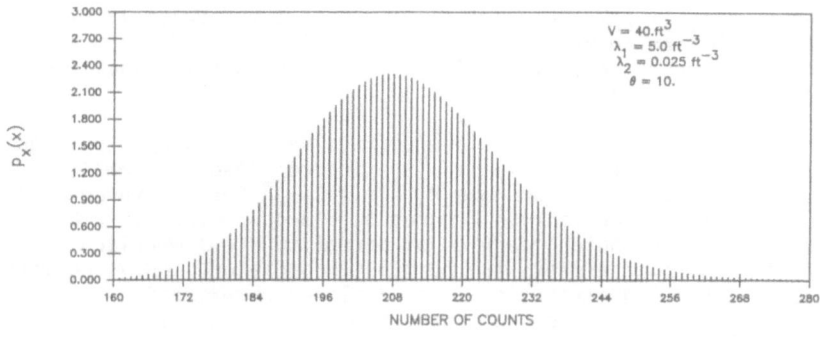

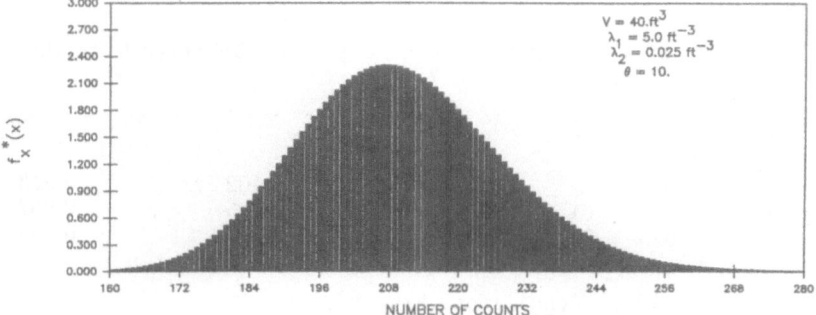

Figure 9. Comparison of the Discrete non-Poisson Distribution with the Continuous Piecewise-constant Companion Distribution.

At this point, the above equations can be substituted into the initial expression for functional approximation, Inequality (18). Replacing the discrete distribution by Equation (21) gives

$$\left| \int_{z_\alpha \sigma}^{\infty} \frac{1}{\sigma\sqrt{2\pi}} \epsilon^{-\frac{1}{2}\left[\frac{x}{\sigma}\right]^2} \left[\frac{m_3}{3!\,\sigma^3} \left[\frac{x^3}{\sigma^3} - \frac{3x}{\sigma} \right] \right] dx \right| < \rho\alpha$$

and carrying out the integrations yields

$$\frac{m_3}{6\sqrt{2\pi}\,\sigma^3} \left| z_\alpha^2 - 1 \right| \epsilon^{-\frac{1}{2}z_\alpha^2} < \rho\alpha \tag{25}$$

Substitute Equation (24) for m_3 and take $\alpha = 0.05$, $z_\alpha = 1.65$ and $\rho = 0.2$. Then we find the desired functional inequality is

$$\frac{\lambda_1 V + \lambda_2 V \theta(\theta^2 + 3\theta + 1)}{[\lambda_1 V + \lambda_2 V \theta(\theta + 1)]^{3/2}} \leq 0.341 \tag{26}$$

When the parameters λ_1, λ_2, θ and V satisfy this inequality, the normal density applies.

$$f_z(z) = \frac{1}{\sigma_z \sqrt{2\pi}} \epsilon^{-\frac{1}{2}\left[\frac{z - \mu_z}{\sigma_z}\right]^2} \tag{27}$$

with mean $\mu_z = \lambda_1 V + \lambda_2 V\theta$ and variance $\sigma_z{}^2 = \lambda_1 V + \lambda_2 V\theta (\theta+1)$ is a good approximation to the discrete distribution $P_Z(z)$ defined by Equation (6). Accordingly, the left hand side of this inequality will be referred to as a normality index (N.I.).

To demonstrate that distributions defined by the non-Poisson counting process which satisfy Inequality (26) are indeed close to normal, we compute the normality index and the $P\{Z > 1.65\sigma_z\}$ for the three discrete distributions plotted earlier in Figures 5, 6 and 7. The distribution parameters and the computed measures of normality are summarized in Table I. The third case meets the normality condition, and as depicted in Figure 6, the normal distribution gives a close approximation to the discrete distribution. Note that cases 1 and 2, which fail the normality condition, also fail the initial criterion on $P\{Z \geq 1.65\ \sigma_z\}$, defined by Equation (18).

Table I. Summary of Parameters for non-Poisson Probability Distributions Shown in Figures 5, 6 and 7.

CASE	λ_1	λ_2	θ	V	σ_z	N.I.	$P\{Z \leq 1.65\sigma_z\}$
1	2.0	0.01	20	1.0	2.49	6.102	0.013
2	5.0	0.10	10	1.0	4.00	2.125	0.067
3	5.0	0.025	10	40.0	17.61	0.277	0.054

3.6 Moments of Z(0,V) by Conditional Expectation

This section presents a third way to determine the mean and variance of the counting process Z(0,V) without assuming specific probability distributions for N, the number of spikes, or for X_i, the number of counts in the ith spike. These equations are derived from an identity in probability theory regarding conditional expected values. Let S(0,V) represent the random sum component of the counting process.

$$S(0,V) = \sum_{i=1}^{N(0,v)} X_i$$

Then the identity [5] states that

$$E\{S(0,V)\} = E\{E\{S(0,V)\,|\,N(0,V)\}\} \tag{28}$$

The inner expected value is on S but conditioned on a random value for N; it defines a function of random variable N. The outer expected-value is with respect to N.

When the random variables X_i are independent and identically distributed with mean μ_x and variance $\sigma_x{}^2$, Equation (28) leads to the following expressions for the moments of S(0,V).

$$E\{S(0,V)\} = \mu_x E\{N(0,V)\}$$

and

$$E\{S^2(0,V)\} = \mu_x{}^2 E\{N^2(0,V)\} + \sigma_x{}^2 E\{N(0,V)\} \tag{29}$$

Hence, the variance of S(0,V) is

$$\sigma_S{}^2 = \mu_x{}^2 \sigma_N{}^2 + \sigma_x{}^2 \mu_N \tag{30}$$

66

where μ_N and $\sigma_N{}^2$ are the mean and variance of N. Now combining the Poisson counting process with the random sum gives the desired moments of Z(0,V)

$$\mu_z = \mu_K + \mu_x \mu_N$$

$$\sigma_z{}^2 = \sigma_K{}^2 + \mu_x{}^2 \sigma_N{}^2 + \mu_N \sigma_x{}^2 \tag{31}$$

Here μ_K is the mean and $\sigma_K{}^2$ the variance of the Poisson counting process. These equations are general relationships which are valid independent of the distributions of N(0,V) and X_i. Note that the variance of Z(0,V) is resolved into three components: the variance of the Poisson process, a term containing the variance of N(0,V) and a term containing the variance of X_i. Typically in particulate testing μ_x is large, so that the second term is dominant.

When the distributions of N(0,V) and X_i are taken as Poisson, the new equations for μ_z and $\sigma_z{}^2$ reduce to the earlier equations. Since the Poisson distribution is well suited for modeling the number of spikes per unit volume, we may want to use its moments in Equations (31) but leave the distribution of X_i undefined. Then

$$\mu_z = \lambda_1 V + \lambda_2 V \mu_x$$

$$\sigma_z{}^2 = \lambda_1 V + \lambda_2 V(\mu_x{}^2 + \sigma_x{}^2) \tag{32}$$

With these equations, it is possible to model the number of counts in a spike by a discrete or continuous distribution. If a non-Poisson model is used, however, the probability distribution of Z(0,V) is no longer given by Equation (6). The normality condition may still hold, but this needs to be re-examined.

4. FORMULATION OF A COMPLIANCE TEST FOR PARTICULATES

There are at least two ways in which the non-Poisson counting process can be used to formulate a compliance test for particulates in process gases. First, we can estimate the values of the parameters λ_1, λ_2 and θ from test data obtained by sampling a fixed volume of gas, substitute these values into the probability distribution and compute the probability of exceeding a specified number of counts. Expressed mathematically, this means

$$P\{Z(0,V) > m\} = 1 - \sum_{l=0}^{m} p_z(l; \hat{\lambda}_1, \hat{\lambda}_2, \hat{\theta}, V) \tag{33}$$

where P_z (1, $\hat{\lambda}_1$, $\hat{\lambda}_2$, $\hat{\theta}$, V) is the distribution function defined by Equation (6) with the parameters replaced by estimates while m is the critical number of counts obtained by multiplying the specification on counts per cubic foot by the sampled volume; $m = \lambda_{SPEC}V$. If this specification is to be met at, say a 95 percent confidence level, then to pass the compliance test the probability stated in Equation (33) must be less than 0.05.

Since m is large (greater than 100) for typical test volumes, severe numerical problems are encountered in evaluating $p_z(1; \lambda_1, \lambda_2, \theta, V)$ for large values of 1. However, these are usually situations in which the normal approximation is valid, and the sum in Equation (33) can be replaced by an integration over a normal density function.

$$P\{Z(0,V) > m\} = 1 - \int_0^m f_Z(\xi, \hat{\mu}_z, \hat{\sigma}_z)\, d\xi \qquad (34)$$

where μ_z and σ_z are given by Equations (7) and (9).

The second approach is to formulate the compliance test directly in terms of the mean or average of the total number (Poisson counts plus spikes) of counts per cubic foot. This formulation uses an estimator of the first moment of the distribution as the decision variable. The estimator is given by

$$\hat{\Lambda} = \frac{Z(0,V)}{V} \qquad (35)$$

where $Z(0,V)$ is the total number of counts observed in volume V. The estimator $\hat{\Lambda}$ is a random variable, because Z is random, and it may be treated as normal when Z satisfies the normality condition. The mean and variance of $\hat{\Lambda}$ are derived from the moments of $Z(0,V)$. Accordingly, the mean is

$$\mu_\lambda = E\{\hat{\Lambda}\} = \frac{1}{V} E\{Z(0,V)\} = \lambda_1 + \lambda_2\theta \qquad (36)$$

and the variance is

$$\sigma_\lambda^2 = E\{(\hat{\Lambda} - \mu_\lambda)^2\} = \frac{1}{V^2} E\{[Z(0,V) - \mu_z]^2\}$$

$$\sigma_\lambda^2 = \frac{\lambda_1 + \lambda_2\theta(\theta+1)}{V} \qquad (37)$$

Since particulate compliance tests typically involve large sampled volumes, the normal approximation is usually valid. In this event, $\hat{\Lambda}$ will be normally distributed with mean

$$\mu_\lambda = \lambda_1 + \lambda_2\theta$$

and variance

$$\sigma_\lambda^2 = \frac{\lambda_1 + \lambda_2\theta(\theta+1)}{V}$$

The most likely situation where the normality condition would not be met is when only a few spikes occur, but each contains a large number of particle counts; i.e., θ is large. Although the probability of exceeding a specification on counts per cubic foot can, in theory, be calculated from the discrete distribution, it is very unlikely to pass a compliance test because λ will have a very large variance.

Referring again to Equation (37), note that when there are no spikes in the test data, $\lambda_2 = 0$, and the equation gives the variance for a Poisson counting process.

Figure 10 shows the density function for $\hat{\Lambda}$ compared to a "spec" of 1.0 counts per cubic foot. The estimate $\hat{\lambda}$ and its standard deviation S_λ are indicated on the plot. The relationship between these statistics and the moments μ_λ and σ_λ^2 will be taken up in the next section. The area under the curve to the right of the $\lambda = 1.0$ counts/ft^3 represents the probability of not meeting the "spec". If the "spec" is to be met at a 95 percent confidence level, then the indicated area must be no more than 0.05.

68

Two different formulations of a particulate compliance test have been presented. The second formulation offers three advantages

1. It is a direct extension of the deterministic calculation for the average number of counts.

2. The effect of the parameters λ_1, λ_2, θ and V on the uncertainty in the estimate $\hat{\lambda}$ is easy to observe in the variance equation.

3. The decision whether to accept or reject a system can be cast into a simple hypothesis test.

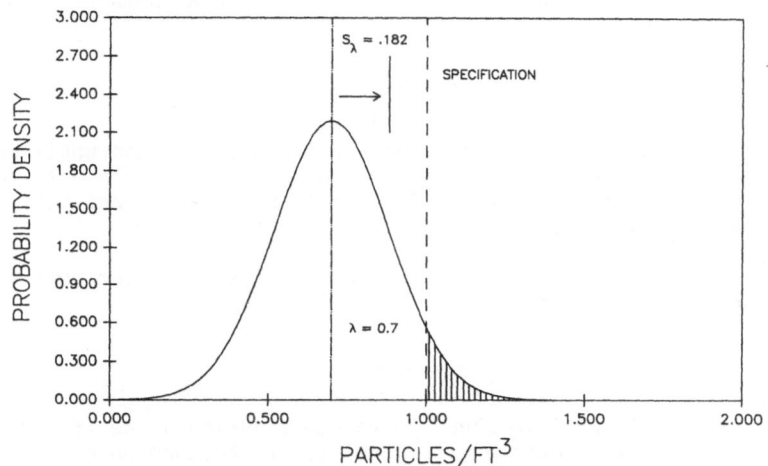

Figure 10. Compliance with a Specification is Determined from the Probability Distribution of $\hat{\lambda}$, the Estimator of Total Counts per Cubic Foot.

5. ESTIMATION OF DISTRIBUTION PARAMETERS AND KEY STATISTICS

The mathematical model of particle counts expressed by Equation (1) is designed to accomodate multiple particle arrivals like those observed in Figure 3. To estimate the parameters in this model from test data, it is necessary to distinguish between Poisson counts and spikes. Ideally, Poisson counts are identified by unit jumps in the cumulative count function as shown in Figure 4. However, the test data obtained with a laser spectrometer gives the number of counts in unit volume intervals. As a result, there are usually several counts in each unit volume even when the counts are generated by distinct unit arrivals. To resolve this problem, a threshold is set, and if the number of counts in a unit volume exceeds this threshold, the event is called a spike. Otherwise, the event is called Poisson. Guidelines on how to set the threshold and the effect it has on the estimated parameters are taken up in the next section. Here, we simply note that in typical tests, spikes contain substantially more counts than Poisson events. And in this situation, the estimates of λ and σ_λ are not sensitive to the threshold if it is set at a reasonable value.

After sorting the data into two classes, spikes and Poisson events, the parameters in the model are estimated. The estimate of the parameter λ_1 is found by dividing the total number of Poisson counts by the sampled volume

$$\hat{\lambda}_1 = \frac{1}{V} \sum_i k_i \tag{38}$$

The estimate of λ_2 is obtained by dividing the number of occurrences of spikes by the sampled volume.

$$\hat{\lambda}_2 = \frac{n}{V} \tag{39}$$

The estimate of the θ is obtained by dividing the total number of counts due to spikes by the number of spikes

$$\hat{\theta} = \frac{1}{n} \sum_j x_j \tag{40}$$

The volume elements classified as spikes are actually a superposition of Poisson counts and a spike. Although the counts in these elements could be reduced by the expected Poisson count, this correction was not used in examples presented in Section 5.1.

The estimate of the total count (Poisson plus spikes) per unit volume is determined from Equation (35)

$$\hat{\lambda} = \frac{z(0,V)}{V} \tag{41}$$

where z is the total particle count in volume V during a specific test. Since the total count is constant, regardless of the threshold, λ is constant for a given data set. Note that the same value is found for λ by substituting the parameter estimates into the Equation (7).

$$\hat{\lambda} = \hat{\mu}_\lambda = \hat{\lambda}_1 + \hat{\lambda}_2 \hat{\theta} \tag{42}$$

An estimate of the variance of λ may be computed by two different methods. The easiest is to substitute the estimated parameters into the Equation (9)

$$S_\lambda^2 = \hat{\sigma}_\lambda^2 = \frac{\hat{\lambda}_1 + \hat{\lambda}_2 \hat{\theta}(\hat{\theta}+1)}{V} \tag{43}$$

The second method is to divide the data set into m nonoverlapping sequential experiments and then apply the well known formula for sample variance

$$S_K^2 = \frac{1}{N_K - 1} \sum_{i=1}^{N_K} (k - \bar{k})^2 \tag{44}$$

to compute estimates for σ_K^2, σ_N^2 and σ_θ^2. These values are then substituted into Equation (31) for the estimate of σ_z^2.

When the data is closely approximated by the assumed Poisson models for K(0,V), N(0,V) and X_j, both estimators converge in probability to the same value. The first approach is used in the examples presented in Section 5.1.

5.1 Simulated Particulate Test Data

As mentioned in the preceding section, to apply the non-Poisson model, the particulate data must be sorted into two groups, Poisson counts and spikes. A threshold T is set, unit volumes containing more than T counts are called spikes while those containing no more than T counts are called Poisson. Estimates of the parameters λ_1, λ_2 and θ will depend on the choice of T. If T is too low, Poisson counts will be assigned to the group called spikes, and on the other hand, if T is too high, some smaller spikes will be assigned to the group called Poisson counts. In either case, the estimates of λ_1, λ_2 and θ will be in error. The best choice for T will be the value which minimizes the probability of making an error in assigning the data to the two groups. An optimal value can be found by formal mathmatical procedures using statistical decision theory.

To investigate the effect of different threshold values, synthetic data were generated by computer simulation using the non-Poisson counting model described previously. The counts generated from this model were then grouped into volumes of 1 cubic foot which simulated the way in which data are collected in an actual test. Values of λ_1, λ_2 and θ were chosen which modeled the actual data set presented in Figure 3. The parameters used were $\lambda_1 = 0.748$ ft^3, $\lambda_2 = 14.48$ ft^3 and $\theta = 21.35$. Several data sets were generated using the above parameters, the only difference between them being the seed numbers used to start the random number generators. The synthetic data sets were analyzed, with a series of different values for the threshold. The effect of differing T values on the recovered estimates for λ_1, λ_2 and θ could then be readily determined by comparing them to the parameters used to generate the data set.

The results of this study for the first synthetic data set are summarized in Table II. In this table, column 6 is $1/\hat{\lambda}_1$, column 7 is $1/\hat{\lambda}_2$ and column 8 is $\hat{\theta}$. One can see that at low or high values of T, the parameters are very sensitive to T and give poor estimates of the initial parameters. For threshold values of 4-15 particles per cubic foot, the parameters are relatively insensitive to the value of T and give good estimates. Since the total number of counts is constant, the estimate of particulate concentration $\hat{\lambda}$ is fixed at 2.60 counts per cubic foot for all values of T. Column 9 gives the standard deviation of $\hat{\lambda}$ computed using Equation (43) and column 10 gives the normality index.

The reason for the relative insensitivity of the parameters to different values of T is the wide separation between the average number of counts in the Poisson process, λ_1, and the average number of counts in a spike, θ. This is readily apparent from the consideration of Figure 11, which is the probability density function generated from the estimated parameters and a cubic foot sample volume. The distribution is bimodal. The left hand peak in the probability distribution is based primarily on the number of counts observed in a cubic foot sample of gas from the underlying Poisson process (including counter background). The right-hand peak is that observed primarily from the spike process. If the two peaks in the probability density do not overlap, then a threshold value set anywhere in the gap between the two will properly discriminate between Poisson and spike counts. The estimates of the parameters will be good, and they will be insensitive to changes in the threshold. If the threshold is set too low then Poisson counts are improperly assigned to the spike process, and if set too high the opposite error occurs. As is expected, the estimates of the parameters suffer in either case.

Table II. Estimates of Parameters λ_1, λ_2 and θ in the non-Poisson Model as Function of Threshold T for Synthetic Data

THRESHOLD COUNTS/FT³	TOTAL VOLUME FT³	NUMBER POISSON COUNTS	NUMBER SPIKE COUNTS	NUMBER SPIKES	AVG. POISSON INTERARRIVAL VOLUME (FT³)	AVG. SPIKE INTERARRIVAL VOLUME (FT³)	AVERAGE COUNTS/SPIKE	S_D PART/FT³	INORM
0	1838.97	0	5170	1374	∞	1.338	3.763	0.085	0.035
1	1838.97	538	4632	836	3.418	2.199	5.540	0.095	0.041
2	1838.97	1324	3846	443	1.388	4.151	8.681	0.107	0.052
3	1838.97	1858	3312	265	0.9897	6.939	12.49	0.117	0.064
4	1838.97	2218	2952	175	0.8291	10.50	16.86	0.127	0.077
5	1838.97	2458	2712	127	0.7482	14.48	21.35	0.137	0.089
6	1838.97	2656	2514	94	0.6924	19.56	26.74	0.146	0.103
7	1838.97	2719	2451	85	0.6763	21.63	28.83	0.150	0.108
8	1838.97	2807	2363	74	0.6551	24.85	31.93	0.154	0.115
10	1838.97	2983	2187	55	0.6165	33.43	39.76	0.165	0.133
15	1838.97	3208	1962	37	0.5732	49.70	53.02	0.180	0.162
20	1838.97	3315	1855	31	0.5547	59.32	59.83	0.185	0.176
30	1838.97	3596	1574	19	0.5114	96.78	82.84	0.200	0.224
50	1838.97	3911	1259	11	0.4702	167.1	114.4	0.210	0.294
100	1838.97	4350	820	4	0.4228	459.7	205	0.226	0.485
200	1838.97	4562	608	2	0.4031	919.4	304	0.237	0.685
300	1838.97	4862	308	1	0.3782	1838.	308	0.172	0.933
1000	1838.97	5170	0	0	0.3557	∞	0	0.039	0.014

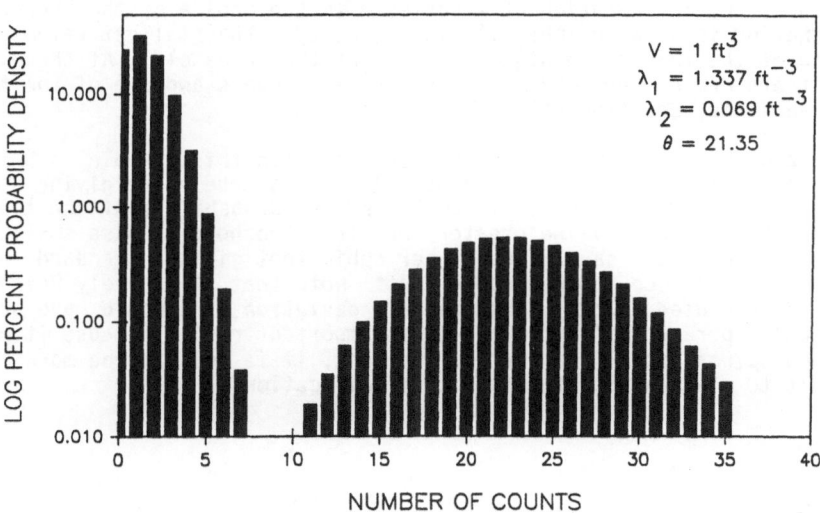

Figure 11. Non-Poisson Probability Density Function for Computer
Simulated Counting Process.

A more difficult case to discriminate occurs when the average counts in the Poisson process and the average counts in the spikes are close together. This narrows or eliminates the gap between the two peaks in the distribution. In this case the choice of a threshold value becomes more critical, and some error in assigning counts as either Poisson or spike becomes unavoidable.

5.2 Analysis of Actual Particulate Test Data

The experimental data set displayed graphically in Figure 3 was used to estimate values for the parameters in the non-Poisson model. Again, a threshold T was set, and the data were sorted into two groups, Poisson counts and spikes. The values of the estimated parameters as a function of the threshold are listed in Table III using the same format as in Table II. The estimate of the particle concentration λ, which is based on the total particle count, is 2.81 counts per cubic foot for all values of T. The standard deviation of λ was calculated using Equation (43) and the results are listed in column 9 of the table. As observed for synthetic data, the standard deviation of λ varies with the choice of the threshold. This behavior is shown graphically is Figure 12. The smallest values were obtained at the lowest and highest values of the threshold. At these points there are either no Poisson or no spike counts and one of the terms in the variance, Equation (43), approaches zero.

A reasonable value to use for the threshold in this example is 8-10 counts per cubic foot. Then the probability of a true spike giving a number less than the threshold is small and the probability of the true Poisson count giving a value greater than the threshold is also small. Using a threshold of, say 10 counts per cubic foot gives a standard deviation of 0.165 counts per cubic foot. Note that if a purely Poisson model had been used (T=∞), the standard deviation of λ would have been 0.039 counts per cubic foot. This is an important result because it means that when spikes occur in a particulate test, it is going to be more difficult to meet an accepted quality specification.

Table III. Estimates of Parameters λ_1, λ_2 and θ in the non-Poisson Model as a Function of Threshold T for Particulate Test Data

THRESHOLD COUNTS/FT³	TOTAL VOLUME FT³	NUMBER POISSON COUNTS	NUMBER SPIKE COUNTS	NUMBER SPIKES	AVG. POISSON INTERARRIVAL VOLUME (FT³)	AVG. SPIKE INTERARRIVAL VOLUME (FT³)	AVERAGE COUNTS/SPIKE	S_D PART/FT³	INORM
0	732	0	1906	731	∞	1.001	2.607	0.113	0.052
2	732	765	1141	94	0.9568	7.787	12.13	0.171	0.107
4	732	930	976	44	0.7871	16.63	22.18	0.210	0.151
5	732	940	966	42	0.7787	17.42	23.00	0.212	0.155
6	732	946	960	41	0.7733	17.85	23.41	0.213	0.156
10	732	953	953	40	0.7681	18.30	23.82	0.214	0.158
15	732	966	940	39	0.7577	18.76	24.10	0.214	0.160
20	732	1127	779	30	0.6495	24.40	25.96	0.203	0.178
30	732	1714	192	5	0.427⁻	146.4	38.40	0.132	0.344
50	732	1906	0	0	0.3840	∞	0	0.060	0.023

The distributions of the estimator $\hat{\Lambda}$ for the Poisson and non-Poisson models are shown in Figure 13. Both distributions are essentially normal because their normality indices are much less than 0.341. Note that the distribution based on the non-Poisson model is broader due to the larger standard deviation. For this data set, the broader distribution has no impact on the compliance with the specification, since the average counts per cubic foot is well below the specification indicated.

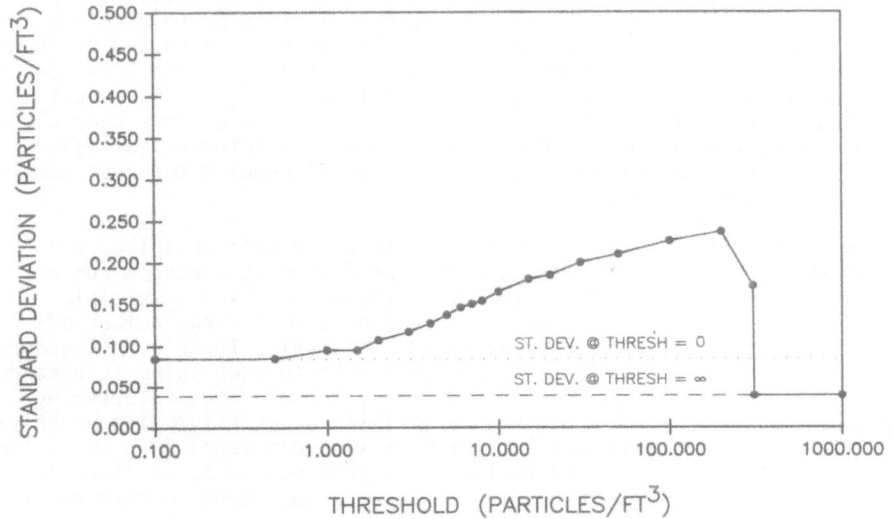

Figure 12. Standard Deviation of the Estimate of Average Counts per Cubic Foot in a Process Gas as a Function of the Threshold.

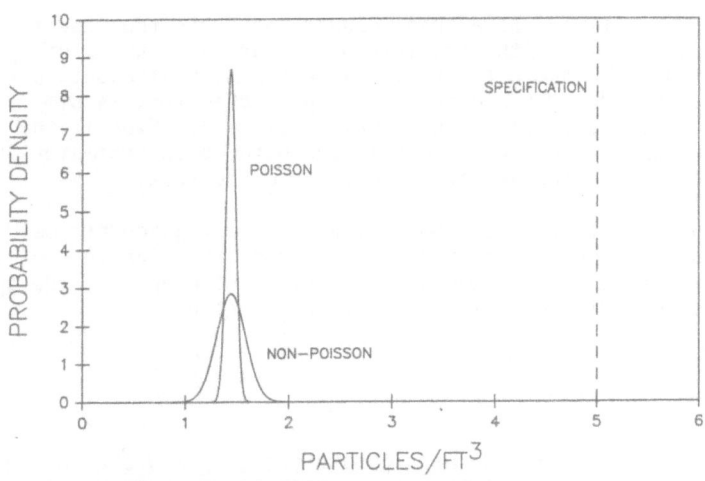

Figure 13. Probability Distributions of the Estimator of Average Counts per Cubic Foot for Poisson and non-Poisson Counting Processes.

6. CONCLUSIONS

Random counting processes provide a theoretical basis for the development of statistical procedures used to determine particulate concentrations in high purity process gases. The research discussed in this report concerns the application of Poisson and non-Poisson counting processes to the characterization of particle counts in these gases. The following conclusions are a direct result of this investigation.

Background counts in the LP101 laser spectrometer are adequately modeled as a Poisson Process. The interarrival times or volumes are closely approximated by an exponential distribution as required by theory. Background data from two spectrometers were examined using a Chi-squared goodness of fit test on the exponential distribution at a 95 percent confidence level. One data set passed the test and the other came close, but failed because of an inordinate number of zero interarrival volumes. In spite of the latter result, the Poisson model remains the best simple model for background counts.

Particulate test data containing multiple arrivals or spikes are adequately modeled by a Poisson counting process plus a random sum of independent and identically distributed random variables. Multiple arrivals or spikes refer to the occurrence of a very large number of particle counts in a small time or volume interval. These events generally occur at random points and the number of counts in each spike is a random number. Their statistics are different from those of the intervening particle counts, and their occurrence violates a key assumption underlying the Poisson model; the probability of more than one count in a small time interval should be negligibly small. This problem can be resolved by treating the spikes as a sum of independent random counts occurring at random points in time.

The estimate of total counts per cubic foot is the same for both the Poisson and non-Poisson counting processes. For the non-Poisson model, the estimate of total counts per cubic foot is given by the equation

$$\hat{\lambda} = \hat{\mu}_\lambda = \hat{\lambda}_1 + \hat{\lambda}_2 \hat{\theta}$$

where λ_1 is the estimate of average counts per cubic foot due to the Poisson component, λ_2 is the estimate of the average number of spikes per cubic foot, and θ is the estimate of average number of counts per spike. For the Poisson model the average counts per cubic foot is simply the total number of counts K divided by the sampled volume V. Even if the Poisson model is used when there are spikes in the data, both estimators give the same numerical value for average counts per cubic foot.

The standard deviation of the estimate of average counts per cubic foot is always larger for the non-Poisson model than for the Poisson model and it is very sensitive to the average number of counts in the spikes. The standard deviation of λ is given by the equation

$$S_\lambda^2 = \hat{\sigma}_\lambda^2 = \frac{\hat{\lambda}_1 + \hat{\lambda}_2 \hat{\theta}(\hat{\theta}+1)}{V}$$

If there are no spikes in the particulate data, $\lambda_2 = 0$ and the equation reduces to the standard deviation for a Poisson counting process.

Since the second term in the above equation represents the contribution of the spikes to the variance, it is always positive. Consequently, the standard deviation of λ for the non-Poisson model is always larger than the standard deviation for the Poisson model.

Under typical particulate test conditions the probability distribution of the estimator of total counts per cubic foot is approximately normal. The probability distribution of   , the estimator of total counts per cubic foot, approaches a normal distribution as the sampled volume increases. An inequality condition determines when the normal distribution is a valid approximation. It is given by

$$\frac{\lambda_1 V + \lambda_2 V \theta(\theta^2 + 3\theta + 1)}{[\lambda_1 V + \lambda_2 V \theta(\theta+1)]^{3/2}} \leq 0.341$$

where λ_1 is the average number of Poisson counts per cubic foot, λ_2 is the average number of spikes per unit volume and θ is the average number of counts per spike. The above condition can always be met for finite values of λ_1, λ_2 and θ by taking the sampled volume V sufficiently large.

Compliance with electronic industry standards can be formulated as statistical hypothesis test. A test for compliance of process gases with electronic industry standards can be formulated as a statistical hypothesis test on the mean of a normal distribution which has unknown standard deviation. After preliminary estimates of λ_1, λ_2 and θ are determined, the sample volume required to achieve a desired standard deviation of the estimate can be predicted. The validity of the normal approximation can also be verified by checking the above normality condition.

If the normality condition is not met in a test, statements about the compliance of a process gas with a particulate specification must be calculated using the non-Poisson distribution function, Equation (6).

APPENDIX

A. MOMENT GENERATING FUNCTION OF NON-POISSON DISTRIBUTION

The moment generating function of Z(0,V) provides an alternate, but mathematically equivalent, formulation of this new counting process. The moment generating function is used here to confirm the equations for the mean and variance of Z(0,V), but more importantly, it is used to prove the asymptotic convergence of the distribution of Z(0,V) to a normal distribution. The distribution function, Equation (6), can also be derived from the moment generating function but for this case it is extremely tedious.

The moment generating function of the Poisson distributed random variables X_i, the number of counts in a spike, is

$$M_{X_i}(t) = E\{\epsilon^{-tX_i}\} = \sum_{l=0}^{\infty} \epsilon^{tl} \left[\frac{\epsilon^{-\theta} \theta^{l}}{l!} \right]$$

$$M_{X_i}(t) = \epsilon^{-\theta(\epsilon^t - 1)} \qquad\qquad (A.1)$$

Similarly, the moment generating function of $N(0,V)$, the random number of spikes, is

$$M_N(t) = E\{\epsilon^{tN(0,V)}\} = \epsilon^{\lambda_2 V(\epsilon^t - 1)} \qquad (A.2)$$

and for $K(0,V)$, the Poisson counting process, it is

$$M_K(t) = E\{\epsilon^{tK(0,V)}\} = \epsilon^{\lambda_1 V(\epsilon^t - 1)} \qquad (A.3)$$

Let $S(0,V)$ represent the random sum in Equation (1). Then the generating function for $Z(0,V)$ is given by

$$M_Z(t) = E\{\epsilon^{t[K(0,V) + S(0,V)]}\} = M_K(t)M_S(t) \qquad (A.4)$$

The generating function for $S(0,V)$ can be derived from the relationship [5]

$$E\{\epsilon^{tS(0,V)}\} = E\{E\{\epsilon^{tS(0,V)} \mid N(0,V) = n\}\} \qquad (A.5)$$

The inner expected-value is the conditional generating function of $S(0,V)$ given that there are n random variables in the sum. Therefore,

$$E\{\epsilon^{tS(0,V)} \mid N(0,V) = n\} = E\{\epsilon^{t[X_1 + X_2 + \cdots X_n]}\}$$

$$= \prod_{i=1}^{n} M_{X_i}(t) = M_{X_i}^{n}(t) \qquad (A.6)$$

Returning to equation (A.5), we find that

$$M_S(t) = E\{M_{X_i}(t)\} = \sum_{n=0}^{\infty} M_{X_i}(t)\, P\{N(0,V) = n\} \qquad (A.7)$$

Temporarily, let

$$M_{X_i}(t) = \epsilon^{\xi n}$$

Then

$$M_S(t) = \sum_{n=0}^{\infty} \epsilon^{\xi n}\, P\{N(0,V) = n\}$$

But, this is the definition of $E\{e^{N\xi}\}$ and since, $\xi = \ln M_{X_i}(t)$, we get

$$M_S(t) = M_N[\ln M_{X_i}(t)] \qquad (A.8)$$

Combining these equations gives the desired generating function

$$M_Z(t) = \epsilon^{-\{\lambda_1 V(1-\epsilon^t) + \lambda_2 V[1-\epsilon^{\theta -\epsilon^t})]\}} \qquad (A.9)$$

The mean and variance of the counting process are obtained from the first and second derivatives of the moment generating function. The mean is derived from [4]

$$\mu_z = \left.\frac{d\ M_z(t)}{d\ t}\right|_{t=0}$$

$$\mu_z = \lambda_1 V + \lambda_2 V\theta \qquad\qquad \text{(A.10)}$$

Similarly, the second moment m_2 is derived from

$$m_2 = \left.\frac{d^2\ M_z(t)}{d\ t^2}\right|_{t=0} \qquad\qquad \text{(A.11)}$$

and hence, the variance is

$$\sigma_z^2 = \lambda_1 V + \lambda_2 V\theta(\theta+1)$$

Note that these equations for the mean and variance are the same as those obtained earlier in Section 3.3.

REFERENCES

1. R.A. Van Slooten, Statistical treatment of particle counts in clean gases, Microcontamination 4(2) 33 (1986).

2. M.L. Malczewski, J.D. Borkman and G.T. Vardian, Measurements of particulates in filtered process gas streams, Solid State Technol. 29(4) 151 (1986).

3. T.T. Soong, "Probabilistic Modeling and Analysis in Science and Engineering", John Wiley, 1981.

4. H. Cramer, "Mathematical Methods of Statistics", Princeton University Press, 1963.

5. W. Feller, "An Introduction to Probability Theory and Its Applications", Volume 1, John Wiley, 1950.

6. W. Mendenhall and R.L. Scheaffer, "Mathematical Statistics with Applications", Duxbury Press, Boston, 1973.

7. A Papoulis, "Probability, Random Variables and Stochastic Processes", McGraw-Hill, 1965.

LIQUID PARTICLE COUNTER COMPARISON

C. Willis

Texas Instruments, Inc.
P.O. Box 655012, MS 9
Dallas, Texas 75265

Liquid laser particle counters from two manufacturers were compared by the simultaneous measurement of particles in deionized (DI) water and in various acids. Good correlation was observed for certain size ranges; for others, the numbers were dissimilar but predictable. The particle distribution was characteristic for each counter type, and consistent for the different fluids examined.

INTRODUCTION

Particles are recognized as one of the major contributors to yield loss in semiconductor manufacturing. Efforts to reduce particles in liquids coming into contact with silicon wafers necessitates the ability to measure their presence. There has often been concern over the accuracy of liquid particle counting, due perhaps to the occasional generation of careless data. The qualification of a semiconductor manufacturing bath is routinely performed indirectly by running pilot wafers through the appropriate steps. A correlation between particle level in process chemicals and wafer surface counts has been demonstrated using a static bath[1]. However, recirculation filtration will achieve liquid particle counts three orders of magnitude cleaner than in that study; due to contributions from other factors, the relation between liquid and wafer particle counts in this region remains obscure.

Investigation into reduction of particles in semiconductor liquids must, of course, start with a reliable measuring tool. The most common such device incorporates a laser to illuminate a stream of liquid and measures the light scattered by passing particles. At present the only two domestic manufacturers of instruments capable of handling hydrofluoric acid (HF) based solutions are Particle Measuring Systems (PMS) and Pacific Scientific, although there are other companies making laser counters compatible with DI water.

Liquid particle counters are routinely calibrated by uniform sized polystyrene latex (PSL) microspheres suspended in water. Differing diameter spheres are used to develop a curve of the voltage response of the sensor versus particle diameter. This serves to calibrate the instrument for particle size but not concentration. A method for number calibration was developed at Millipore Corporation[2] involving counting of the PSL beads caught on analytical discs using a scanning electron microscope. This remains the only documented method of number calibration, but is rather complicated and time consuming. It possibly involves more effort in cost and time than some liquid particle counter users are willing to devote.

As users of electronic grade chemicals have become more interested in the particulate level of their liquid chemicals, suppliers have begun to institute particle specifications. As with specifications limiting the level of metallic contaminants in these materials, users and suppliers must develop the ability to correlate their analyses. Hence there exists the necessity to know the comparative response of different particle counters. The question of the accuracy and precision of these tools was the primary reason for the present research. The results are simply many comparisons of the numerical values outputted by instruments of the two primary liquid counter manufacturers.

EXPERIMENTAL PROCEDURE

Particle counter sensors used were the latest models offered by each company for batch analysis of corrosive liquids. All sensors used had been size calibrated within the previous six months using PSL microspheres. The Pacific Scientific sensor was Hiac-Royco model 346BCL. Three different units were utilized in the study; they are referenced in the figures as Hiac #2, #3, and #4. These sensors were size calibrated at flows of 25 and 100 ml/min to register particles with diameters of 0.5 microns and larger. Their associated model 4100 counters assign their signals into six adjustable size bins, generally set to >0.5, >0.6, >0.8, >1, >2, and >5 microns to correspond to those of the PMS. The millivolt settings for the 0.8 and 1.0 micron channels were interpolated from the 0.72 and 1.1 micron microspheres responses. Hiac instruments #2 and #4 were calibrated by Pacific Scientific with beads of nominal sizes 0.5, 0.6, 0.72, 1.1, 2.0, and 5.0 microns. Hiac #3 was calibrated at Texas Instruments using a pulse height analyzer by verifying that responses of three or four of the sizes were unchanged from factory settings.

The PMS sensor utilized was model IMOLV (Integrated Micro-Optical Liquid Volumetric); PMS #3 refers to one with a sapphire capillary capable of handling hydrofluoric acid and PMS #2 refers to one with a borosilicate capillary capable of handling any other acid. They are sensitive down to 0.3 microns, and the companion LLPS-X counter is calibrated to size the counts into 15 bins. These instruments were calibrated at the factory using five PSL microsphere sizes below 1 micron and four above 1 micron; the channel thresholds were obtained by interpolation. This calibration was performed at the midpoint of the specified flow rate range of 20 to 100 ml/min. As a particle passes through the sensor, the PMS electronics examines the height of its voltage signal, so that the length of time which a particle spends traversing the sensing volume is not a factor in its sizing. The Hiac technology, on the other hand, uses the signal area, so that the millivolt threshold values vary with flow rate.

The schematic of the setup used is shown in Figure 1. The configuration of the sensors being in series rather than in parallel was chosen in the belief that the maintenance of identical flow rates through each sensor was more critical than the possible release of particles from surfaces between the sensors and the unequal pressures at the sensors. Most measurements were made with the Hiac upstream of the PMS, but some were also made with the PMS upstream. To minimize the locations available for trapping and later re-entrainment of particles, the sensors were

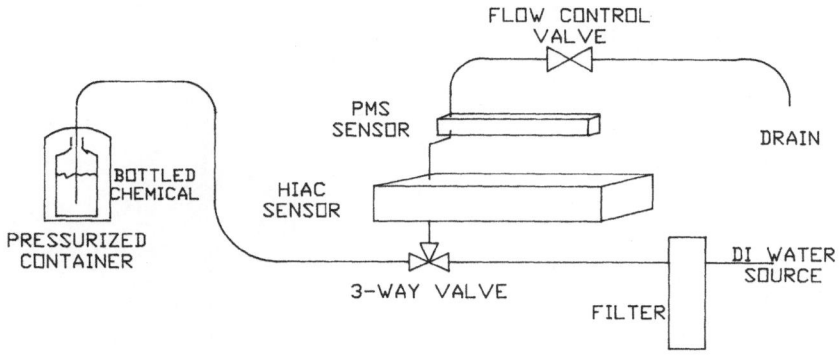

Figure 1. Particle counter comparison system schematic.

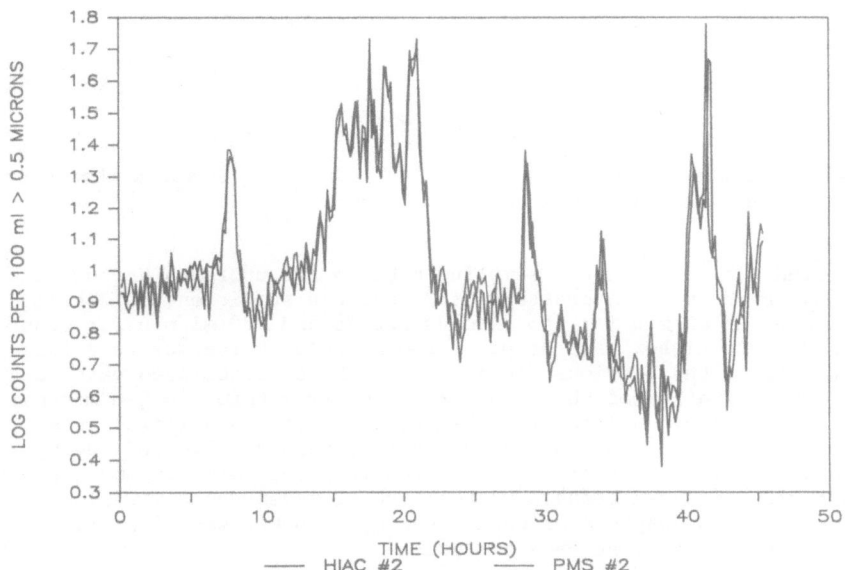

Figure 2. DI water counts over time, PMS and Hiac. PMS is the upstream
sensor; ten minute sample intervals.

coupled with at most one 1/4" tube union and 2" of tubing. All tubing
and fittings used were made of PFA Teflon[©].

     The measurements of DI water were performed at a flow rate of 100
ml/min, bypassing the point-of-use filter. That filter was used for
flushing out the system between bottled chemical measurements. The level
attainable by a few hours of flushing was less than 1 total count per 100
milliliters. However, a background this low was necessary only for the
cleanest of the bottled chemicals. A background of less than 1% of the
subsequent bottle count was enough to be insignificant, and typically was
achieved by total counts of close to 100 per 100 ml. Flushing entailed
turning the three-way valve upstream of the sensor so that the rinse-water
would flow both through the sensor and through the tube leading toward
where the sample would be placed. Measurements of chemicals were taken
from individual bottles placed inside of a pressurized chamber. A fitting

[©] Registered trademark of E.I. DuPont de Nemours & Co., Inc.

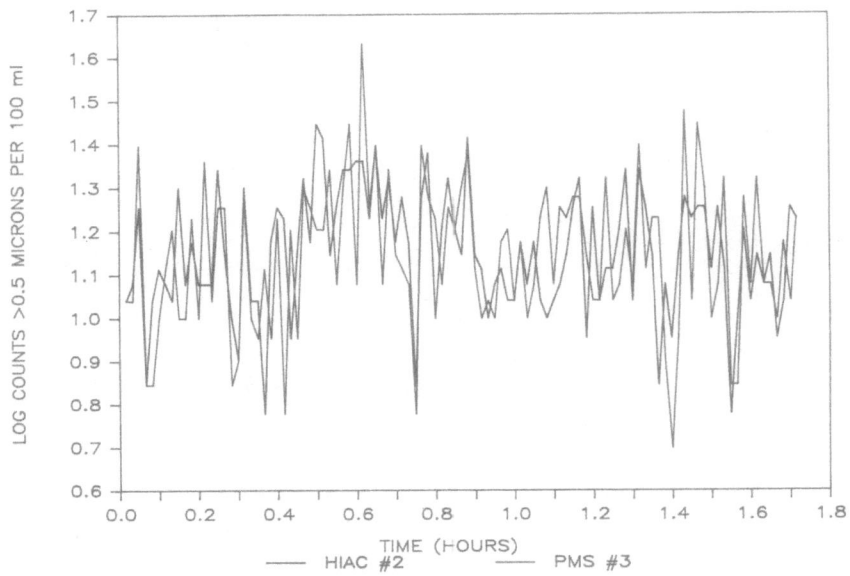

Figure 3. DI water counts over time, PMS and Hiac. Hiac is the upstream sensor; one minute sample intervals.

was drilled out to allow the sampling tube to run unbroken through the wall into the bottle. A pressure of 20 psi was sufficient to obtain a flow rate of 25 ml/min for the viscous fluids and 100 ml/min for other chemicals. The higher rate is preferred because it results in a faster displacement of the previous fluid along with its associated particulate level. Stabilization of the counts required from three to ten minutes of uninterrupted flow; data recorded before the counts stabilized were discarded. Numbers reported are averages of between five and twenty one-minute runs. In describing these results, the term 'counts' as opposed to 'particles' is used because of the lack of independent verification of their number. The capture of each count by computer was essential in performing the subsequent data analysis.

MEASUREMENTS OF DI WATER

Figure 2 is a graph of >0.5 micron counts reported by each instrument operating on a DI water source over a period of 45 hours. The counts are plotted using a log scale to show more detail in the lower counts. It shows excellent correlation throughout. Data for this graph were collected with the PMS being upstream. Figure 3 is a similar graph covering a much shorter duration using a different Hiac sensor and with the PMS being downstream. Although the counts exhibit a seemingly random variability, the overall averages are the same. Each data point in Figure 2 represents counts registered over ten-minute intervals and is converted to the 100 ml basis, while Figure 3 incorporates data from single-minute time spans. The more precise agreement reflected in Figure 2 arises in part because of this difference in sampling time, adhering to the statistical tenet that the smaller the population, the higher the variation. This lack of coincidence of each observed particle event could be due to some of them being caught between the sensors and so delayed in 'eluting' through to the downstream sensor.

A condition of which to be wary entails failing to obtain a clean

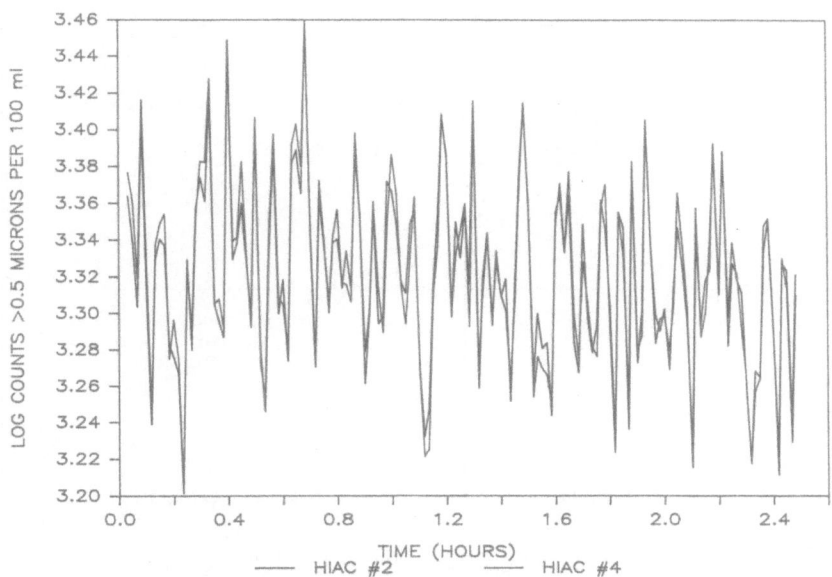

Figure 4. DI water counts over time, Hiac vs Hiac. One minute sample intervals.

system prior to data collection. One illustration of this occurred upon introducing a sensor into a previously clean system (indicated by its unvarying counts). Even though the initial counts outputted were low (63 per 100 ml), fourteen hours of flushing at 100 ml/min were required before the counts of the downstream sensor diminished to the same level of 10-15 steadily reported by the upstream sensor. Apparently, particles caught on the surfaces between the two sensing zones were gradually being washed away. Cleanliness of the system is critical in obtaining the true measure of the desired count.

Although only a single PMS data collection unit (LLPS-X) was available, each Hiac had its own 4100 unit and so two counters could be operated simultaneously. Figure 4 is data collected by two Hiac instruments from moderately particle laden DI water. Although these data are from one-minute runs, the plots are coincident on account of the large population of particles for each data point.

The Hiac records particle counts for six size ranges and the PMS for fifteen ranges. How do the numbers for other sizes compare? Figure 5 was calculated from the same data used in Figure 3. The counts greater than each reported size were averaged and are shown on the y-axis as log cumulative counts with the x-axis being log size. During the timespan in question the average count >0.3 microns measured by the PMS was 66; the number >0.5 microns measured by the Hiac was 13. The PMS counts result in a somewhat concave-down configuration while the Hiac counts form a concave-up curve. The two curves touch at 0.5 microns and cross near 2 microns; these points correspond to sizes at which the two sensors report identical cumulative counts. Between 0.5 and 2 microns the PMS reports more particles while above 2 microns the Hiac registers a greater number. The accuracy of the PMS curve at the large sizes can be inferred from its approximate linearity. The >12 micron data point is a count of three particles in ten liters; this low of a sensitivity was attained on account of the large volume (25 liters) of fluid that was analyzed over the four hours.

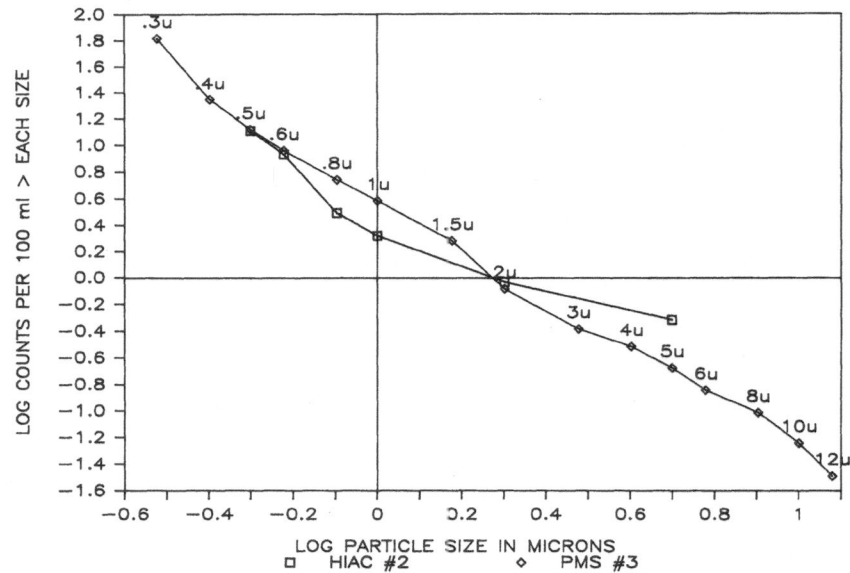

Figure 5. DI water cumulative count distribution of Figure 3 data.

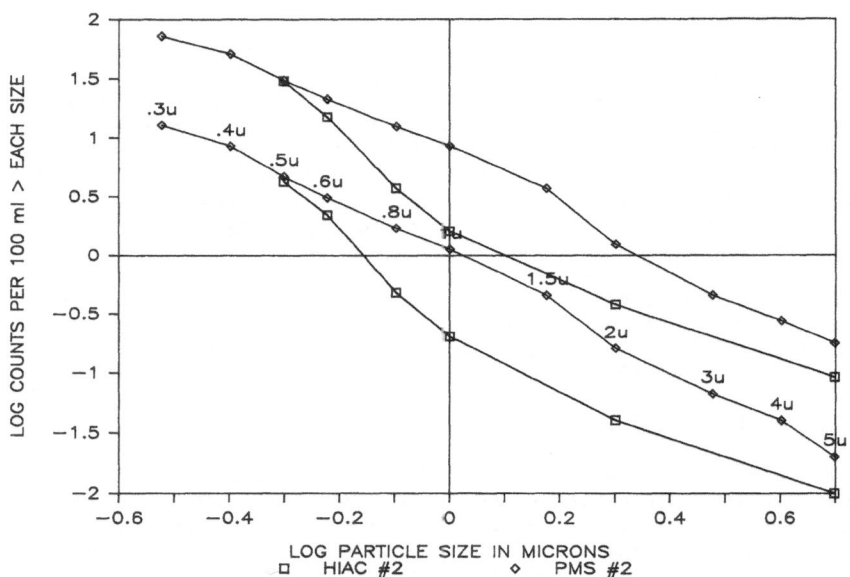

Figure 6. DI water cumulative count distribution, PMS vs Hiac, at two
particulate levels of Figure 2 data.

Figure 6 is the cumulative count distribution from different time
periods of the data from Figure 2. The upper pair of curves originates
from the period of relatively high counts (T=15-20 hours), and the lower
pair is from a region of low counts (T=35-40 hours). These two sets of
curves are strikingly homologous in shape and relative position. They do
vary from the plot in Figure 5 in that the Hiac curves do not 'overtake'
those of the PMS at the high end of the size range. It is significant
that this is exhibited by both pairs of curves, leaving open the

86

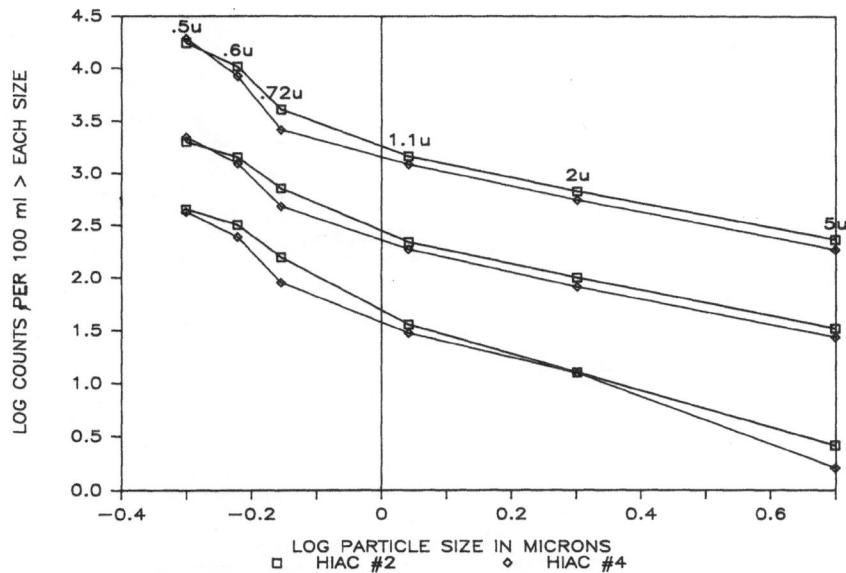

Figure 7. 4.9% HF cumulative count distribution, Hiac vs Hiac; three
bottles at 100 ml/min.

possibility that the discrepancy lies in some physical characteristic of
the water rather than in lack of reproducibility of the counters. In some
other DI water distribution relationships recorded, the Hiac line crosses
that of the PMS at close to 0.6 microns, so that the Hiac reads a greater
number below this value. No variation was observed dependent on the order
of the sensors. Note that the lowest point in Figure 6, the Hiac >5
micron, was contributed to by one count per ten liters.

MEASUREMENTS OF ACIDS

     The acids tested were hydrofluoric, sulfuric, nitric, and
hydrochloric. Other aqueous solutions run were two phosphoric based
etches, a buffered oxide etch, and hydrogen peroxide. The sulfuric acid
and phosphoric etches were run at a flow rate of 25 ml/min while 100 ml/min
was standard for the others. Some of the low viscosity chemicals were
examined at 25 as well as 100 ml/min. The flow rates are indicated on
each chemical graph.

Hydrofluoric Acid

     Figure 7 displays measurements by two Hiac counters of three
bottles of 4.9% HF having >0.5 micron counts of 17,500, 2000, and 450
per 100 ml. The shapes are similar to those obtained from water.
Hiac #4 reads a somewhat lower number at >0.72 microns; this same
relation showed up in each comparison of these two Hiac counters and
is illustrated in Figure 8. The >0.5 micron points are all on the
equal count line, while the 0.72 micron points are offset from it a
consistent distance. This phenomenon is likely an error in the size
calibration of the instruments or in their counting abilities.
Nevertheless, the fact that this was observed in every case spanning
3.5 orders of magnitude of counts using three fluids reflects the
individual reproducibility of each instrument.

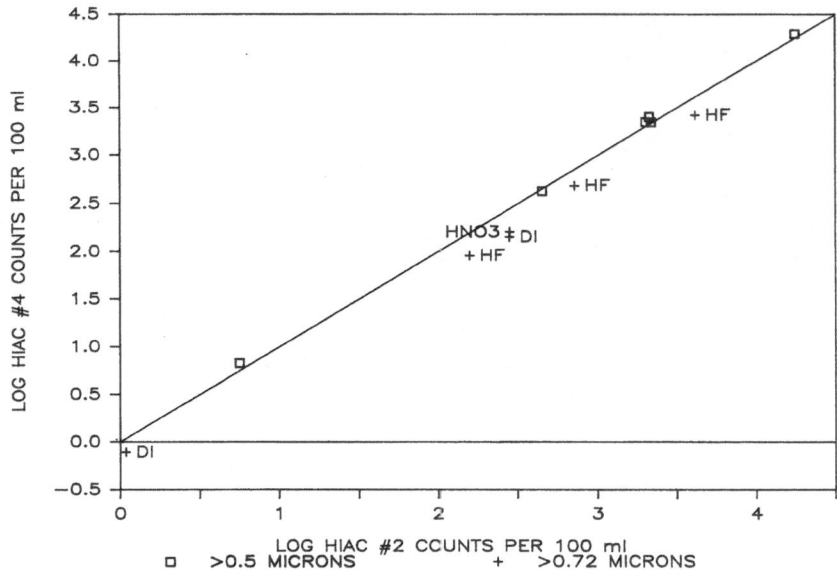

Figure 8. Hiac counter comparison at two particle sizes for DI water,
nitric and HF acids at 100 ml/min. Diagonal line indicates
equal counts.

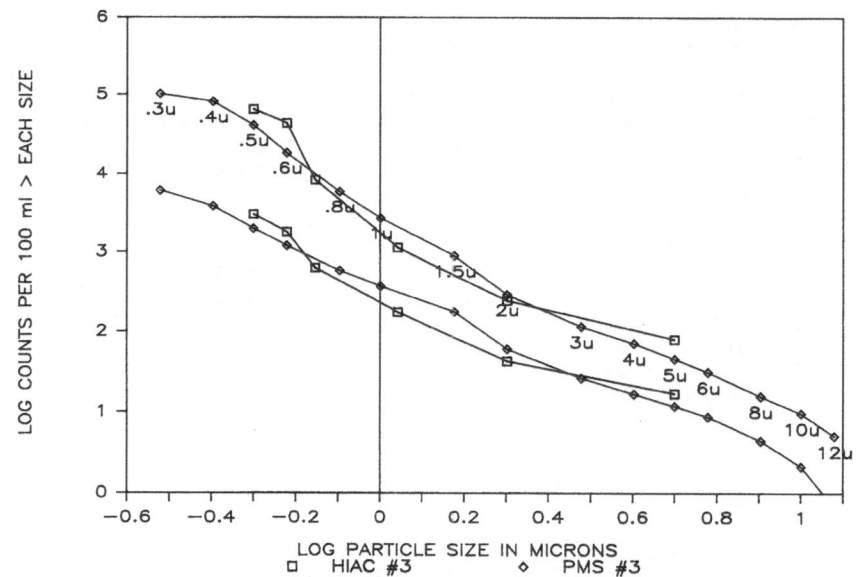

Figure 9. 4.9% HF cumulative count distribution, PMS vs Hiac; two
bottles at 100 ml/min.

Figure 9 shows two typical plots of 4.9% HF run on a PMS and Hiac.
In the majority of cases with this chemical, the curves cross close to
0.7 microns and again somewhere between two and five microns, so that the
PMS records more particles between about 0.7 and 3 microns and the Hiac
reports more outside this range. Note that the convergence of the PMS
curves towards the right reflects a slight difference in their overall
slopes, and is apparent also in the Hiac curves.

88

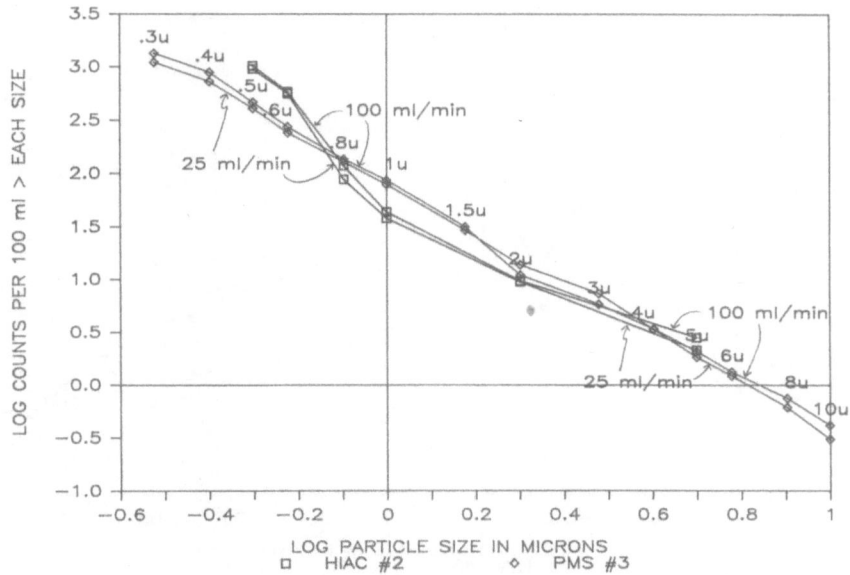

Figure 10. 4.9% HF cumulative count distribution, PMS vs Hiac; one bottle at two flow rates.

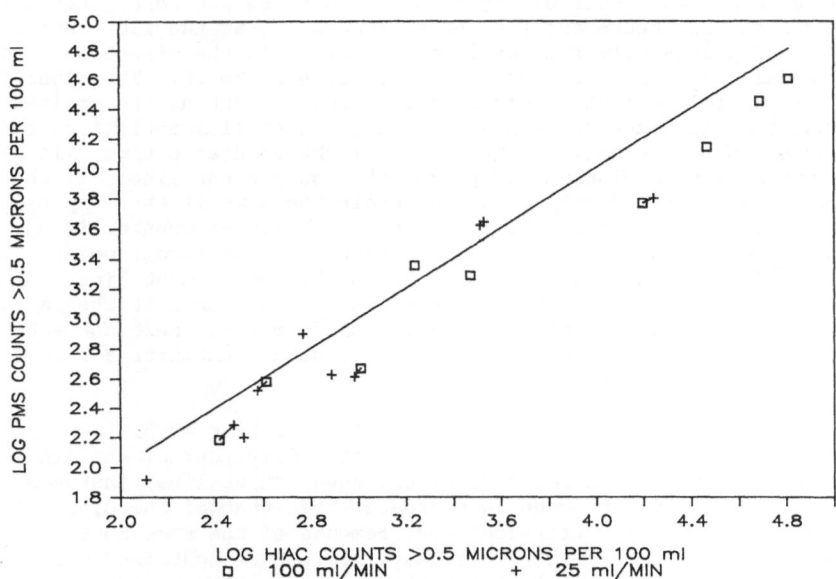

Figure 11. PMS and Hiac comparison, 4.9% HF bottles. Major line indicates equal counts; connected points are of same bottle at different flow rates.

Figure 10 displays measurements of the same bottle at 25 ml/min and 100 ml/min flow rates. The curves correspond well for both instruments, in spite of there being 5% uncertainty in setting the lower rate of flow.

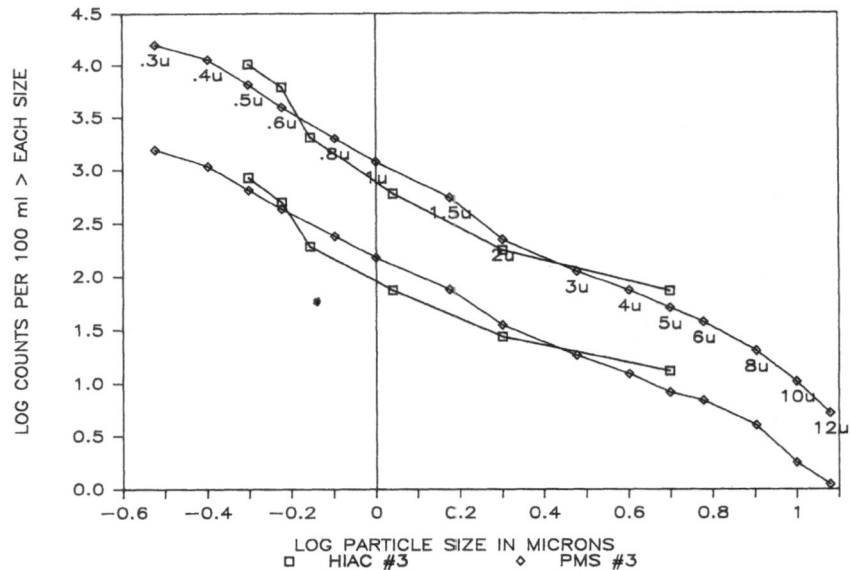

Figure 12. Hydrochloric acid cumulative distribution, two bottles
at 100 ml/min.

    Figure 11 displays the comparative counts >0.5 microns for all samples
of 4.9% HF measured. Four of the 25 ml/min crosses are associated with a
100 ml/min square; these are four bottles measured at two flow rates.
The pairs of points differ in position primarily in the direction parallel
to the equal count line, rather than perpendicular to it. The significance
of this is as follows: there may be a 5% error in setting the desired flow
rate, but the flow rate through each counter is still identical on account
of being assembled in series. The ratio of PMS to Hiac counts will place
each point a certain distance away from the equal count line. If the flow
rate is changed, this distance should remain the same if the response of
each counter is consistent. If the response of either counter is not
consistent, the count ratio will be different and the second point will
be displaced some distance perpendicular to the equal count line. Since
this was not the case with the four bottles, the reason that the pairs of
points do not overlay exactly is because the flow rates were not set
accurately, not because one or both of the counters exhibits a flow-rate
dependency.

    In the majority of bottles of Figure 11, the Hiac >0.5 micron count
was quite close to the PMS >0.4 micron count. Only four of the nineteen
points are located on the PMS side of the equal count line, instances in
which the PMS >0.5 micron count was greater than that of the Hiac. Three
of these points actually represent measurements of the same bottle run
different weeks. It is conceivable that this bottle contained particles
of a different elemental composition than did the others, and that the
PMS was better able to sense them. Such a phenomenon would account for
the data spread of Figure 11.

Other Acidic Chemicals

    Figure 12 shows two samples of 37% hydrochloric acid (HCl). They
are virtually identical to plots obtained from HF. The >0.5 micron counts
of these and various other fluids examined are compiled in Figure 13.

90

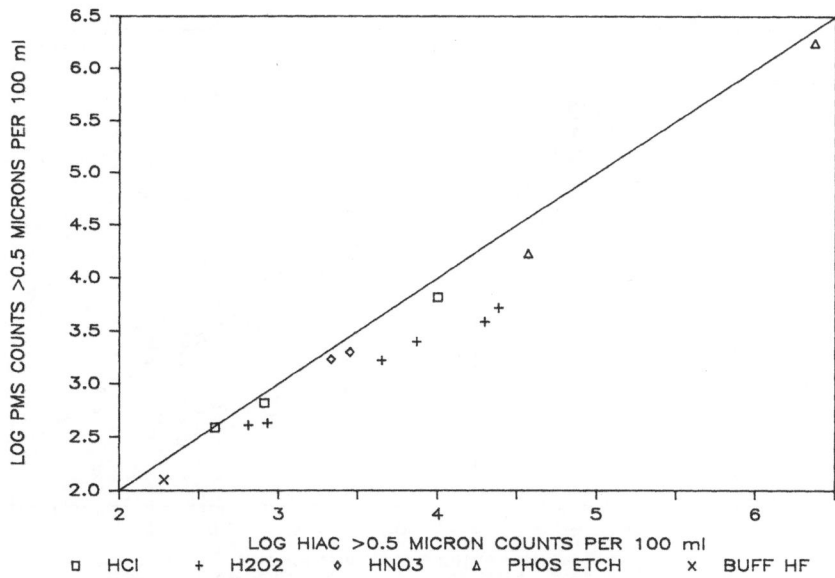

Figure 13. PMS and Hiac comparison with miscellaneous acids. The
diagonal line indicates equal counts.

These data also entail different instruments and flow rates and span five
orders of magnitude. The highest point, a phosphoric based etch, is the
only instance where the concentration is high enough for particle
coincidence effects to be relevant. These follow the generalization of
the Hiac outputting a larger number at >0.5 microns.

Hydrogen peroxide is a difficult chemical in which to count particles,
as certain individual bottles form bubbles even under as high a pressure
as 35 psi. Bubbles show up predominantly in the highest channel, and are
characterized by greatly fluctuating counts. Hydrogen peroxide was a
chemical affected by the sensor order: due to its lower pressure the
downstream sensor would have an increased tendency to form bubbles.
Formation of occasional bubbles most likely was responsible for the
frequently greater number of large-sized counts of the downstream sensor
compared to the upstream one. The difference was not enough to
significantly alter the total counts, however.

Figure 14 shows two samples of 96% sulfuric acid. The consecutive
one-minute (25 ml) runs of this chemical exhibited more variation than
other acids (see error section); as a consequence, the distribution
plots were not as consistent in their shape as were others. (This can
be seen in the jaggedness of the right hand 'tails' of the curves in
the figure). In contrast to results from the other fluids, the PMS
reported more particles at all sizes than did the Hiac. A compilation
of the thirteen sulfuric samples analyzed are displayed in Figure 15.
Four of them are just slightly below the equal count line but the
remainder are considerably above it.

Refractive Index Correction

The amount of light scattered by a particle of a certain diameter
and reaching the photodetector depends on the index of refraction of the
fluid as well as that of the particle. In general, the higher the fluid's

91

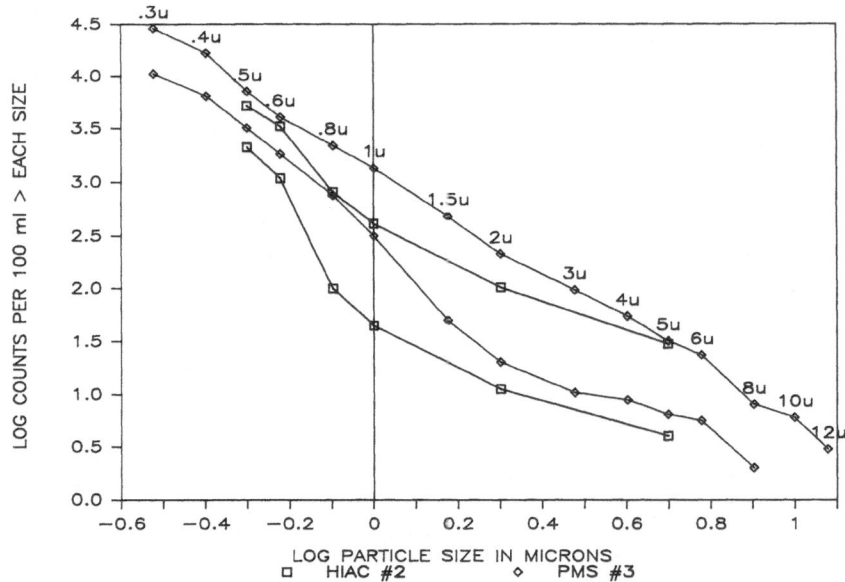

Figure 14. Sulfuric acid cumulative count distribution, PMS and Hiac; two bottles at 25 ml/min.

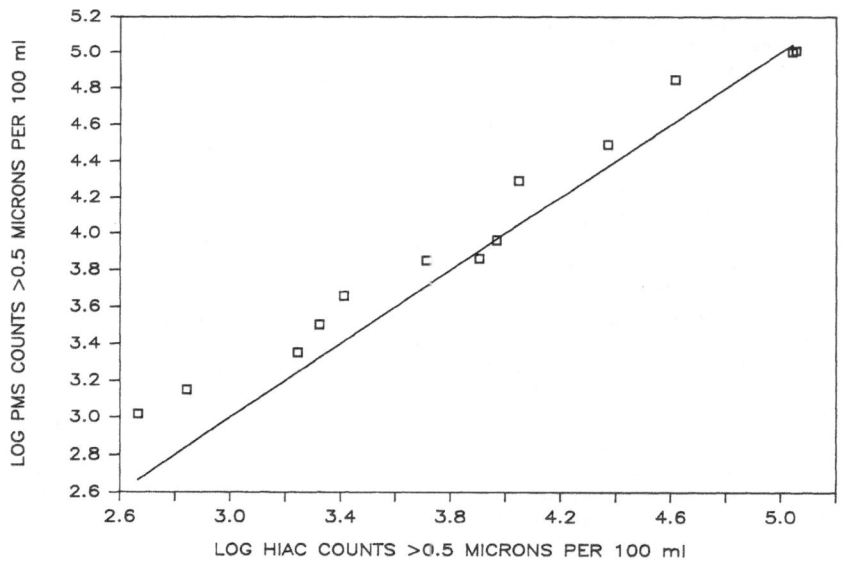

Figure 15. PMS and Hiac comparison, sulfuric acid bottles at 25 ml/min. Diagonal line indicates equal counts.

refractive index, the harder it is to detect particles. The reason that gas counters are so much more sensitive than their liquid counterparts is because many gases have a refractive index of 1, while water's is 1.29, much closer to that of solids. Liquid counters are calibrated assuming that the refractive index of PSL, 1.59, is representative of that of real particles. In reality, particles will have a continuum of refractive indices whose distribution varies with the source. A recent article[3]

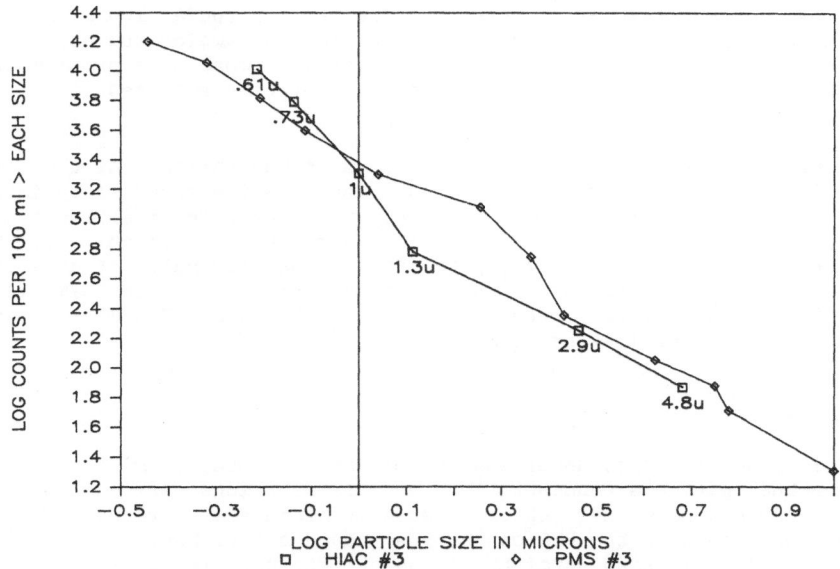

Figure 16. Refractive index correction for upper set of curves of HCl in Figure 12.

describes theoretical corrections to be applied to the sizing of particles in fluids having refractive indices varying from that of water. The adjustment for a given liquid is a complex function of particle diameter. In addition, it varies slightly between the PMS and Hiac instruments due to a difference in the angles over which they gather the scattered light. The correction factors were calculated for spherical particles of refractive index equivalent to polystyrene latex;[4] real particles are of varying refractive indices and are also far from spherical. This results in a correction that is closer to the truth but still an approximation.

On account of its high water content, counts from 4.9% HF will require no correction; but the size adjustment for hydrochloric and sulfuric acids is substantial. Assuming a value for the average particle refractive index equal to that of PSL, a 0.5 micron 'count' in HCl correlates with a 0.61 micron particle. Figure 16 shows the plots resulting from corrections applied to the higher particle bottle of HCl in Figure 12 (none of the other graphs has been corrected for refractive index). The curves are shifted to the right and somewhat compressed. In addition to increasing the disparity at the mid-sized counts, the PMS representation is altered from a smooth to a disjoint curve. In view of this and of the congruence of the uncorrected HCl curves of Figure 12 with the already-correct 4.9% HF curves of Figure 9, it is difficult to accept the refractive index size correction without corroboration. In any case, to avoid this ambiguity the term 'counts' rather than 'particles' has been used throughout the present paper.

Uncertainty

As stated previously, some of the initial readings of each chemical were discarded waiting for the signals to stabilize. The counts almost always diminished during this period and were undoubtedly caused by light scattered from the mixing fluids. The surprisingly long time required for this is due to the absence of plug flow through the ten feet of 1/4" tubing between the sample and sensors. The samples run at 100 ml/min equilibrated usually within 2-3 minutes, while the samples run at 25 ml/min required more time. The highly viscous acids were more of a problem than the others run at the same flow rate in terms of time required for stabilization.

Flow rate dependence was also reflected in the variance of the data from which the averages were calculated. The relative standard deviation (RSD) of a given sample increased smoothly with particle size: the smaller the data population, the higher the relative run-to-run variability. For those chemicals run at 100 ml/min, the RSD was usually below 4% at >0.5 microns and below 12% at >5 microns for both instruments. The same fluid run at 25 ml/min, however, resulted in up to 15% RSD at >0.5 microns and ranging to well over 100% at >5 microns.

CONCLUSIONS

This study makes no determination of which counter is more accurate, as the actual particle concentrations and distributions were unknown. The Hiac can not be deemed better because it often sees more particles >0.5 microns, just as the PMS can not be judged better because of its inherently more pleasing straight line distribution. In order to make an independent analysis, the particle concentrations must be corroborated in some other fashion such as capturing them on an analytical disc for subsequent microscopic counting. This can be accomplished using uniformly sized latex microspheres, but might not be feasible with real particles.

Liquid particle counter users with access to more than one instrument can set them up in series on a DI water system for a straight-forward method of examining their numerical counting accuracy; if the instruments do not yield counts which coincide, then the instruments are not operating correctly.

The two models of liquid particle counters discussed can be made to appear either coincident or dissimilar depending on the size considered. It is advantageous to be aware of the distribution across a range of sizes; an abnormal distribution is probably indicative of a problem (bubbles or electronic noise). Cleanliness is critical when measuring extremely clean fluids; it is recommended that the system be qualified by first running a clean bottle of water. Using water supplied through a filter and overflowing a test bottle, a count of <10 particles >0.5 microns per 100 ml can be achieved. The faster sampling rate of 100 ml/min is preferred because of its more repeatable measurements and faster removal of the spurious 'mixing' counts.

Based on the Hiac 346BCL and PMS IMOLV instruments utilized in this study, the Hiac more often reports a somewhat higher value at >5 microns. However the number generated by one is at times twice that of the other. Although part of this variability is a consequence of the small number of particles >5 microns in semiconductor fluids, the ratio of counts nevertheless remains consistent for a given sample. The greatest instrument disagreement exists in the region around 1 micron, with the PMS typically reading two to five times the count of the Hiac. Part of this discrepancy is due to the fact that the millivolt thresholds of the 0.8 and 1.0 micron bins were obtained by interpolation: the Hiac instruments from 0.72 and 1.1 micron microspheres and the PMS from 0.72, 0.945, and 1.27 micron spheres. For the semiconductor industry these shortcomings are not critical, since by concentrating on reducing submicron particles the larger ones will be made extremely scarce. In addition, large particles can be cleaned off of wafers much more easily than can smaller ones.

Fortunately, as the size range is lowered to the half-micron level the instruments trend toward parity. At >0.5 microns the Hiac usually generates a higher number in all fluids examined except sulfuric acid, for which the Hiac count is lower at all sizes. Since the two counters are approximately equal at >0.5 microns, particle specifications should be set at this size rather than at >1 micron, where the true concentration is in greater dispute. It is surprising that more consistent agreement is observed at smaller sizes, since the signal generated by a particle diminishes by the sixth order of the radius. The discrepancy at >0.5 microns, usually under 20%, is adequate for user-supplier specifications as long as a certain amount of leeway is observed. It is doubtful whether particle counting can ever attain the precision now routine in metallic contaminant measurements: counts can vary with time and storage conditions due to various particle phenomena both increasing the count (release of particles by the container walls, breakup of large particles, crystallization by ammonium fluoride containing etches, formation of microbubbles), and serving to decrease the count (attraction to container surfaces, dissolution in strong acids, attraction to the liquid surface, particle agglomeration).

Upon consideration of the entire size distribution, it is clear that each instrument is internally uniform. The shape of its cumulative count curve becomes a characteristic fingerprint of that counter and appears to be independent of fluid and level of contamination. The relative vertical placement of the curves is remarkably consistent for individual sensors counting bottles of the same chemical as well as for bottles of different chemicals. This justifies the use of liquid particle counters as quantitative analytical instruments. It is the opinion of the author that differences in particle counts are an accurate reflection of the state of the liquid flowing through the cell rather than of variation internal to the counter. This is, of course, exclusive of electronic noise which elevates the lowest channel count and bubbles which elevate the large counts.

REFERENCES

1. T.A. Milner and T.M. Brown, in "Proceedings of Microcontamination Exposition and Conference" held in 1986, pp. 146-154.
2. S.L. Peacock, M.A. Accomazzo and D.C. Grant, J. Environmental Sci., 29(4), 10 (1986).
3. D.C. Grant and W.R. Schmidt, J. Environmental Sci., 30(3), 28 (1987).
4. R. Knollenberg (1988), personal communication

PARTICLE COUNTING OF LIQUID SYSTEMS USING A SCANNING ELECTRON MICROSCOPE

W. A. Zorn

IBM General Technology Division
Essex Junction, Vermont 05452

A procedure to filter particles with diameters as low as 0.2 μm, and statistically quantify them using a scanning electron microscope (SEM) was developed and found to be limited with liquid reagents. Counting particles with this procedure indicated that large differences existed between automated counters and the SEM technique.

INTRODUCTION

Submicron defects caused by particles impact semiconductor yields as product dimensions get smaller. The intense competition in the production of low-cost megabit chips requires maximum yield, thus control of size and number of particles is receiving a great deal of attention.

Statistically valid methods are required to monitor particle levels in solution. The ideal method should be reproducible, accurate, easily calibrated, rapid, and have detection limits at or below 0.1 μm. Particle counting of liquid reagents is currently accomplished with automated laser particle counters. While these instruments are fast and compatible with corrosive liquids, the lower level of particle size detection is usually 0.5 μm [1]. In addition, particle counters cannot distinguish between nonparticulates, such as bubbles, and thus can generate erroneous data. The filtration/SEM method, however, has the ability to detect much smaller-sized particles, but the accuracy and limitations of the technique have not been defined. The purpose of this study was to determine the applicability of the technique to various chemicals, and to determine the correlation with optical counters.

This investigation employed 0.2 μm as the lower limit of particle sizes studied; however, particles much smaller than 0.2 μm can be counted using smaller filter pore sizes. An obvious disadvantage of this technique is that SEM counting takes a great deal of time.

The methodology for counting filtered particles with an SEM is not new [2]. An SEM method has been used for many years at the IBM facility near Burlington, Vermont, to quantify particulate and bacteria levels in deionized (DI) water [3]. A big advantage in the analysis of DI water is that large volumes (hundreds of liters or more) can be filtered, greatly reducing the multiplication factors. In analyzing liquid chemicals which are more expensive and difficult to dispose of, the technique is limited to much smaller volumes (less than 100 ml). Fortunately, because many chemicals contain higher levels of particles, than those present in DI water, the reduced sampling volumes are adequate for statistical validity.

STATISTICAL CONSIDERATIONS: THE NEED FOR RANDOM FIELD SELECTION

Since the results of the filtration/SEM method might be used to provide data for future chemical specifications, it was prudent that the data be statistically valid. Non-random deposition during filtration can result in clusters of particles on filters. Analysis and error predictions would be inaccurate with non-random sampling of clustered particles. To eliminate counting errors, the filter area displayed on the SEM cathode ray tube screen, henceforth called field, was selected randomly. This was accomplished by generating a series of random numbers in which each number designated a specific SEM field coordinate. Two SEMs equipped with automatic staging suitable for stage programming were used. A computer program was written that automatically moved the SEM stage to coordinates determined by a random number generator. The program had the ability to generate sampling sequences for the number of fields desired.

DATA CALCULATIONS

When a volume of fluid was passed through a filter, particles greater than the pore size of the filter were collected. The assumption was made that, in all filtrations, the particles removed were independent of properties of the fluid and distribution of particles across the filter was not necessarily uniform.

The filter, viewed under a known magnification of the SEM, was mathematically sectioned into N fields of view. The number N was very large; only a smaller number (n) of the total N could be counted in a reasonable period of time. When the filter was scanned for particles, the following applied: x was the average number of particles per SEM field for n fields counted out of the total N fields that constituted the filter surface. For the blank, y (see note) was the particle average for an independent n out of N on a blank not exposed to any fluid. The counts per unit volume of sample, C, were then given by the following equation. (The actual operating parameters, volumes used, and field sizes are discussed in the SEM Standard and Chemical Survey sections).

$$C = \frac{[(\overline{X}) - (\overline{Y})]N}{V}, \text{ where V was the volume in liters.}$$

Note: In practice, the background count was obtained using the same volume of filtered DI water as used for the sample volume.

ERROR IN COUNTING

Using the above formula for determining particles/liter, it is possible to calculate the relative error in counts from the following equation:

$$\text{Error} = \frac{\pm 2N}{V} \left[\frac{S_X{}^2 + S_Y{}^2}{n} \right]^{1/2} \text{ (for 95\% confidence limit)}$$

Where N= total number of fields

V= volume in liters

S_X = standard deviation for sample (particles/field)

S_Y = standard deviation for background (particles/field)

n = number of fields counted out of N

For determination of error in counts from multiple filters, the following equation was used:

$$\text{Error} = \frac{\pm t(S)}{\sqrt{K}}$$

S = standard deviation of samples
k= the number of samples compared

Where t is dependent upon the number of samples for a 95% confidence interval:

Sample Size	t-Value
2	12.706
3	4.303
4	3.182
5	2.776
10	2.262
20	2.093
30	2.045
infinity	1.960

Determination of a Suitable Field Size

The application of error calculations is illustrated for the example of determining a suitable field size to scan for our particle counting program. Isopropyl alcohol taken from our bulk delivery line is filtered and packaged in metal containers ("repacked"). One hundred milliliters from this container were filtered and the filter counted using 50, 100, and 200 fields, randomly chosen ("random walks"). For a 200 field random walk this represents 0.057% of area sampled (200/348,100; refer to SEM STANDARD section). The error equation for a single filter yielded the following values:

Field	M̄ (millions of particles/liter)	Error	% Error
50	35.9	± 4.4	12.3%
100	55.1	± 10.8	19.6%
200	45.9	± 5.8	7.9%

Although the % error decreased as the number of fields increased, the 50 field random walk routine was used for our study because it reduced counting time and operator fatigue from over two hours (200 field) to under 30 minutes (50 field) for a single complete random walk, without significantly sacrificing counting error.

Error in Counts: Single Bottle, Multiple Filters

Three separate filtrations, from the same repacked can of isopropyl alcohol used in the field size determination experiment, were counted using a 50 field random walk. The results for the single filter and multiple filter errors are:

Filter	Count	Error	% Error	Multiple Filters	Error	% Error
A	35.9	4.4,	12.3%			
B	42.9	4.1,	9.6%	38.1	10.3	27.0%
C	35.5	9.5,	26.8%			

This data illustrates the variability of particle levels within just one bottle. For bottle-to-bottle or lot-to-lot comparison, one can expect the same or even higher variation.

SEM STANDARD

To determine the exact screen size for the magnification used and verify this size against the values obtained from the SEM, a standard was used, consisting of 0.25 µm lines and spaces etched on a silicon wafer and mounted in the SEM. The number of lines was counted in the x and y axes. The area of the screen size at 8,600 magnification was found to vary slightly for each SEM (For one SEM the field size was determined to be 144 microns^2). The 0.25 µm lines and spaces also allowed verification of the average pore size of the filters. Exact verification of the screen size was necessary to minimize errors in calculation, since the number of screens were multiplied by a very large factor N, which was typically on the order of several hundred thousand for a 13-mm polycarbonate filter. The actual number from one SEM was determined to be 348,100. If one chose to use larger size diameter filters, N would grow correspondingly bigger (4,840,000 for a 25 mm filter).

PARTICLE COUNT VERIFICATION

Mixtures of submicron-sized latex spheres in an aqueous solution were employed to substantiate particle count data. Sphere sizes in 0.5 µm, 0.6 µm, and 1.0 µm were obtained from Filtech Corporation. The original concentration of these spheres was on the order of millions of spheres per cubic centimeter (cc).

Antistatic and antibacterial growth agents were added by Filtech in a proprietary blend to minimize clustering of spheres and bacteria growth so that these solutions could be used for several months. A detailed explanation of this technique is given in the Latex Sphere Standardization Addendum Section in the Appendix.

FILTRATION/SEM SEQUENCE

The following flow chart depicts the sequence used for the filtration/SEM technique:

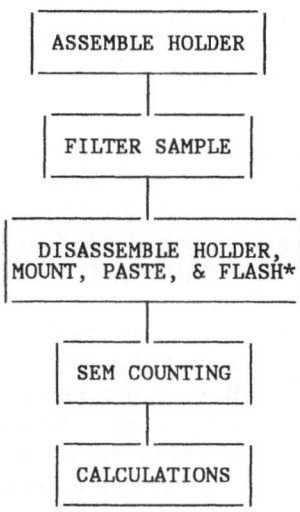

*Mount/Paste/Flash: After filtering, the filters were fixed on carbon planchets using a colloidial carbon/isopropanol based adhesive, and dried by evaporation. A thin film of gold was applied (flashing) and the sample was mounted in the SEM for counting. Carbon planchets were selected because they were flat, which was desirable for focusing, and more importantly; carbon could be obtained in high purity, and is a low atomic number element. Elements with atomic numbers below 23 are invisible to EDX elemental analysis. Thus, the planchets with filters could later be used in elemental analysis when the need arose.

CORRELATION OF SEM/HIAC-ROYCO AUTOMATED OPTICAL COUNTERS

A substantial effort was made to understand the relationship between SEM data and a Hiac-Royco automated optical counter currently used for chemicals [4]. Experimentation was divided into two areas: ideal situations using known concentrations of latex spheres, suspended in DI water, and counting of actual reagents. A narrow window was found in which the particle levels could be compared for the two techniques. If the particle concentration was too high, the counter was saturated with too many particles, and the numbers reported were erroneous. However, if the particle levels were too low, the data collected from the SEM method would be erroneous as there would be too small a sample population of particles to be statistically represented.

Ideal Case

Filtering and sample preparation were done in a clean room environment, where great care was exercised to minimize outside particle contamination. For example, 0.2 µm point-of-use filters on the DI water lines were employed to purge the counter lines between readings. Samples for SEM counting were obtained by collecting and subsequent filtering the sampling lines leaving the counter. The lowest level for background counts of filtered water was 90 total particles/cc.

The optimum window was 600 total particles/cc to 2,000 total particles/cc, using 0.55 µm size latex spheres. Keeping the concentration of spheres within this window gave an average correlation of 1:2.5 for SEM: optical counter. This window is valid only for submicron particles and could change for another SEM/counter system or for other volumes. It is listed only to demonstrate the limited opportunity for comparison of the two techniques. One reason for the two-threefold difference in correlation seen was that the SEM technique allowed visual discrimination between spheres and particles, whereas the optical systems counted everything--spheres, particles, and bubbles. Also, the 0.55 µm size spheres were at the lowest diameter size detectable by the optical counter. A logarithmic break begins below 1.0 µm for correlation coefficient and submicron particle sizes, regardless of which optical counter is employed. While our efforts concentrated on spheres less than 1.0 µm, this trend is well established for various counters and sphere sizes by Filtech Corporation. Above 1.0 µm, approximately a 1:1.2 relationship existed for many optical counters (Figs. 1 and 2).

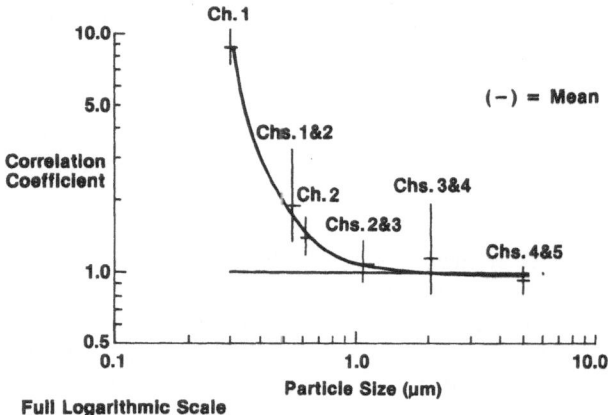

Figure 1.   Calibration curve for PMS counter

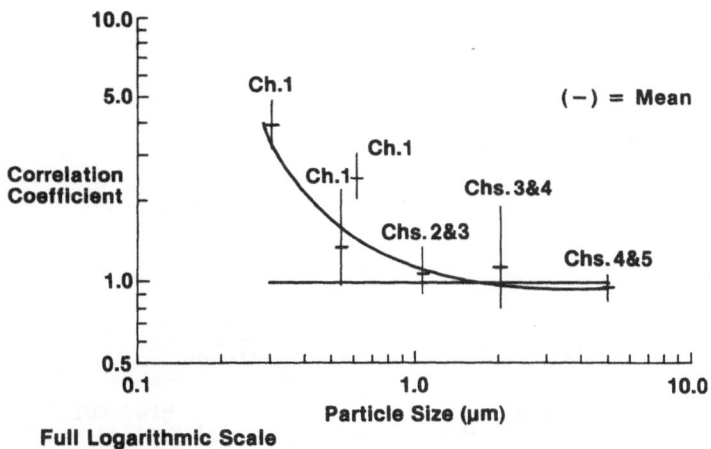

Full Logarithmic Scale

Figure 2. Calibration curve for Hiac/Royco counter

Actual Chemicals

Correlations between the filtration/SEM technique and optical counters were determined for acetic acid, ammonium hydroxide, and isopropyl alcohol. Ammonium hydroxide was chosen because it was one of the few bases compatible with our filters and had sufficient particles present to populate the filter area. The same clean room environment and procedure was used as in the ideal case except that chemicals filtered were not taken from the effluent line of the Hiac/Royco because of potential contamination from the piping. Earlier attempts to filter chemicals using 0.2 µm pore sized filters revealed that large quantities of particles less than 0.5 µm were clustering together. Consequently, the filter's pore size increased to 0.45 µm in an attempt to reduce aggregation effects. Since the optical counter couldn't detect these smaller size particles, it was believed that, by reducing the collection of these smaller particles, less aggregation would occur, giving a closer correlation. During the survey of Burlington reagents, however, 0.2 µm pore filters were used, which involved no optical particle counting, to assess the contribution of the particles less than 0.5 µm.

The results listed in Table I show that the SEM: counter correlation for acetic acid is 33:1, ammonium hydroxide 4:1, and isopropyl alcohol 39:1. One reason for the large differences observed could be that, during drying on the filter, particles may have been crystallizing out of solution. SEM pictures of the isopropyl alcohol filter indicated swelling of the pores which would reduce the flow and trap more particles smaller than 0.5 µm. These particles could aggregate to form larger sized particles. In a separate experiment, rinsing with DI water showed particle levels decreased in acetic acid but increased in ammonium hydroxide. Thus, rinsing the filter was determined not to be a viable technique for reducing aggregation and crystallization. Other possibilities for errors could have been SEM operator error in deciding whether or not a shape was a particle or part of the surface of the filter. Unlike the ideal case where every particle counted was a perfect sphere, all shapes and shades had to be considered.

Table I.   SEM vs counter correlation

	SEM			HIAC/ROYCO		
	0.5-1.0μm	>1.0μm	Total	0.5-1.0μm	>1.0μm	Total
Acetic Acid	7.00 M	3.00 M	10.0 M	293 K	5.20 K	298 K

Acetic Acid Correlation: 33:1 (SEM:Counter)

	SEM			HIAC/ROYCO		
Ammonium Hydroxide	35.1 M	14.1 M	49.0 M	12.8 M	0.90 K	12.8 M

Ammonium Hydroxide Correlation: 4:1 (SEM:Counter)

Isopropyl Alcohol*

	SEM			HIAC/ROYCO		
Bulk Feed #10	1.70 M	350 K	2.05 M	98.1 K	0.50 K	98.6 K
Bulk Feed #11	4.90 M	1.00 M	5.90 M	104 K	0.70 K	105 K
		average=	3.98 M		average=	102 K

Isopropyl Alcohol Correlation: 39:1 (SEM:Counter)

*Samples taken from bulk feed lines located inside Chemical Distribution Building.

Notes: Results in particles/liter (M = million, K = thousand); numbers reflect average counts for both techniques; background counts were subtracted for the SEM method.

FILTRATION

Filtering was performed in a class 1000 or better hood to minimize exogenous particle introduction. A tremendous difference in the levels of airborne contamination was found to exist between different hood systems (see Table II). The need for such a hood is well documented[5,7]. Details of our filtration steps are given in the Appendix section.

Table II. Hood comparison in terms of particles per ft^3

A. Wet Laboratory (Inorganic Area)

	0.2 μm	0.5 μm	5.0 μm
Position A	237,907	25,700	127
Position B	231,021	17,877	86

B. Class 1,000 Hood (Light Microscopy Laboratory)

Position A	8	2	1
Position B	3	1	1

C. Class 100 Wet Station (Site Contamination Laboratory)

Position A	32	2	0
Position B	313	33	4

FILTERS

The 13-mm diameter filters supplied by Nuclepore Corporation were available in polycarbonate or polyester material with very smooth surfaces that facilitated particle identification. This diameter was selected to increase the number of particles per unit area of the filter. The materials, however, were not compatible with all of the reagents tested. Teflon* filters could not be used because their surfaces were too fibrous to allow easy identification of small particles (Fig. 3). Consequently, dilutions were employed to reduce chemecal attack on the filter.

---

* Trademark of E. I. duPont de Nemours and Company

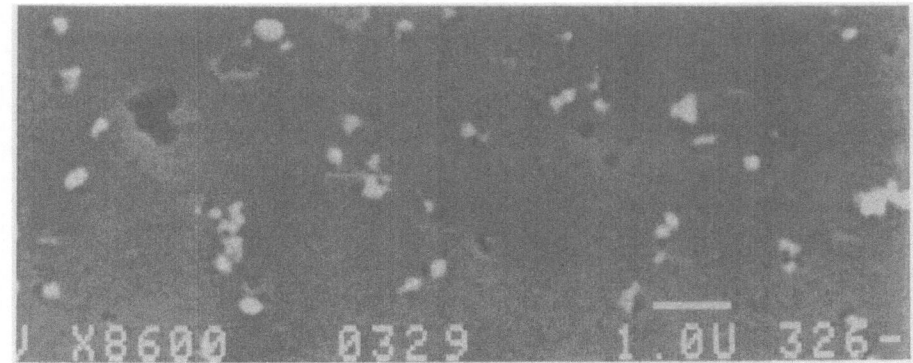

0.2 µm Pore Size Polyester Type Filter (at X8600)

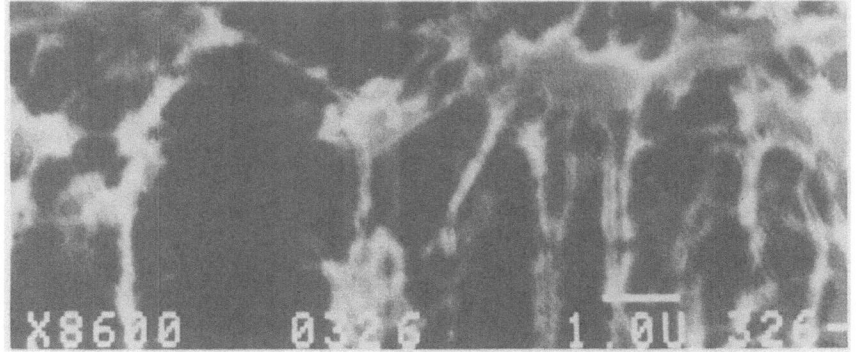

0.2 µm Pore Size Teflon Type Filter (at X8600)

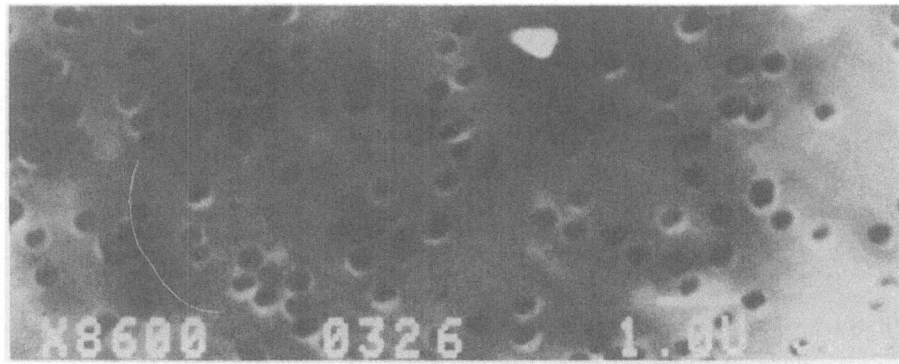

0.2 µm Pore Size Polycarbonate Type Filter (at X8600)

Figure 3.   Comparison of SEM filters at high magnification

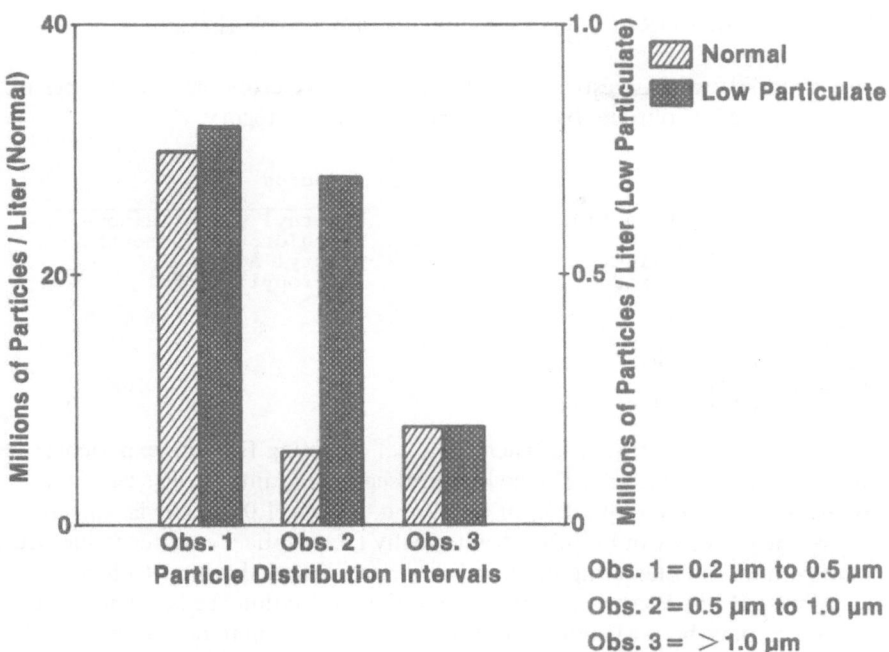

Obs. 1 = 0.2 µm to 0.5 µm
Obs. 2 = 0.5 µm to 1.0 µm
Obs. 3 = > 1.0 µm

Figure 4.   Comparison of normal vs low particulate buffered oxide etch

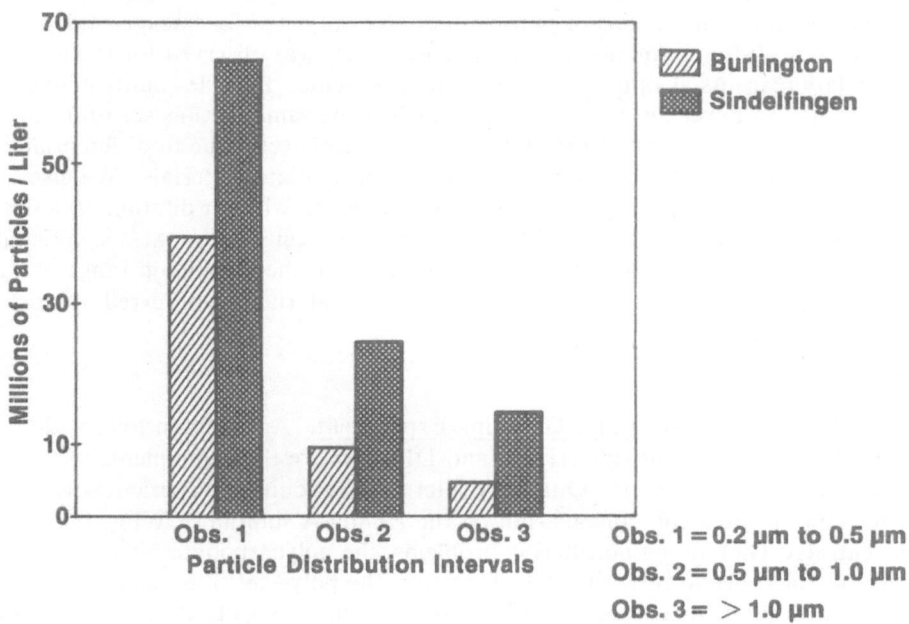

Obs. 1 = 0.2 µm to 0.5 µm
Obs. 2 = 0.5 µm to 1.0 µm
Obs. 3 = > 1.0 µm

Figure 5.   Particulate comparison of Burlington vs Sindelfingen: NH_4OH

SURVEY OF IBM BURLINGTON PROCESS CHEMICALS

These materials were investigated as a representative cross section of types of wet chemicals used at our site by their respective class category:

Inorganics	Organics
Buffered Oxide Etch (10:1)	N-methyl pyrrolidinone
Huang A Reagents	Trichlorotrifluoroethane
Hydrogen Peroxide	N-Butyl Acetate
Ammonium Hydroxide	Isopropyl Alcohol
Nitric Acid	
Hydrofluoric Acid	
Sulfuric Acid	
Hydrochloric Acid	
Phosphoric Acid	
Potassium Hydroxide	

Following is a brief description of each material depicting filtering experiences and the particle levels found. The counts are segregated into three size categories: 0.2 µm to 0.5 µm, 0.5 µm to 1.0 µm, and > 1.0 µm. It is expected that the particle levels would differ dramatically from optical counter values due to crystallization and clustering of submicron particles on the filter to form larger aggregated particles, as noticed during the correlation study. Therefore, these numbers may be misleading and are presented for comparison only.

Inorganics

Buffered Oxide Etch (10:1). Vendor-supplied buffered oxide etch with a ratio of 10 parts hydrofluoric acid to 1 part ammonium fluoride (both normal and low particulate reagents) were examined. Ten milliliters of etchant were filtered and counted, using a 50 field random walk for each case. Figure 4 depicts the results, which are in millions of particles per liter, adjusted for background particles. A 20-fold reduction in total particle levels was observed for the low particulate reagents as opposed to the normal reagents. Particle counts might have been even lower but the counts approached the same level as seen for our background filter, indicating the inability of the SEM technique to differentiate between particle levels when used to count low particulate materials. We also observed that this reagent could be filtered as received without dilution on a 0.2 µm pore size polycarbonate type filter in a stainless steel holder. Also significant was that the largest number of particles was found in the submicron range of 0.2 µm to 0.5 µm. This trend was not unique to this material but occurred in most chemicals surveyed.

Huang A Reagents and Wafer Gettering Experiment. Ammonium hydroxide (NH_4OH), hydrogen peroxide (H_2O_2), and DI water are the components of the Huang A preclean reagents. Our early filtering and counting experiences were on the concentrated components--not on the Huang A solutions. While concentrated H_2O_2 posed no filtering problems, the polycarbonate filters were unusable for concentrated NH_4OH. However, the polyester filter could be employed to filter the concentrated NH_4OH. Comparisons of NH_4OH from both IBM Burlington and IBM Sindelfingen, West Germany vendors were also made because of the availability of reagents at that time from Sindelfingen. Results of the two samples of NH$_4$OH (Fig. 5) indicated fairly large particle levels in both. Figure 7 shows the particle levels found in 30% H$_2$O$_2$. along with the results for diluted nitric acid. It is obvious that, from a review of data in Figs. 5, 6 and 7, dilution of the components in making the Huang A reagent decreased the levels of particles. In some cases, increased dissolution of certain types of particles may have occurred.

Also important was the effect of wafer gettering. A separate experiment was conducted with materials sampled in a manufacturing line. A reduction in particle levels was observed for chemicals after processing wafers through the tanks. Since foreign material counts taken at the bright-light inspection station on the wafers themselves were not high, the DI water rinse was apparently effective in removing particles from the wafers (Fig. 6).

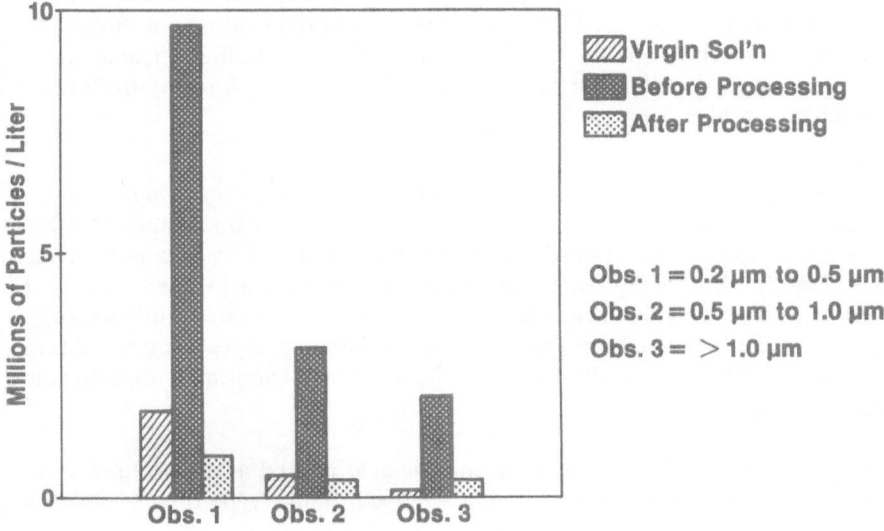

Figure 6. Comparison of Huang A Station: Before / after wafer processing

All of the concentrated acids surveyed had to be diluted in order to reduce attack on the filter and portions of the holder. The Kel-F* filter holder had to be employed as well. This procedure was followed for nitric, hydrofluoric, sulfuric, hydrochloric, and phosphoric acids. Extrapolation to the concentrated reagent will not necessarily reflect the true number.

---

* Trademark of 3M Company.

Nitric Acid. Concentrated nitric acid (HNO_3) is frequently used in the front-end-of the line (FEOL) to clean wafers prior to various processing steps. A 5:1 dilution of HNO_3. with DI water was sufficient to reduce filter attack, when used with the Kel-F holder and polycarbonate type filter. The total level of particulates was found to be quite high even after dilution, on the order of a hundred million particles per liter (Fig. 7).

Hydrofluoric Acid. Concentrated hydrofluoric acid is used in FEOL at certain process etch steps. A 4:1 dilution was made and filtered through a polyester type filter. The counts for this dilution showed an average of 8 million total particles per liter (Fig. 8).

Sulfuric Acid. Concentrated sulfuric acid is used as a preclean for certain FEOL processing steps. Sampling of this material required a 5:1 dilution with DI water and a polycarbonate type filter. The results from two one-gallon containers revealed a total of 83 million particles per liter in the first container and 29 million particles per liter in the second container--almost a threefold difference between two bottles each from the same lot number. The average total count was 46 millon particles per liter shown in Fig. 8 for hydrofluoric and hydrochloric acid.

Hydrochloric Acid. Concentrated hydrochloric acid is a component of the Huang B preclean station along with H_2O_2 and DI water for certain FEOL processing steps. A 4:1 dilution was filtered and counted using a polyester type filter (Fig. 8). An average of 19 million total particles per liter was found. It should be noted that this acid was the most corrosive on our stainless steel supports, located inside both types of filter holders. Best results were obtained by limiting contact of the diluted acid with the holder and immediate flushing with DI water.

Phosphoric Acid. Concentrated phosphoric acid is used as a metal preclean in wafer processing. A 4:1 dilution using a polycarbonate type filter yielded an average of 49 million particles per liter present (Fig. 9).

Potassium Hydroxide. Potassium hydroxide is used as a developer in certain resist apply levels on different product lines. A ten milliliter aliquot of 0.2 normal potassium hydroxide was filtered from one-gallon containers using a polycarbonate filter. The material was tested as received, with no dilution required. Total particle counts averaged 1.2 million/liter (Fig. 9).

Organics

Swelling of the filter was observed to occur for most of the organics, the exception being trichlorotrifluoroethane. This could result in a reduction of the pore sizes if the problem was severe enough.

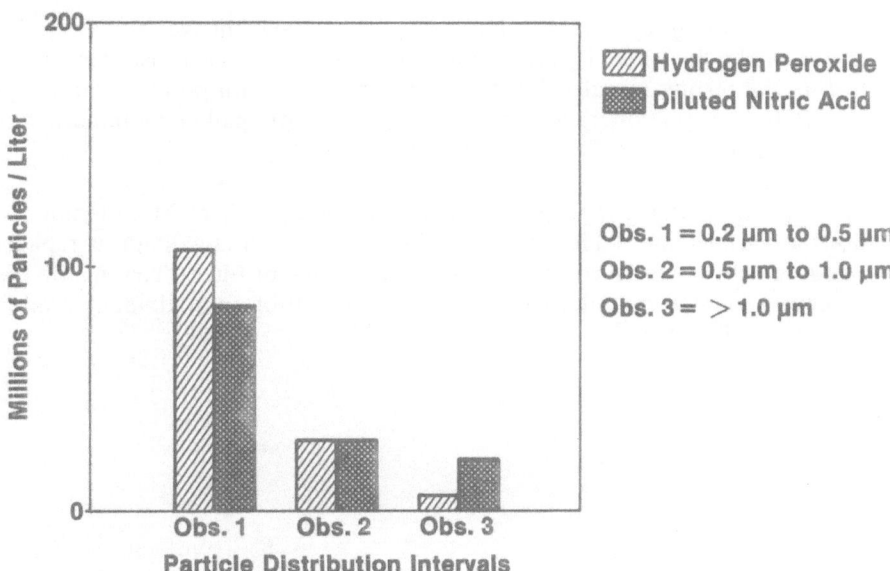

Figure 7.   Particles in hydrogen peroxide and diluted nitric acid

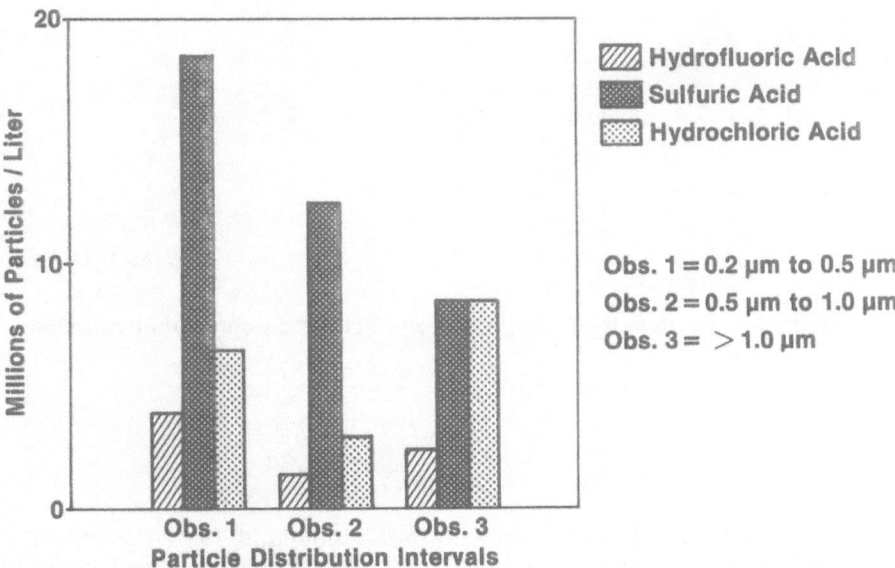

Figure 8.   Comparison of particles in hydrofluoric, sulfuric, and hydrochloric acids

N-Methyl-Pyrrolidinone (NMP). NMP is used primarily in the back-end-of-the-line (BEOL) processing to strip polyimide or resist from the wafer. Initial testing revealed NMP was incompatible with polycarbonate type filters. Polyester type filters used to filter particles from gallon containers revealed total particle levels of 30 million/liter (Fig. 10).

Trichlorotrifluoroethane (Freon TF). Freon TF is used for resist adhesion stations in various lines. This material was sampled from a gallon-size repacked can and was shown to be compatible with either type of filter. Ten mls filtered on a polycarbonate-type filter gave a fairly low count of approximately 3 million particles/liter (Fig. 10).

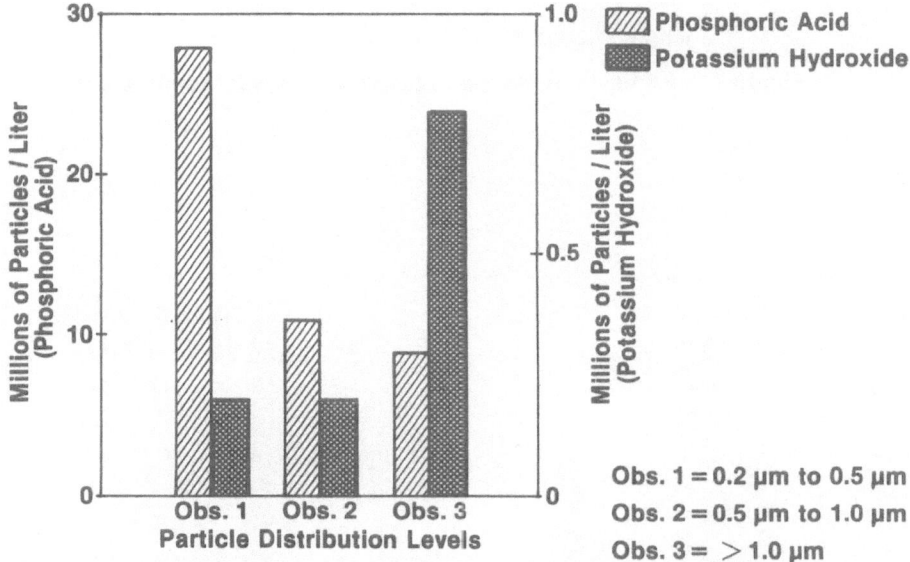

Figure 9.   Particle levels in phosphoric acid and potassium hydroxide

Isopropyl Alcohol (IPA). IPA is used in resist stripping processes. Sampling was done from the repack cans and the bulk tank wagon. Ten mls of samples were filtered without problems using polycarbonate type filters. The particle levels in the repack cans averaged 3.6 million/liter. As expected, the tank wagon samples were found to be much higher--103 million particles/liter (Fig. 11).

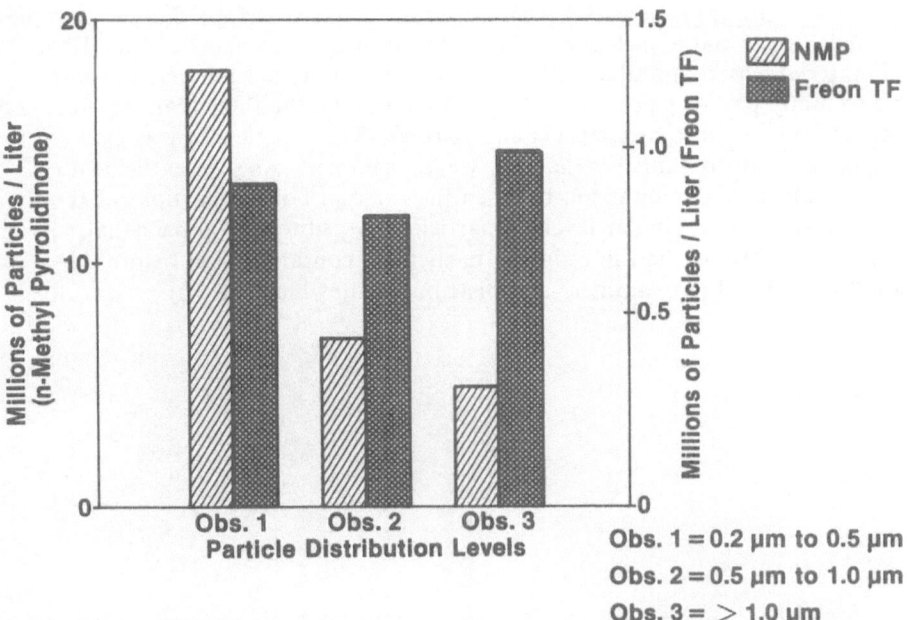

Obs. 1 = 0.2 µm to 0.5 µm
Obs. 2 = 0.5 µm to 1.0 µm
Obs. 3 = > 1.0 µm

Figure 10. Particle levels in N-methyl pyrrolidinone and trichlorotrifluorethane

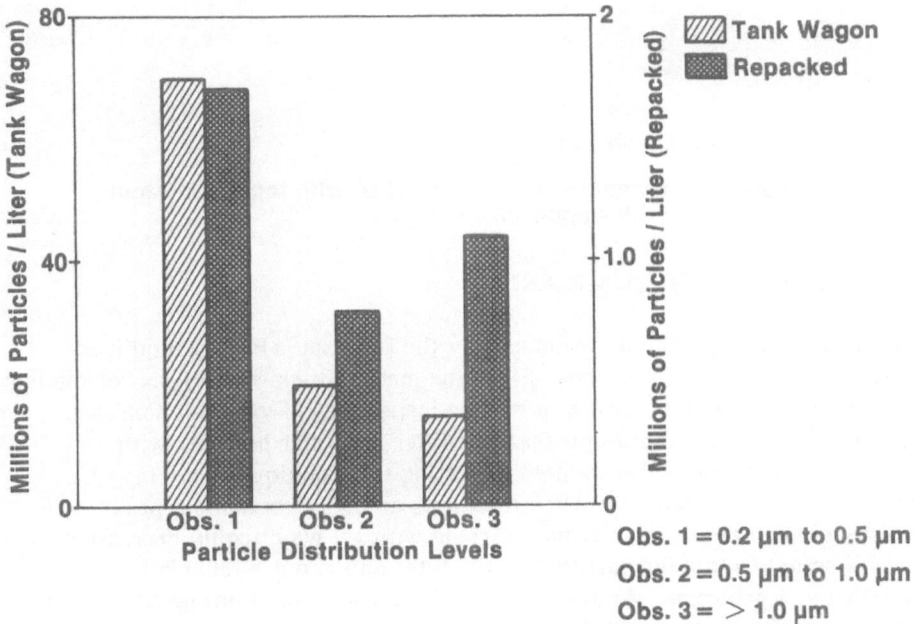

Obs. 1 = 0.2 µm to 0.5 µm
Obs. 2 = 0.5 µm to 1.0 µm
Obs. 3 = > 1.0 µm

Figure 11. Comparison of repacked IPA with tank wagon IPA

<u>N-Butyl Acetate (NBA).</u> NBA is also used in the resist-stripping area. Samples were taken from the repack cans and bulk tank wagon, as was done for IPA. The material was compatible with the polycarbonate-type filter. Since there were so many particles present (> 500 million/liter), the filter scan was reduced from 50 fields to 10 to minimize counting time. The sample size was also changed from 10 to 5 mls for the tank wagon samples. As seen earlier with isopropyl alcohol, a comparison between the top and bottom portions of the tank wagon revealed similar levels in particles and shows the impact that point-of-use filtration had in reducing particulate contamination before the material was filled in containers and distributed sitewide (Fig. 12).

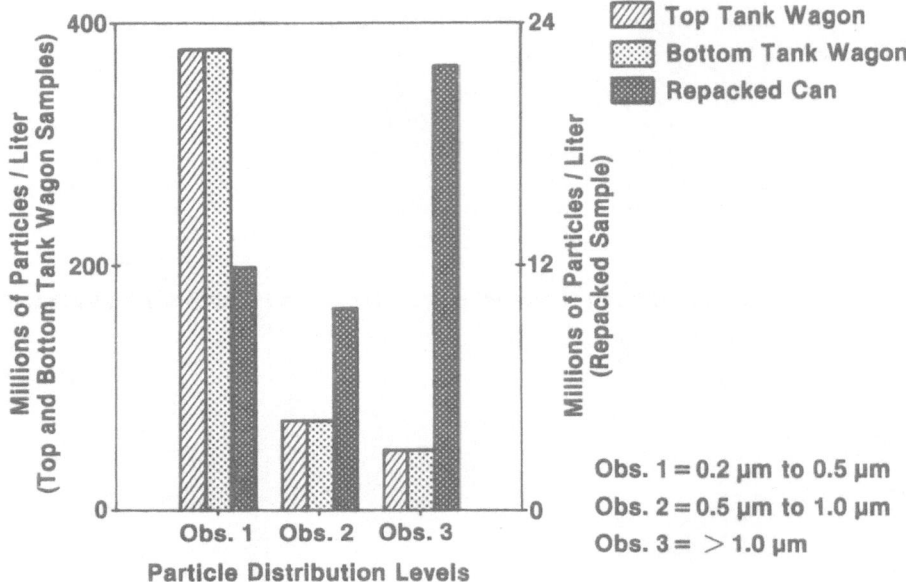

Figure 12. Comparison of repack NBA with top and bottom tank wagon NBA samples

CONCLUSIONS AND SUMMARY

Particle counting of liquid chemicals by the filtration/SEM method was demonstrated to have limitations. Since the method is slow and time consuming, it would not be practical for use in routine inspections of various chemicals. Only those chemicals compatible with the filter and filter housing, with no dilution required, should be counted. In using the technique, there may be additional particles generated when the filters are dried, as some types of particles crystallize out of solution. Rinsing with DI water could decrease or increase actual counts for particles on the filter and is not a reliable cure for crystallization problems. Another source of error may be counting of filter surface imperfections as particles. Yet, the most serious error is due to the clustering of submicron-size particles to form larger particles on the filter.

To minimize errors in counting, the particle levels should be sufficiently high in order to populate the filter area. Counting of low particulate reagents is not recommended because these materials are too low in both particulate matter to be statistically represented and in numbers required to exceed background counts.

Precision for the SEM technique was found to increase as particle levels/unit volume increased. A good example was the comparison between repacked isopropyl alcohol (IPA) and tank wagon IPA. For repacked IPA (after point-of-use filtration) with an average of 3.6 million particles/liter, the error rate was 32% for three samples. For tank wagon IPA with an average of 103 million particles/liter, the error rate dropped to 17%. Obviously, chances of randomly finding particles were better when the filter was more populated with particles.

Potential applications for the SEM method would be to verify concentrations of latex spheres in solution, which could be used as calibration standards since consistent and close correlation was found to exist. The differences seen for the three types of chemicals during the correlation experiment helped explain why discrepancies between the SEM method and past analyses using optical counters were so large. Had it been possible to use volumes greater than one liter without plugging up the filter, it is believed that the differences now seen would be significantly reduced, as the multiplication factors would be much less.

Bulk and cylinder gases, although not tested in this study, would be excellent materials to characterize by SEM. The volumes required to offset the multiplication factor and collect enough particles would be quite high; tens, hundreds, maybe thousands of liters could be used and, since the viscosity of most gases is less than comparable liquids, resistance when filtering would not be a problem. The SEM technique is also the only method available which can quantify large concentrations of particles in fluids, as optical counters cannot handle high particle counts.

Keeping the working environment as clean and particle-free as possible is a key factor in reducing exogenous particle introduction. The container types and the seals/caps used are critical in maintaining clean chemicals [5]. Ground glass containers should not be used as a collection vessel for any liquid, if one intends to quantify particles from this container.

In general, regardless of the material or grade, the largest concentration of particles resided in the submicron range (0.2 μm to 0.5 μm). As chip-size dimensions shrink, the impact that the submicron region and smaller particle sizes will have on our products will undoubtedly increase. Presently, SEM techniques offer the only method of detecting and quantifying these smaller particles.

ACKNOWLEDGEMENTS

The author wishes to thank the following people for their support in this project: Glen Enos, Brian Burnor, and Jim Chappel for building certain SEM staging units and the Kel-F holder; David Paul, for providing the SEM standards, computer-generated random walk program, and help in setting up our SEMs for random walk capability; Roy Hango, for sharing his experiences in quantifying particles for our site's deionized water system; Wendell Carr, for providing statistical guidance and stressing the need for complete randomness of our sample base selection; Jack Hedman, for sharing his calibration experiences and data; Robert Suslavich of Filtech Corporation, for his advice in using and developing the latex sphere calibration procedure; Hal Sullivan, for his efforts to obtain ammonium hydroxide from IBM's Sindelfingen plant located in Germany; John Kehley, for arranging and facilitating the sampling of chemicals in the manufacturing line (Huang A Station); Eric Vanslette, for his help in counting particles from the various chemicals using the SEM technique; Larry Austin, for his help in counting chemicals using the Hiac/Royco counter; Larry Doane, for his technical ideas and support; and Ken Racette, for his support, constructive criticism and help in reviewing this paper for publication.

REFERENCES

1. K. Dillenbeck, Measuring particulates in processing chemicals: The need to standardize, Microcontamination, 23, November 1985.

2. J. R. Millette, P. J. Clark, M. F. Pansing, Sizing of particulates for environmental health studies, Scanning Electron Microscopy, Vol. 1, 253, 1978.

3. R. Hango, DI water: A common sense approach, in "Proceedings of the Third Annual Semiconductor Pure Water Conference," 1984.

4. S. Peacock, Quantitative calibration of light scattered particle counters, J. Environmental Sci., July/August 1986.

5. K. Dillenbeck, Characterization of particle levels in incoming materials, Microcontamination, 58, December 1984.

6. J. T. Przybytek, K. L. Calabrese, Measuring low level particle count in solvents, Microcontamination, 53, June 1985.

7. ASTM F-60, Detection and enumeration of microbial contaminants in water for microelectronic devices, 1974.

APPENDIX

Additional Comments Regarding Filtration

Filtering was done with either a 10 ml or 50 ml syringe with a Luer type adapter and a stainless steel (ss) or Kel-F filter holder (Fig. 13). The 50 ml syringe was used when larger volumes were required because of dilutions or the need to increase the particle concentration on the filter (refer to individual reagents). The syringes forced measured aliquots of the reagents through a filter supported by a holder. Due to the small size of the pores, a moderate degree of pressure was required to effect a reasonable flow of reagent through the filter. The plunger seat was made of Teflon to minimize back leakage in the syringe body during filtration. A plastic syringe with a rubber seat was employed for chemicals that attacked glass. The filter itself was supported on a ss screen with a 10 mm polyester filter disc sandwiched between. The disc cushioned the filter and eliminated indentations caused by the underlying screen support, which induced particle channeling to areas immediately above the screen pores.

Pore expansion during filtration was not observed to be a problem. SEM measurements showed pore dimensional integrity was maintained after filtration, assuming inelasticity of the filter material. We were confident, therefore, that no significant amount of particles above 0.2 μm diameter escaped through the filter pores as a result of pore expansion from syringe pressure.

A small filter mount using 13 mm diameter filters was employed to minimize the amount of filter area to scan for particles. Although larger diameter filters were available and would have greatly reduced filtering time, especially in situations where larger volumes were needed, it was decided to stay with the 13 mm filters because the number of fields needed would increase proportionly to the area. The savings in filter time would not offset the longer SEM time required for scanning.

As mentioned earlier, ss and Kel-F filter holders were used. Since ss is less inert to many of the concentrated inorganic reagents than the filters themselves, a fluorocarbon substitute mount of Kel-F was designed. Teflon was not used because of the better structural integrity of Kel-F. The cold flow properties of Teflon made dimensional retention impossible. The Kel-F holder was machined from a solid rod of Kel-F by our model shop, since it was not available commercially. Experiments conducted to verify the filter efficiency of both holders using known quantities of 0.62 μm size latex spheres in aqueous solution revealed the ss holder retained > 99.9% of the original spheres, while the Kel-F holder initially removed only 90% of the spheres. Modifications to the Kel-F holder, using larger Teflon 'O' rings, increased the filter efficiency to 99.8%.

Latex Sphere Standardization Addendum

The filtering procedure for latex sphere calibration was very technique-dependent. Slight variations from this technique caused severe discrepancies in the calibration results. The major problem to avoid was clustering of spheres. The exact sequence outlined below gave minimal clustering and best overall results. An ultrasonic bath was needed to sonicate the solutions a minimum of one hour prior to filtering, to break-up any clustering of spheres that might have occurred on standing.

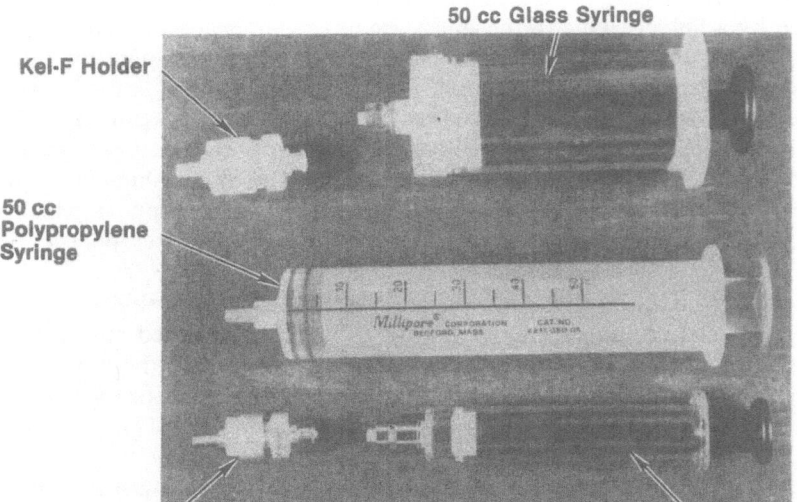

Kel-F Holder

50 cc Glass Syringe

50 cc
Polypropylene
Syringe

Stainless Steel Holder

10 cc Glass Syringe

Stainless Steel Holder
(Top)

Round Sided Teflon 'O' Ring

10 mm Polyester Filter Discs

Stainless Steel Screen

Flat Sided Teflon 'O' Ring

Stainless Steel Holder
(Bottom)

Figure 13. Types of syringes and holders used

<u>Pre-DI Wash.</u> With new filters (polycarbonate or polyester) and 2-3 filter discs in filter holders, holders were rinsed with 7-10 mls of DI water. An extra holder was dedicated as a control--one unexposed to any latex sphere solution.

<u>Latex Solution.</u> The syringe was rinsed with the latex solution, filled to the volume required, and inverted to remove air bubbles. A filter was attached to the holder and syringe in the inverted position. The solution was slowly pushed until drops started forming at the syringe tip. The syringe was turned to its normal position, and the pushing continued until the desired volume was removed and 1-2 mls of solution remained. It was important that no air bubbles were introduced in this step. The holder was removed from the syringe and placed upright. The syringe was cleaned with DI water by refilling with 4-6 mls of DI water. The holder was attached and inverted, then slowly pushed until drops reappeared at the tip. The syringe was turned back to the normal position and water slowly pushed through the holder until 1-2 mls of water were left. The same sequence was repeated two more times, with the same volume of water used for all the filters involved, including the blank. This washing procedure removed surfactants that might reside on the filter and cause clustering during the following drying step.

<u>Mounting.</u> The holder was carefully removed from the syringe and taken apart. The actual filter was removed and mounted on a carbon planchet with the carbon based adhesive. This was done within the class 100 hood. The planchet was covered with a clean, dry, glass beaker and the filter dried overnight before flashing with gold. Note: Attempts to hasten the drying process using an oven or heat lamp were found to insure clustering of spheres and/or ejection of spheres from filter area.

IMPROVED METHODOLOGY FOR MEASUREMENT OF PARTICLE CONCENTRATIONS IN SEMICONDUCTOR PROCESS CHEMICALS

Donald C. Grant

Millipore Corporation
80 Ashby Road
Bedford, Massachusetts 01730

Particle concentrations in semiconductor process chemicals are usually measured using light scattering instruments calibrated for use in water. These chemicals frequently have high vapor pressures, high particle concentrations, and refractive indices which are significantly different from that of water. All of these characteristics can lead to errors in data interpretation. This paper describes methodology developed to reduce the magnitude of these errors.

INTRODUCTION

The processes used to produce state-of-the-art microcircuits require fluids with extremely low levels of particles. Ideally, no particles should be present. Particles and associated impurities can open or short circuits, affect photolithographic reproduction, change electrical properties, and introduce crystal structure defects in microcircuit devices. Any of these effects can result in failure or diminished reliability of the finished device.

Particles can originate from numerous sources including the manufacturing environment, process gases, water, chemicals and processing equipment. Although the first three sources are usually controlled to meet process needs, chemicals and processing equipment are often sources of numerous particles. The environment can be controlled by manufacturing in a class 10 clean room (<10 particles/cubic foot >0.5μm), while process gases and water are filtered through microporous membranes to significantly reduce particle concentrations. Chemicals are either unfiltered or filtered and bottled prior to use. Under these conditions, typical particle concentrations in fluids which contact microcircuits during manufacture are:

Fluid	Particles/liter >0.5μm
Clean room air	< 0.4
Process gases	<10 (>0.01μm)
Process water	<10
Process chemicals	>10,000

Particle concentrations on wafers have been shown to depend on concentrations in the fluids.[1-3] Therefore, as process chemicals are a major source of particles, accurate particle concentration measurement techniques are highly desirable.

Particle concentrations in chemicals from various sources have been reported by several authors.[4-7] The data were obtained using optical particle counters (OPCs) which determine particle size by measuring the amount of light scattered by individual particles passing through a fine capillary. However, the concentrations reported are often subject to errors inherent in the test methodology used. These errors arise from counting gas bubbles, particle coincidence, OPC sensor inefficiency, and differences in index contrast (the ratio of the particle refractive index to that of the medium). This paper describes these sources of errors and presents methodology for reducing their magnitude.

Data are presented for two OPCs:

1. PMS (Particle Measurement Systems, Boulder, CO) IMOLV.3 sensor with a Model LLPS-X Counter, 0.3-15µm sensitivity, 5mW TEM_{00} HeNe laser, 100ml/min flow rate, 1.25×10^{-5}ml sensing volume, sapphire sensing chamber.

2. H/R (Hiac/Royco, Silver Springs, MD) 346BCL Sensor with Model 4100 Counter, 0.5-25µm sensitivity, 5mW Multimode HeNe laser, 100ml/min flow rate, 2.5×10^{-5} sensing volume, sapphire sensing chamber.

POTENTIAL SOURCES OF COUNTING ERRORS

Counting of Gas Bubbles

Gas bubbles are often included in particle counts measured in process chemicals. Until recently, particle concentrations were determined using "batch sampling" techniques. In this method fluid is drawn through the OPC sensor at a controlled rate by vacuum. Solutions with high vapor pressure tend to outgas when subjected to reduced pressures, hence, bubble formation is likely to occur in a number of semiconductor process chemicals. Once bubbles are formed, they are counted efficiently by OPCs. Recent publications have indicated that maintaining a positive pressure on chemical solutions reduces the apparent particle concentration measured.[8-10]

Particle Coincidence

The data output from OPCs assumes that each pulse of light generated results from a single particle and that each particle produces a pulse. However, at high particle concentrations OPCs measure less than the actual particle concentration due to particle coincidence errors.

Two types of coincidence occur: optical and electronic. Optical coincidence is a result of the simultaneous presence of more than one particle in the OPC sensing volume. When multiple particles are in the sensing volume simultaneously, they are counted as one larger particle. Electronic coincidence (or saturation) occurs because time is required by the counter to process the signal generated by a particle passing through the sensor. If a second particle passes through the sensor during the time that the signal is being processed, the pulse generated by the second particle is not counted. Both types of coincidence increase with increasing particle concentration.

Two types of errors result from coincidence: counting and sizing. Electronic coincidence results in only counting errors. Optical coincidence results in both counting and sizing errors since multiple small particles are counted as one larger particle.

The coincidence occurring in the two instruments used in this study was determined using a method developed at Millipore.[11] In this method polystyrene latex (PSL) beads are injected into a flowing stream of water, passed through the OPC sensor and collected on a filter which is subsequently examined by scanning electron microscopy (see Figure 1).

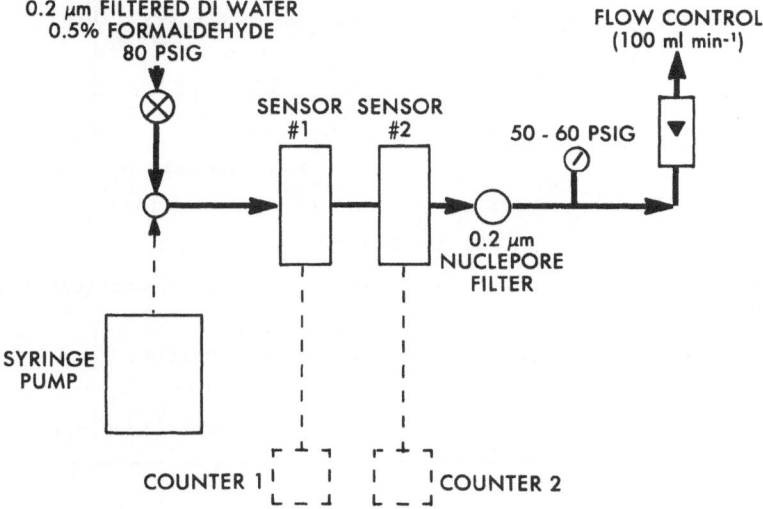

Figure 1.   Calibration test system schematic.[11]

Performance of the two sensors examined is shown in Figure 2. Errors due to coincidence occur at lower concentrations in the PMS instrument than in the H/R instrument; however, the errors are small for concentrations below 1,000,000 particles/liter. The errors start to become significant in the H/R instrument at about 10,000,000 particles/liter.

The coincidence occurring in the PMS instrument is electronic rather than optical and is a result of the signal processor used rather than the sensing cell.[12] The signal processor was designed to process signals from a variety of sensors, some of which require long signal processing times. If the processor were modified to accept signals from only the IMOLV.3 sensor, then the coincidence would be similar to that of the H/R instrument.

Refractive Index Effects

The index contrast is of major importance in determining the amount of light scattered by a particle. It is defined as follows:

$$\text{Index Contrast} = \frac{\text{Particle refractive index}}{\text{Medium refractive index}}$$

Refractive indices for some commonly used semiconductor processing fluids are shown in Table I. Examination of the table reveals that a large variation in index contrast can result, depending upon the nature of the contaminant particles and the process fluid undergoing measurement.

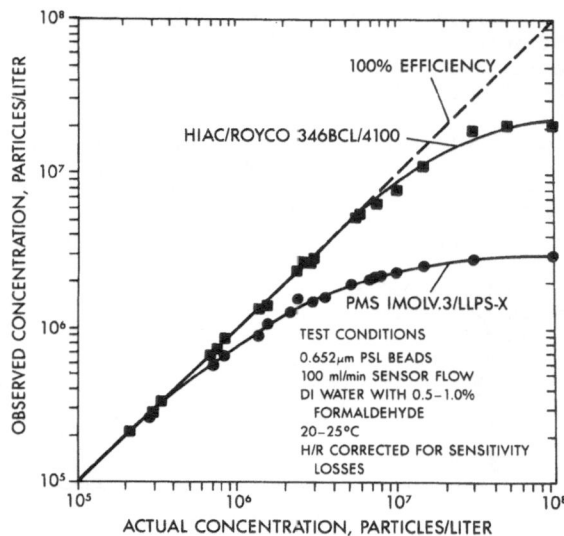

Figure 2. The effect of coincidence on observed particle concentrations.[8]

Table I. Fluid Refractive Indices.

Fluid	Refractive Index
Gases	
Nitrogen	1.00
Oxygen	1.00
Argon	1.00
Hydrogen	1.00
Liquids	
49% Hydrofluoric Acid	1.29
100% Water	1.33
29% Ammonium Hydroxide	1.33
30% Hydrogen peroxide	1.36
37% Hydrochloric acid	1.41
96% Sulfuric acid	1.45
Solids	
Silicon dioxide	1.46
"Average" aerosol particle	1.55
Polystyrene latex sphere	1.59
Germanium	3.50
Silicon	3.90

OPCs usually are calibrated using PSL spheres in water. Instrument determination of contaminant particle size assumes that particles in the medium being measured have the same index contrast as PSL beads in water.

Typical atmospheric aerosol particles have refractive indices which are close to that of PSL beads.[13] Hence, if one assumes that the sources of chemical and atmospheric contamination are similar, then index contrasts in chemicals can be calculated, as follows, using a particle refractive index of 1.59:

Liquid	Refractive index	Index contrast
49% Hydrofluoric Acid	1.29	1.23
100% Water	1.33	1.19
29% Ammonium hydroxide	1.33	1.19
30% Hydrogen peroxide	1.36	1.18
37% Hydrochloric acid	1.41	1.12
96% Sulfuric acid	1.45	1.09

The theoretical effects of these changes in index contrast on light scattering are shown in Figures 3 and 4. These figures present the effective scattering cross section of particles of various diameters in media with various refractive indices determined using Mie scattering theory. The effective scattering cross section is defined as the cross section having an area such that the power flowing through that area is equal to the power of the light scattered. Figure 3 presents calculated performance of the PMS IMOLV.3/LLPS-X instrument, while Figure 4 shows the same calculations for the H/R 346BCL/4100 OPC.

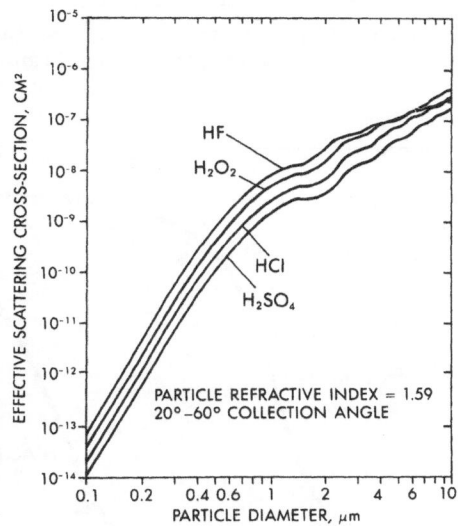

Figure 3. The effect of liquid media refractive index on scattering intensity--PMS IMOLV.3 (Data courtesy of R. Knollenberg--PMS, Boulder, CO).[8]

Examination of Figures 3 and 4 reveals that changes in index contrast result in sizing errors. The closer the index contrast is to 1.00, the larger the particle required to scatter the same amount of light. For example, a 0.8µm particle in 49% hydrofluoric acid scatters the same amount of light as a 2.3µm particle in 96% sulfuric acid. Hence, if corrections are not made, dramatic errors can result. In particular, particle sizes in sulfuric acid are grossly underestimated.

It should be pointed out that the amount of light scattered by a particle depends upon how far the index contrast deviates from 1.00 rather than the absolute magnitude of the index contrast. For example, air bubbles in water (index contrast=0.75) scatter somewhat more light than PSL spheres of an equivalent size (index contrast=1.19).

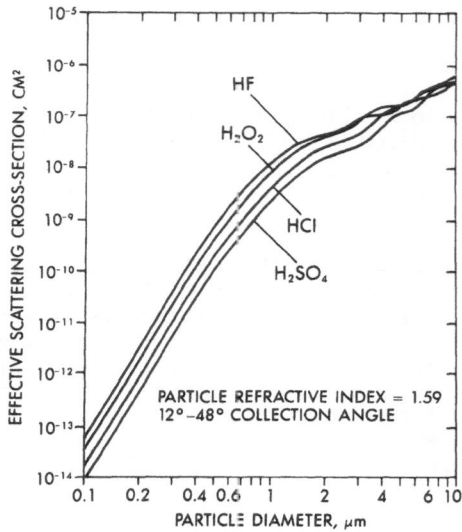

Figure 4. The effect of liquid media refractive index on scattering intensity--H/R 346BCL (Data courtesy of R. Knollenberg--PMS, Boulder, CO).[8]

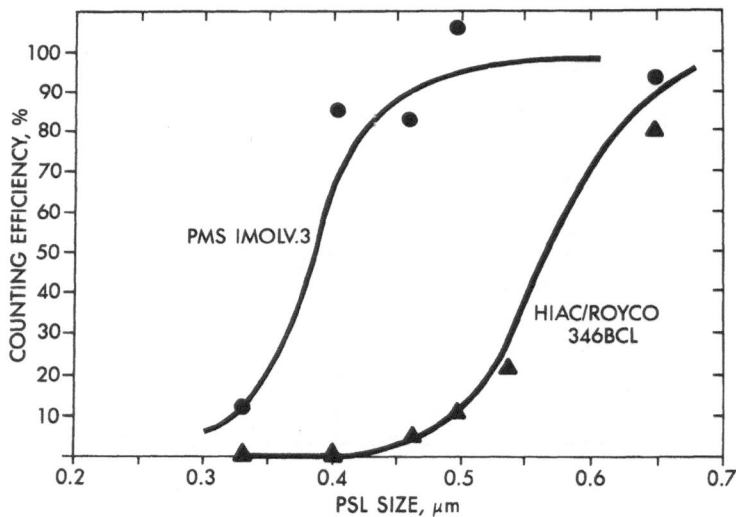

Figure 5. Instrument counting efficiency versus particle size.

Sensor Efficiency Losses

The sensitivity of OPC sensors decreases rapidly near the minimum detectable particle size as the amount of light scattered is proportional to approximately the sixth power of the particle diameter. The loss in efficiency for the two sensors described above is given in Figure 5.[11] Both sensors count 10-15% of the particles at the rated minimum detectable particle size (0.5μm for the H/R sensor, 0.3μm for the PMS sensor). Near one hundred percent efficiency is achieved at approximately 0.1-0.2μm above the minimum size rating.

126

DESCRIPTION OF IMPROVED METHODOLOGY

The sources of errors described above were reduced by performing all measurements under positive pressure and by making corrections for index contrast, OPC sensor counting efficiency losses, and coincidence.

Pressure sampling was performed using the system shown in Figure 6. Particle concentrations in bottled chemicals were measured using the configuration shown in Figure 6A, while filter removal efficiencies were measured using that shown in Figure 6B. Bottles were fitted with a perfluoroalkoxy (PFA) cap which had a PFA dip tube and a polytetrafluoroethylene (PTFE) gas inlet filter and placed inside a pressurized PFA container used routinely for chemical storage in a centrifugal spray acid processing system. Regulated gas pressures of 30-40psig were used to force chemicals through the OPC sensor at flow rates of 30-100ml/min by adjusting a downstream valve/flow meter. The gas inlet filter prevented contamination of the sample by the pressurized gas.

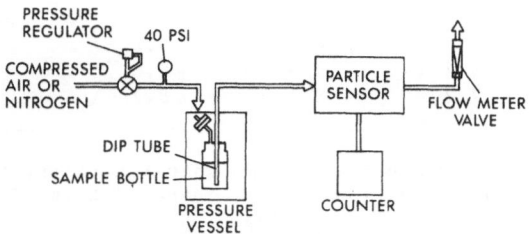

A. BOTTLE SAMPLING MODE OF OPERATION

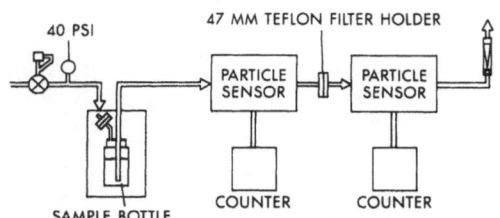

B. FILTER EFFICIENCY MEASUREMENT MODE OF OPERATION

Figure 6. Pressurized test systems.

The use of this system has been shown to result in stable, repeatable measurements of particle concentrations in a number of chemicals including 30% hydrogen peroxide and 29% ammonium hydroxide.[8]

Refractive index corrections were made by determining the theoretical sizing error which occurred in each fluid using Mie scattering calculations. First, the instruments were calibrated using PSL spheres in water. Then Mie scattering calculations were used to determine the particle sizes which would scatter the same amount of light in the test fluids. A particle refractive index of 1.59 was assumed. In other words, no instrument calibration was performed in the test fluids. Rather, the data obtained using an instrument calibrated in water were plotted using the theoretical particle diameters with the same scattering cross section determined by using Mie scattering theory.

The corrected particle sizes for the H/R and the PMS instruments for several index contrasts are shown in Table II and Table III. Substantial corrections are indicated for sulfuric acid.

Table II. Particle Diameters (in μm) with Equivalent Scattering Cross Sections in Various Media—IMOLV.3 (20 to 60 deg Collection Angle).

H_2O NH$_4$OH	HF	H_2O_2	HCl	H_2SO_4
0.30	0.28	0.31	0.36	0.40
0.40	0.37	0.42	0.48	0.55
0.50	0.46	0.52	0.62	0.72
0.60	0.54	0.63	0.77	0.93
0.80	0.70	0.85	1.1	1.8
1.00	0.87	1.1	1.8	2.3
1.5	1.2	1.7	2.3	2.7
2.0	1.8	2.1	2.7	3.8
2.5	2.2	2.7	3.9	4.9
3.0	2.6	3.3	4.2	5.7
4.0	4.0	4.2	5.6	7.3
5.0	5.2	5.0	6.0	7.9
8.0	7.2	8.1	8.7	>10.0

Table III. Particle Diameters (in μm) with Equivalent Scattering Cross Sections in Various Media—346BCL (12 to 48 deg Collection Angle).

H_2O NH$_4$OH	HF	H_2O_2	HCl	H_2SO_4
0.50	0.46	0.51	0.61	0.69
0.60	0.55	0.63	0.73	0.85
0.72	0.66	0.74	0.90	1.07
1.0	0.90	1.0	1.3	1.7
2.0	1.8	2.1	2.9	3.6
5.0	5.0	4.7	4.8	6.1

The validity of the corrections has been demonstrated by two experiments. In the first experiment PSL beads were injected into fluids with different refractive indices and the response of the H/R 346BCL sensor was determined. Three fluids were used:

1. Water

2. Simulated hydrochloric acid, a mixture of glycerin and water with the same refractive index as hydrochloric acid

3. Simulated sulfuric acid, a mixture of dibutyl phthalate and isopropyl alcohol with the same refractive index as sulfuric acid

Instrument response is shown in Figure 7 where the OPC signal produced as a function of particle size is shown for each of the three fluids tested. The shapes of the curves in Figure 7 are similar to the theoretical curves calculated for this sensor (Figure 4).

In the second experiment, particle concentrations in a bottle of sulfuric acid were measured, then the acid was diluted tenfold with DI water containing <100 particles/liter, and the concentration remeasured. Dilution of the acid to this extent reduced the refractive index to essentially that of water, thereby eliminating the sizing error. Figure 8 compares the corrected and uncorrected particle size distributions with that of the diluted acid (multiplied by 10 to account for the dilution). The size corrected data agree quite well with the diluted data. Similar results have been obtained by Knollenberg.[13]

Corrections for the sensor counting efficiency were made by multiplying the particle concentrations in the two most sensitive channels of the OPC by correction factors. The correction factors are a function of the size distribution of the particles which are being counted and are determined by calculating the instrument response to an assumed particle size distribution using the counting efficiency curve shown in Figure 5.

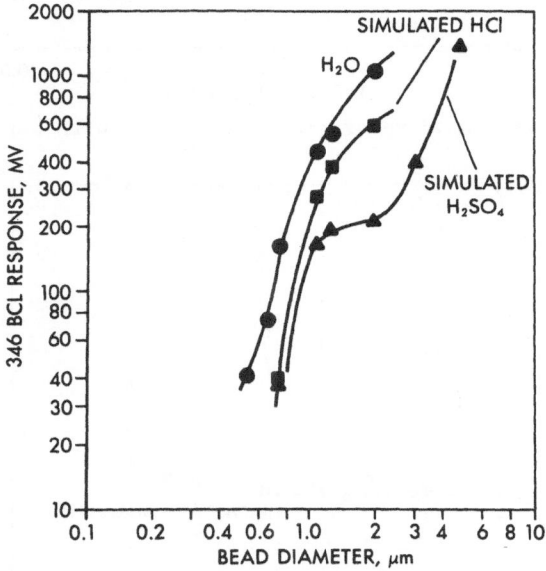

Figure 7. H/R 346BCL response to PSL beads in liquids with different refractive indices.

Particle size distributions can often be described accurately using a log-log distribution. Typical slopes of the distribution are -2 to -3 and the correction factors are fairly insensitive to changes in this range. Therefore, a slope of -2.5 was assumed in this study, resulting in the following correction factors:

Particle size range, μm	Correction Factors Hiac/Royco	PMS
0.3-0.4	–	3.1
0.4-0.5	–	1.3
0.5-0.6	2.2	1.0
0.6-0.72	1.2	1.0

Correction factors for coincidence counting were determined by dividing the indicated particle concentrations shown in Figure 2 by the actual concentration. The results are shown in Figure 9. No corrections for sizing errors were included.

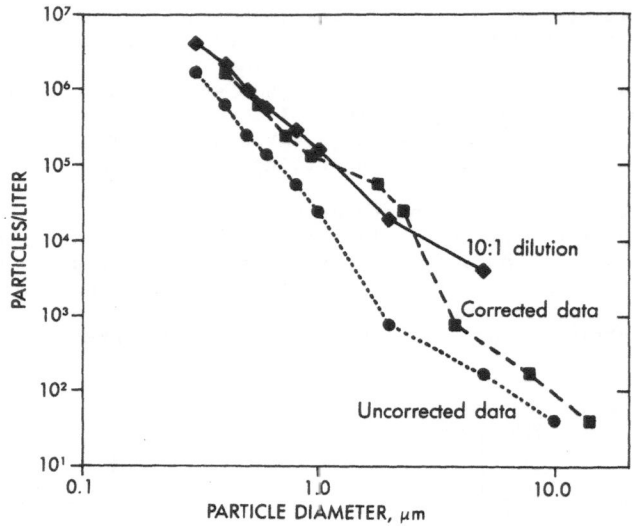

Figure 8. Comparison of theoretical correction with actual data--H_2SO_4.[8]

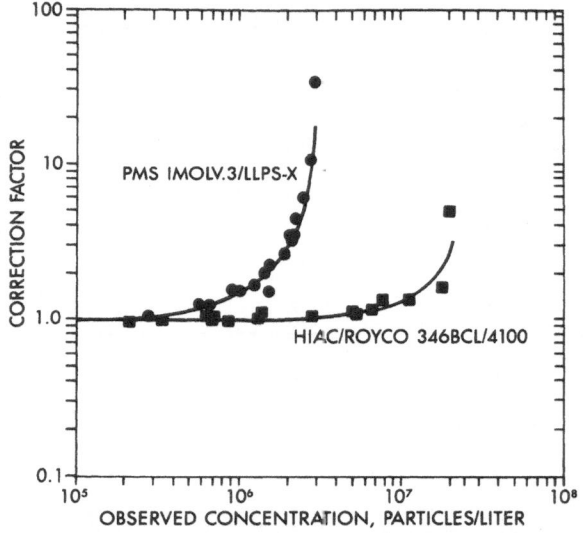

Figure 9. OPC coincidence correction factors.

CORRECTION FACTOR MAGNITUDES

Numerous bottles of low-mobile-ion grade chemicals were measured using this methodology. Examples of the data obtained are shown in Table IV in which the raw data and the effect of each correction factor on the data are given.

The magnitude of the correction factors varies significantly with the largest correction factors due to either coincidence or refractive index. The correction factor for OPC sensor inefficiency is about 30-40%. The coincidence factor increases as concentration increases. Some of the data shown required essentially no correction for coincidence, with the largest correction a factor of 3.2.

Table IV. Examples of correction factor magnitudes.

Chemical	Uncorrected	Corrected for Efficiency Losses	Corrected for Coincidence and Efficiency Losses	Corrected for Refractive Index, Coincidence and Efficiency Losses
			Particles/L >0.3μm	
30% H_2O_2	250,000	350,000	370,000	380,000
29% NH_4OH	2,400,000	3,400,000	11,000,000	11,000,000
37% HCl	200,000	280,000	290,000	469,000
96% H_2SO_4	1,100,000	1,500,000	2,500,000	8,200,000
49% HF	250,000	320,000	320,000	290,000

*>0.5μm

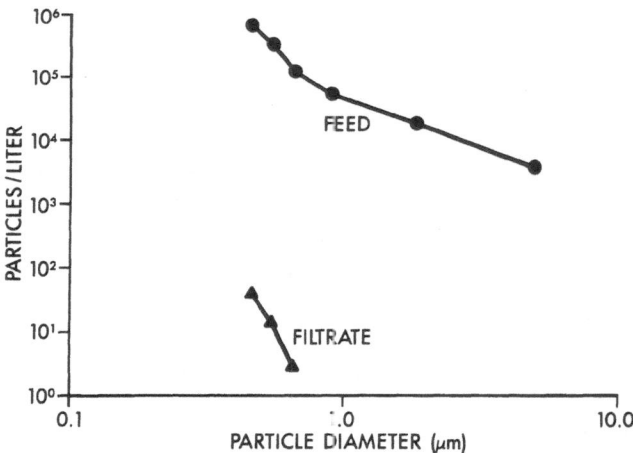

Figure 10. 0.2μm Filter perfcrmance in hydrofluoric acid.

The refractive index correction also varied significantly. The smallest correction factor was for 49% hydrofluoric acid (0.91) and the largest for 96% sulfuric acid (3.3).

These data indicate that significant errors can result if these correction factors are not used. The overall sulfuric acid correction was close to an order of magnitude.

UNRESOLVED ISSUES

There are still unresolved issues involved in particle counting in chemical solutions despite the corrections performed on the data. Evidence of these issues has been obtained in experiments performed to determine the effectiveness of 0.2μm Wafergard[(R)] PF-40 microporous membrane filters in removing particles from chemicals.

One indication of counting problems involves the shape of the particle size distributions in chemical filtrates. In fluids with relatively low vapor pressure and viscosity such as water, 49% hydrofluoric acid and 37% hydrochloric acid, the filtrate particle size distribution is much steeper than that of the feed. An example is shown in Figure 10. This difference in shape is expected since membrane filters are known to have a fairly sharp retention efficiency cutoff. Measured filtration efficiencies in these fluids are also very high. In the example shown, the filter removed >99.99% of the particles >0.3μm.

However, in fluids with high vapor pressure or high viscosity, the shapes of the particle size distributions of the feed and filtrate are similar. An example of a high vapor pressure solution (30% hydrogen peroxide) is shown in Figure 11 and one of a high viscosity solution (96% sulfuric acid) in Figure 12. Measured filtration efficiencies in these fluids are less than those in low vapor pressure, low viscosity fluids. The indicated removal efficiency of particles >0.3μm in 30% hydrogen peroxide was 99.8% while in sulfuric acid it was only 98%. One would expect to see similar particle size distributions and filtration efficiencies in the different solution types.

In another experiment, a series filtration was run with a 30% hydrogen peroxide feed. The first filter had a pore size rating of 0.2μm, the second 0.05μm. The feed contained 92,000 particles/liter >0.3μm, while both of the filtrates contained 3000-6000ppl >0.3μm. It seems highly unlikely that the first filter would remove 95% of the particles while the second filter with a smaller pore size rating would remove no particles.

These data indicate that there are additional sources of error which result in artificially high counts. Further work needs to be done to identify these sources.

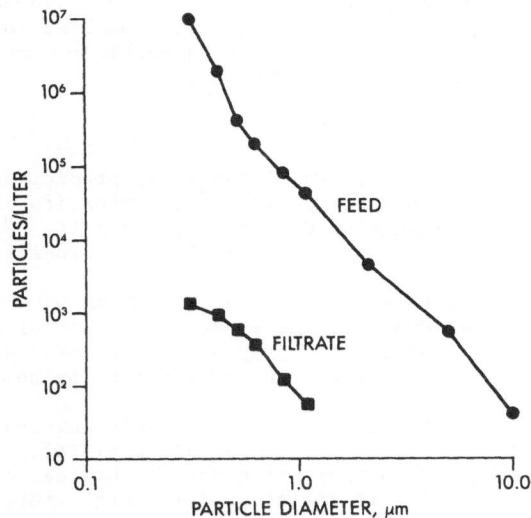

Figure 11. 0.2μm Filter performance in 30% hydrogen peroxide

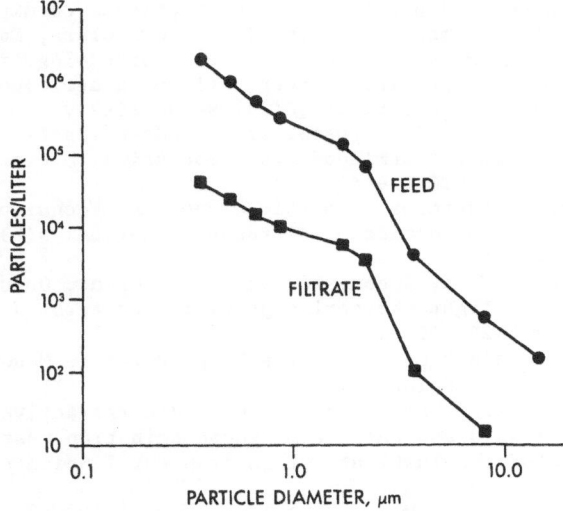

Figure 12. 0.2μm Filter performance in concentrated sulfuric acid.

SUMMARY

Test methodology has been developed to measure particle concentrations accurately in semiconductor process chemicals. The methodology uses elevated pressures in the particle counter sensor to reduce counting of gas bubbles, and makes corrections for sensor inefficiency, particle coincidence, and liquid medium refractive index.

The magnitude of the corrections was found to be almost a factor of 10 in some cases. Therefore, these corrections are necessary when comparing particle concentrations in different chemicals. Also, a general standard cannot be written for all chemicals unless the corrections are performed. Instead, a standard for each chemical is needed.

There are still some unidentified sources of error. It appears that overcounting occurs in solutions with either high vapor pressures or high viscosities. This issue needs to be resolved to further increase the accuracy of chemical solution particle concentrations measured using light scattering techniques.

REFERENCES

1. T. A. Milner and T. M. Brown, A model for predicting the effect of a processing bath on wafer particle contamination, Proceedings of the 2nd Microcontamination Conference, Nov 1986.
2. A. Hiratsuka, Problems of chemicals in VLSI process, VLSI, July 1986.
3. W. R. Schmidt, C. Becker, J. Mehta, R. Novak, D. C. Grant, K. A. Foster, and C. P. Myhaver, The effects of point-of-use chemical filtration in a centrifugal spray system on wafer surface particle counts, Millipore Microelectronics Technical Symposium, Santa Clara, May 1987.
4. K. Dillenbeck, Characterization of particle levels in incoming chemicals, Microcontamination, 2(6):56-62 (1984).
5. I. Bansal, Particle contamination during chemical cleaning and photoresist stripping of silicon wafers, Microcontamination, 2(4):34-39, 90 (1984).
6. P. Wood, Pushing impurities to 1 ppb: Japanese chemical industry leads the way, Solid State Technology, 28(11):69-70 (1985).
7. D. Jones, Submicron particle levels in process chemicals, Millipore Microelectronics Technical Symposium, Santa Clara, May 1986.
8. D. C. Grant and W. R. Schmidt, Improved methodology for determination of submicron particle concentrations in semiconductor process chemicals, J. Environ. Sci., 30(3):28-33 (1987).
9. N. J. Csikai and A. J. Barnard, Jr., Counting particles in mobile liquid chemicals for semiconductor processing, Microcontamination, 4(11):44-50, 128-130 (1986).
10. K. Dillenbeck, Advances in particle-counting techniques for semiconductor process chemicals, Microcontamination, 5(2):30-38,65 (1987).
11. S. L. Peacock, M. A. Accomazzo, and D. C. Grant, Quantitative count calibration of light scattering particle counters, J. Environ. Sci., 29(4):23-27 (1986).
12. Personal communication, R. Knollenberg, Particle Measurement Systems, Boulder, Colorado.
13. R. G. Knollenberg, The importance of media refractive index in evaluating liquid and surface microcontamination measurements, Proceedings of the Institute of Environmental Sciences, pp. 501-511, 1986.

CALIBRATION OF THE PHOTO-SEDIMENTOMETER LUMOSED

WITH A POWDER OF KNOWN PARTICLE SIZE DISTRIBUTION

M. Hangl and G. Staudinger

Abteilung für Apparatebau und Mechanische Verfahrenstechnik
Technische Universität Graz
A-8010 Graz, Austria

The photo-sedimentometer LUMOSED measures the particle size distribution of a powder by determining the light extinction of the suspended powder while it is sedimenting in a cuvette. By using three light beams the time required for one measurement can be reduced by a factor of 100 compared with instruments having only one stationary light beam. The three signals are combined into one single transmission / time curve.

The conversion of a transmission / time curve into a cumulative distribution by volume / particle size curve is possible if the extinction coefficient of a particular material at particular optical conditions is known. The LUMOSED has incorporated a routine, which calculates the extinction coefficient from a measurement of a powder with known particle size distribution for all particle diameters and stores this data in a library.

INTRODUCTION

The LUMOSED is a photo-sedimentometer for particle size analysis, which has eliminated the long measuring times required by other sedimentation instruments (of up to 12 hours for 1 µm particles) by placing not only one light beam at the full liquid depth h_1 but three light beams at the distances of 1.5, 15 and 150 mm below the liquid level as shown in the schematic of Figure 1. The time required for one measurement is thus reduced by a factor of 100. [1]

There are three light sensitive sensors which give three signals proportional to light intensity. All three signals principally contain the same kind of information, however with different accuracy. Intensity I is easily converted to transmission T by Equation 1.

$$\frac{I(x)}{Ir} = T(x) \tag{1}$$

The selection of the most accurate parts of the three transmission curves and the combining of them into one curve as shown in Figure 1 was described by Staudinger et al. [1] The changing of the light trans-

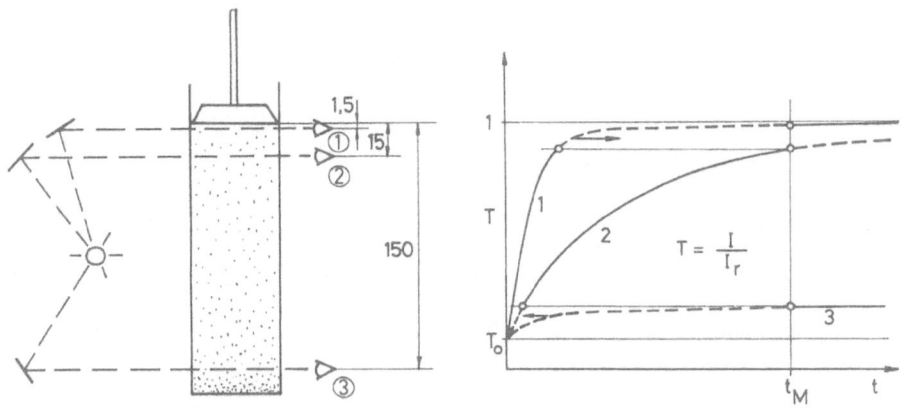

Figure 1. Schematic of the photo-sedimentometer LUMOSED.

mission with time allows the calculation of the particle size distribution using the methods of Rose [2,3] and Telle. [4] Further investigations were carried out by Johne [5], Kurz and Johne [6], Allen [7,8] and Heidenreich and Schuldt [9]. The attenuation of a beam of light by suspended particles was described recently by Leschonsky and Boeck. [10]

In the above mentioned calculation methods the relation between transmission T, the number concentration of particles in suspension C_n and the particle diameter x is determined by using Lambert-Beer's law (Equation 2) with L being the length of the

$$\ln T(x) = C_n L K(x) \frac{x^2 \pi}{4} \qquad (2)$$

light beam's path through the suspension and K(x) the extinction coefficient. K(x) depends not only on particle size and shape, but also on the refractive index of the liquid and on equipment parameters such as the aperture of the photodetector, the wave length distribution of the light beam and the wave length sensitivity distribution of the sensor. Calculation procedures on the basis of the wave theory of light are available; however, the correlation with reality is insufficient. K(x) can only be determined through measurements on the given system, i.e. the LUMOSED itself.

DETERMINATION OF EXTINCTION COEFFICIENT

The cumulative distribution by volume $Q_3(x)$ is defined by Equation (3)

$$Q_3(x) = \frac{\int_0^x \frac{x^3 \pi}{6} q_1(x) \, dx}{\int_0^{x \max} \frac{x^3 \pi}{6} q_1(x) \, dx} \quad . \qquad (3)$$

Differentiation of Equation (2) and introduction into Equation (3) results in a relation between the measured transmission T(x) and $Q_3(x)$.

$$Q_3(x) = \frac{\displaystyle\int_0^x \frac{x}{K(x)} \; \frac{d \ln T(x)}{dx} \; dx}{\displaystyle\int_0^{x \; max} \frac{x}{K(x)} \; \frac{d \ln T(x)}{dx} \; dx} \qquad (4)$$

To determine the calibration factor $K(x)$ the following procedure is followed: A powder of which the cumulative distribution by volume $Q_3^w(x)$ is known as pairs of data $Q_3(x_i)/x_i$ is processed in the LUMOSED and the measured transmission converted to a distribution $Q_3^l(x)$ with $K(x) = 1$ by using Equation (4) (Figure 2). A factor $F(x)$ is defined as the ratio between the measured distribution $Q_3^l(x)$ and the true distribution $Q_3^w(x)$ as

$$F(x) = \frac{Q_3^w(x)}{Q_3^l(x)} \qquad (5)$$

At the data points x_i is $F(x_i) = Q_3^w(x_i)/Q_3^l(x_i) = F_i$. In between the data points $F(x)$ is derived from linear interpolation:

$$F(x) = F_i + \frac{F_{i+1} - F_i}{x_{i+1} - x_i} \; (x - x_i) \qquad (6)$$

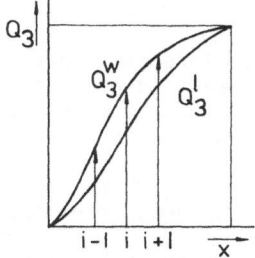

Figure 2. True distribution $Q_3^w(x)$ and measured distribution $Q_3^l(x)$

It was shown by Boeck [10] and it is also apparent from the integrals in Equation (4) ($\frac{x}{K(x)} \frac{d (\ln T(x))}{dx} = q_3^w(x)$)   that

$$K(x) = \frac{q_3^l(x)}{q_3^w(x)} \qquad (7)$$

with $q_3^l(x)$ and $q_3^w(x)$ being the derivatives of $Q_3^l(x)$ and $Q_3^w(x)$ respectively. $q_3^w(x)$ is obtained from the differentiation of Equation (5):

137

$$q_3^W(x) = F(x) \ \frac{dQ_3^1(x)}{dx} + Q_3^1(x) \ \frac{dF(x)}{dx} \tag{8}$$

$\frac{dQ_3^1(x)}{dx} = q_3^1(x);$ and $\frac{dF(x)}{dx}$ is derived from the differentiation of Equation (6):

$$\frac{dF(x)}{dx} = \frac{F_{i+1} - F_i}{x_{i+1} - x_i} \tag{9}$$

Since $F(x)$ was interpolated linearly between the data points the gradient $\frac{d\,F(x)}{dx}$ will change its value at each data point. For better approximation the gradients from both sides of the data points are averaged and introduced into Equation (8) to replace $\frac{d\,F(x)}{dx}$.

$$q_{3i}^W = F_i \ q_{3i}^1 + Q_{3i}^1 \ \frac{1}{2} \left(\frac{F_{i+1} - F_i}{x_{i+1} - x_i} + \frac{F_i - F_{i-1}}{x_i - x_{i-1}} \right) \tag{10}$$

THE CALIBRATION FACTOR

It is now possible to calculate the extinction coefficient K_i according to Equation (7) for each pair of data $Q_3(x_i)/x_i$:

$$\frac{1}{K_i} = F_i + \frac{Q_{3i}^1}{q_{3i}^1} \ \frac{1}{2} \left(\frac{F_{i+1} - F_i}{x_{i+1} - x_i} + \frac{F_i - F_{i-1}}{x_i - x_{i-1}} \right) \tag{11}$$

However, it must not be forgotten that K comprises not only all material data but also all specific properties of the instrument. Therefore, for clarity, the expression "calibration factor" shall be introduced for K(x) in connection with its application to the LUMOSED.

With the calibration factor $K_{(x)}$ known, the distribution of the powder can be calculated from the measured transmission by equation (4).

In order to calibrate the LUMOSED with a particlular powder this powder is processed as in a normal measurement[1], but the option "calibration" is selected on the computer and the pairs of data $Q_3(x_i)/x_i$ of the known "true" distribution are typed in , together with an identification of the powder to name the file, where K is stored. For a later measurement K can be called from the file. There is space for the K's of twenty powders.

EXAMPLE

A calibration run was performed using a quartz powder the cumulative mass distribution by volume $Q_3^W(x)$ of which was determined first with a sedimentation balance. In Figure 3 the small dots show the uncalibrated size distribution $Q_3^1(x)$. Both the uncalibrated and the true

particle size density distributions by volume of this powder are shown
in Figure 4. Evaluation of Equation (11) with the measured data yields
K_i which, as can be seen from Figure 5, has a curve similar to those
found in the literature on the extinction coefficient[10]. For other
powders the curve is somewhat different, but still remains principally
the same.

CONCLUSION

For measuring particle size distribution of powders with an optical method like a photo-sedimentometer, calibration is required because
of the change in the optical properties with particle size. The photo-
sedimentometer LUMOSED has incorporated a routine which makes it possible
to calibrate itself by determining the calibration factor with a powder
of known particle size distribution and storing it in the memory of the
computer.

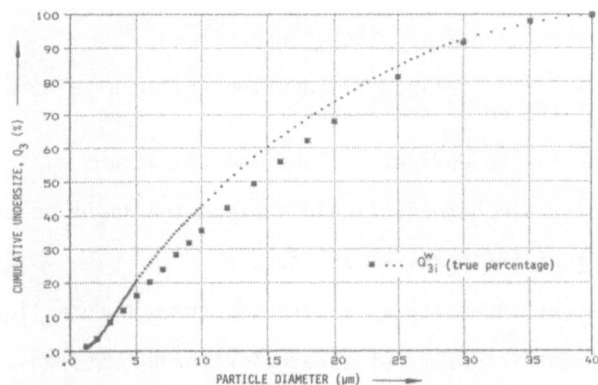

Figure 3   True particle size distribution Q_{3i}^w (large dots) and particle
size distribution as measured with K(x)=1 (small dots).

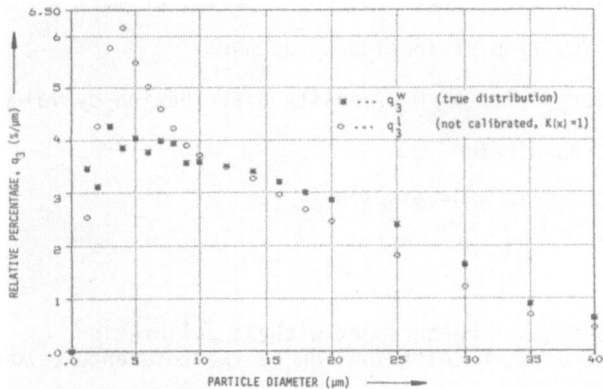

Figure 4   True particle size density distribution q_{3i}^w (dots) and
particle size density distribution q_{3i}^1 (circles) as measured
with K(x)=1.

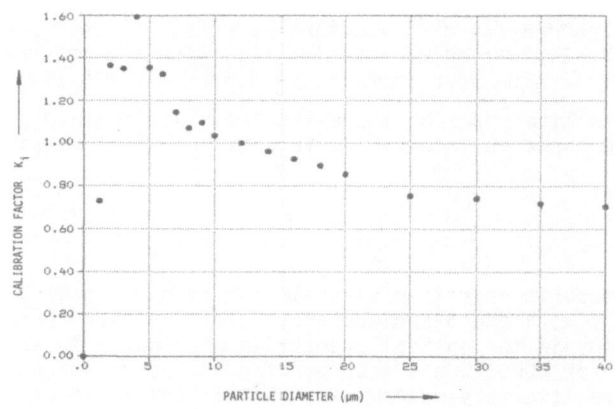

Figure 5 Calibration Factor K_i for quartz powder.

SYMBOLS

C_n .... number concentration of particles

F_i .... factor as defined by Equation (5) at a value x_i of the reference powder

$F(x)$ ..... factor as defined by Equation (5) at any x.

h_1, h_2, h_3 distance between liquid surface and light beams

I .... light intensity at a detector

I_r .... light intensity at a detector through pure liquid

$I(x)$ ..... light intensity at a detector through suspension

$K(x)$ ..... extinction coefficient for particle size x

l .... length of light beam through suspension

Q_3 .... cumulative distribution by volume or mass

q_1 .... density distribution by number

q_3 .... derivative of Q_3, density distribution by volume or mass

T .... transmission

x .... particle equivalent diameter

Indices

1 .... denotes a measurement without calibration
w .... denotes the distribution of the reference powder

REFERENCES

1. G. Staudinger, M. Hangl and P. Pechtl, Quick optical measurement of particle distribution in a sedimentation apparatus, Particle Characterization 3 158-162 (1986).

2. H.E. Rose, Determination of the extinction coefficient-particle size relationship for spherical bodies. J. Appl. Chem. 2 80 - 88 (1952)

3. H.E. Rose, Einige Fragen der Bestimmung von Teilchengrößen sehr feiner Stäube durch einfache meist optische Methoden. VDI-Berichte, 7 35 - 48 (1955)

4. O. Telle, Ein verbessertes lichtelektrisches Sedimentometer. VDI-Berichte, 7 31 - 33 (1955)

5. R. Johne and M. Ramanujam, Genauigkeit der Kornanalyse mit dem Photosedimentometer. Staub 23 269 - 278 (1963)

6. H.P. Kurz and R. Johne, Über die Anwendbarkeit der photometrischen Oberflächenbestimmung auf breite Korngrößenverteilungen und auf Pulver mit hohen spezifischen Oberflächen. Powder Technol. 3 83 - 91 (1969/1970)

7. T. Allen, Determination of the size distribution and specific surface of fine powders by photoextinction methods. I. Theoretical estimate of variation in extinction coefficient with particle size using a white light source. Powder Technol. 2 133 - 140 (1968/1969)

8. T. Allen, Determination of the size distribution and specific surface of fine powders by photoextinction methods. II. Comparison between wide angle and narrow angle photosedimentometers and experimental determination of extinction coefficients. Powder Technol. 2 141 -153 (1968/1970)

9. E. Heidenreich and U.Schuldt, Fotometrische Ermittlung der mengenspezifischen Oberfläche mit dem Spekol EK 5. Silikattechnik 25 267 - 270 (1974)

10. K. Leschonsky and T. Boeck, Photometric On-line measurement of surface area of powders. Particle Characterization 2 81 - 90 (1985)

PARTICLE CONTAMINATION CONTROL AND MEASUREMENT IN ULTRA-PURE VLSI GRADE INERT GASES

R. M. Thorogood, A. Schwarz, and W. T. McDermott

Air Products and Chemicals, Inc.
Allentown, PA 18195

Particle microcontamination in the inert gases, such as nitrogen, used in the manufacture of integrated circuits greatly affects the yield loss of semiconductors. With the continuously increasing complexity of integrated circuits, there are no particles so small that their presence in the inert gas supply systems can be safely ignored. It is, therefore, necessary to accurately measure and control the level of particle contamination in gas delivery systems in order to supply virtually particle free gas. The key factors in production and supply of particle-free gas are ultra-high efficiency membrane filters, total integrity of the filter installations, and absolute cleanliness of all surfaces in contact with supplied gas.

To develop the capability to supply particle-free gas, one must have sophisticated experimental facilities and specialized measurement capabilities. This paper describes the key aspects of a multifaceted development program to ensure the highest quality of VLSI grade inert gases. Among topics of interest are selection and verification of advanced particle counters, and development of accurate sampling systems and methods for generating pressurized test aerosols. Experimental results are presented for filter cartridges and multicartridge filter installations obtained in a large-scale laboratory pressurized test loop and at field locations. Performance data indicate extremely low concentration levels in the filter effluent gas down to one particle per standard cubic foot greater than 0.02 μm.

INTRODUCTION

Particulate contamination as a cause of reduction of product yield in the microelectronics industry has become a familiar concern in the last few years. Inert gases are used in substantial quantities in the manufacturing process of integrated circuits. Nitrogen, in particular, is used in many manufacturing steps: Nitrogen, therefore was the first gas to receive the attention of SEMI[1] in preparing a specification for an acceptable particle concentration. This specification, termed "Particulate Specification for VLSI-Grade Nitrogen and Argon Delivered as Pipeline Gas," recommended a maximum

. concentration of 20 particles per standard cubic foot for all particles larger than 0.20 μm.

At the time of its preparation in 1985, this specification recognized both the requirements of the user and the practical limitations of particle counting equipment. Reflecting the needs of the integrated circuit manufacturers, the 0.2 μm particle size conformed to a line width of 2 μm which might be considered appropriate for a 64K to 128K memory device. A concentration of 20 p/sft³ at 0.2 μm also provided a reasonable certainty for reliable measurement above instrument background concentration.

However, industry progress towards 256K, 1M and 4M memory devices with smaller line geometries and larger chip sizes have imposed the dual need for both smaller particle size and concentration. This has coincided with the availability of more sensitive counting equipment and improved techniques such that current microelectronics industry expectations for gas quality far exceed the SEMI specification requirements. 10 p/sft³ >0.1 μm and 20 to 50 p/sft³ >0.02 μm now occur with regularity. As a recognition of this, it has recently been proposed that the term VLSI in the SEMI specification be replaced by "20/0.2."

The sizes of the features on integrated circuits continue to decrease. Table I shows the sizes of representative features in the state-of-the-art Dynamic Random Access Memory (DRAM) circuits from the early 1970's to the present, as well as the projected future sizes into the 1990's. Along with feature sizes, gate oxide thickness becomes progressively smaller and smaller. The minimum lateral feature size, the gate oxide thickness and the increasing chip area are responsible for the trend towards even smaller particle sizes and number concentrations. There are no particles so small that their presence in the inert gas supply systems can be safely ignored. It is, therefore, necessary to develop capabilities to measure and control the level of particle contamination in the inert gas supply systems.

Table I. DRAM technology trends

year	72	74	76	78	80	82	84	86	88	90	92	94	96
capacity (bits)	1k	4k		16K	64K	256K			1M		4M		16M
minimum lateral feature size (μm)	11	8		5	3	2			1.3		0.8		0.5
chip area (mm²)	14.5			15	21	42			57		90		116
junction depth (μm)	1.3			0.8	0.5	0.35			.02		0.15		0.1
gate oxide thickness (μm)	.12			.10	.075	0.05		.035	.025		.02		.015

144

PARTICULATE REMOVAL IN THE PRODUCTION OF NITROGEN

The key factors in the production and supply of virtually particle-free gas are ultra-high efficiency membrane filters, total integrity of the filter installations, and absolute cleanliness of all surfaces in contact with supplied gas. The production of ultrapure nitrogen for the microelectronics industry has been described previously[2]. The principal source of particulates is the atmospheric air from which nitrogen is separated. Air contains from 10 to 100 x 10⁶ p/sft³ >0.1 μm. These particles are removed in several stages within the process plant. The air entering the plant feed compressor is filtered to remove the bulk of the larger particulates (down to 0.2 μm). Further removal takes place during air cooling and by liquid scrubbing in the distillation column. By these means the particle concentration is reduced to be in the typical range of 100 to 5,000 p/sft³ >0.2 μm. It should be noted that this concentration range applies to both gaseous nitrogen and liquid nitrogen produced at the plant.

Surfaces in process equipment, both static (vessels, piping, etc.) and dynamic (valves, compressors, etc.), are potential particle generators. However, these do not contribute to a measurable degree in the plant product particle concentration. Thus, the challenge is the reduction of particle concentrations from up to 5,000 p/sft³ to single digit levels before it reaches the semiconductor manufacturers process equipment.

GAS FILTRATION

Nitrogen is distributed to the point of use in two stages: by bulk transport in a pipeline or liquid tanker and then by pipeline distribution in the semiconductor plant. Filtration occurs at several locations in a typical system:

1. At exit from the plant to the bulk pipeline.

2. At the point of supply to the customer.

3. At the point of use.

On-site gas producing plants and liquid nitrogen sources will only have steps 2 and 3. The filters currently used have microporous polymeric membranes with actual pore openings from 0.3 to 0.05 μm. Particle removal occurs during gas passage through the membrane by direct trapping and inertial or diffusional transfer to the surface, thus particle capture is not limited to the actual pore size. Experiments have shown that the most penetrating particle size for membranes with large pore openings (>0.5 μm) is the 0.05 to 0.1 μm range.

The standard format for bulk gas filter elements is a cartridge diameter of approximately 2.5 inches with length from 10 inches to 40 inches in increments of 10 inches. Each incremental cartridge has a membrane area of 7.5 feet². The membrane pores are typically liquid rated at 0.04 to 0.24m. The membrane is supported between two fibrous support layers, pleated, and formed as a cylinder contained within perforated plastic cages. One end is fused to a sealing device; the other is either closed or fused to adjacent elements. Various seal types are used to connect the element to the housing tube plate. The filters are typically operated at a superficial velocity of about 0.03 ft/sec.

The point of this description is that one has to be concerned with far more than the microporous membrane performance. Particulate-free gas can be obtained only in the absence of leakage (membrane to seal and seal to housing joints must be perfect), and the system downstream of the membrane (support layer, housing outlet, and piping) must be absolutely clean. It is, therefore, very important to measure performance under conditions which simulate the real industrial operating condition, i.e. at the appropriate pressures, flows, and temperatures. From the research viewpoint, it is also important to be able to separate the sources of particulates during experimental measurements.

MEASUREMENT OF ULTRA-LOW PARTICLE CONCENTRATIONS

The key factors in the achievement of accurate low particle concentrations are the accuracy, sensitivity, and reproducibility of the particle counters and gas sampling devices.

Two different types of particle counters have been used in our studies: Laser Optical Particle Counters (LOPC) manufactured by PMS (Particle Measuring Systems, Inc., Boulder, Colorado 80301) and Condensation Nucleus Counters (CNC) manufactured by TSI (TSI, Inc., St. Paul, Minnesota 55164). Three different models of LOPC (LAS-X, LPC-101, and LPC-101-HP) and two different models of CNC (TSI 3020 and 3760) have been tested. Their characteristics are shown in Table II. All three laser counters have a minimum size sensitivity of 0.1 μm. The low sample rate of the LAS-X makes it unsuitable for use outside the laboratory. The high pressure version of the LPC-101 is preferred for process gas sampling and has been found to give greater reliability in use and lower background count levels, probably due to the absence of the sheath gas recirculation pump. All three laser counters are sensitive to electrical disturbances which can cause significant spurious count levels especially in the lowest size channels. Operation at temperatures below 50°F has also been found uncertain.

Table II. Particle counter characteristics

Counter Type	Minimum Sensitivity μm	Time Required To Sample 1 sft^3 of Gas
Laser Counter		
PMS LAS-X	0.09	8 hr
PMS LPC-101	0.1	10 min
PSM LPC-101-HP	0.1	10 min
Condensation Nucleus Counter		
TSI 3020	0.02	1 2/3 hr
TSI 3760	0.02	20 min

The CNC has been rated with a 0.02 μm sensitivity[s]. Two disadvantages of the Model 3020 are its low sampling rate and susceptibility to liquid bath disturbance by movement. This limits its capabilities as a portable instrument. The newer Model 3760 has greater convenience as a portable instrument, and we have found it to have counting performance similar to the Model 3020 in both laboratory and site use. Operating temperatures of both instruments can be difficult to maintain in outdoor situations.

146

The Condensation Nucleus Counters usually operate at ambient pressure and require reduction of sample gas pressure as do the laser counters in most cases. The means of pressure reduction requires careful consideration. Any reduction of pressure by a ratio of two or greater will induce sonic gas velocity. High gas velocity causes particulates to shed from surfaces and prolongs cleanup time or causes false high particle count concentrations. Alternatively, high velocity can cause particulate loss by impaction giving false low values. It is important to provide a smooth flow path through the pressure reduction device, to minimize surface area and for it to be readily cleaned.

Theoretical and experimental studies[6] were performed to determine the extent of particle deposition in commonly used pressure reduction devices, including an electropolished stainless steel orifice device currently used at Air Products, a bellows valve, and a diaphragm valve. Stainless steel bellows valves have been used by others as pressure reducers in ultraclean gas sampling systems. This work showed that such valves can significantly reduce the measured value of particle concentration through a process of internal particle deposition. Particle concentration was measured downstream of each device using TSI Model 3020 condensation nucleus counters as shown in Figure 1.

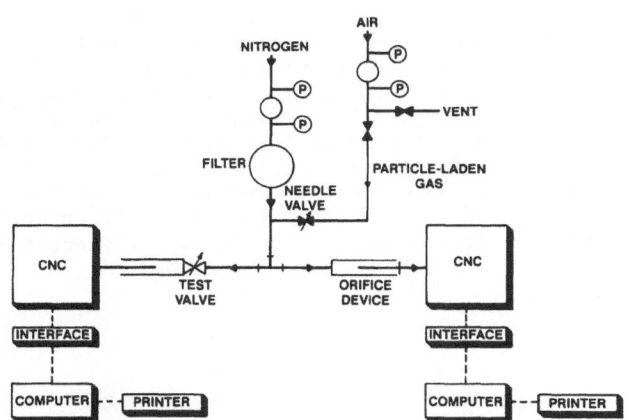

Figure 1. Schematic diagram of experimental apparatus.

Measurements were obtained for subsonic and sonic flow conditions at various challenge rates. Substantial differences were observed in the rate of particle deposition between these various devices.

Subsonic turbulent flow through each device was also studied using a finite difference numerical technique to solve the Navier-Stokes equations. Turbulence was simulated using the k-ε turbulence model. Turbulent particle deposition and impaction were simulated by tracking individual particles in a Lagrangian frame of reference. The results of the computer simulation were used to predict particle removal in each sampling device. The results compared favorably with the experimental data.

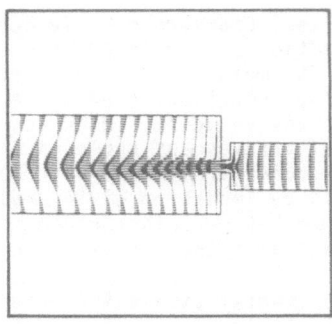

Figure 2. Calculated mean velocity in the orifice device.

Figure 2 shows a vector plot of the calculated mean velocity for
the orifice device. The flowfield is characterized by a high velocity
jet downstream of the orifice and zones of flow circulation lateral to
the jet. The progressive broadening of the jet in the direction of
flow is apparent in the figure. The circulation pattern can be seen
more clearly in the contour plot of stream function shown in Figure 3.
One hundred particles were introduced at the flow inlet of the orifice
device. Figure 4 shows the calculated tracks of 10 particles through
the device. These tracks are typical of those calculated for all 100
particles. Figure 5 shows a vector plot of calculated mean velocity
for the bellows valve geometry. This solution is characterized by a
region of high velocity near the valve seat location and a circulation
zone in the lower part of the valve. Particle tracks were calculated
for the bellows valve design in a manner similar to that for the
orifice design. Capture typically occurred within the upper stagnant
zone and the lower circulation zone of the device (Figures 6 and 7).

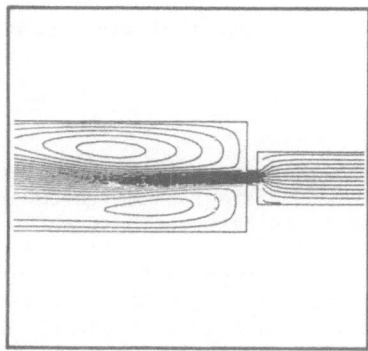

Figure 3. Calculated stream function in the orifice device.

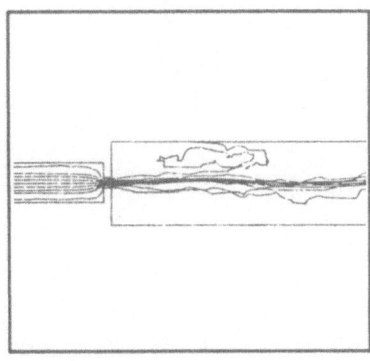

Figure 4. Calculated particle tracks in the orifice device.

The results of comparison tests performed for the orifice and bellows valve devices are shown in Figure 8. The particle concentrations measured downstream of the bellows valve were less than those measured downstream of the orifice device. This is believed to be a result of particle capture within the bellows valve device. The numerical simulation studies demonstrated a possible mechanism by which particles can be captured within stagnant and circulating flow zones within the valve. The rate of particle capture was relatively constant over the range of measured concentration values.

The results of these studies demonstrate the importance of a proper gas sampling system design in obtaining an accurate measurement of particle concentration in a pressurized system.

Removal of the gas sample from the pipeline source, transport to the orifice and delivery into the particle counter with the minimum of flow disturbance (velocity or direction) is also critical for accurate measurements. Isokinetic sampling through axial probes is preferred, with the shortest possible sampling distance. Ideally, no valves should be allowed in the sample train, although experience has shown that a fully open diaphragm valve does not remove or contribute

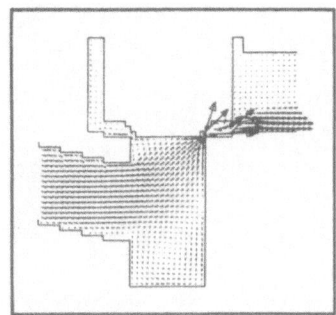

Figure 5. Cross-section of bellows valve showing calculated mean velocity vectors for turbulent flow pressure reduction.

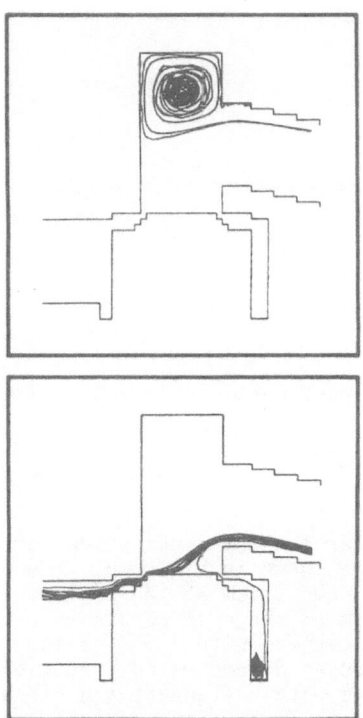

Figures 6 and 7. Calculated particle tracks through bellows valve.

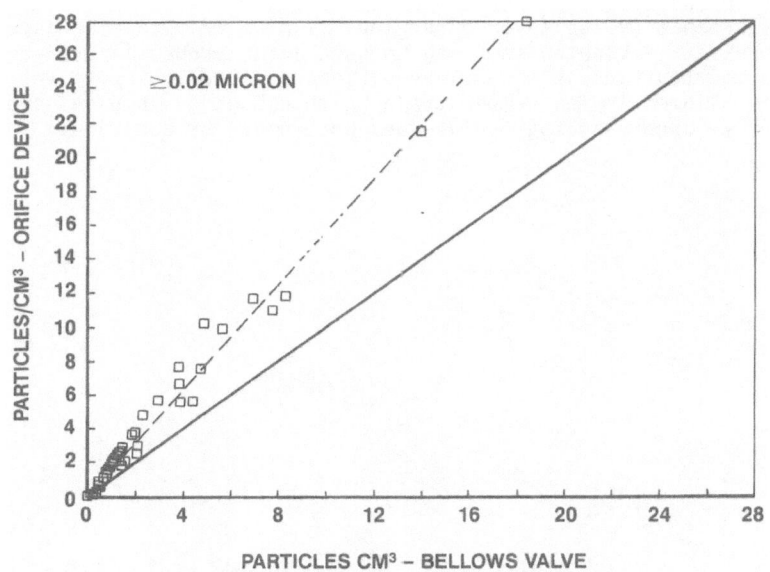

Figure 8. Comparison of particle concentrations measured at outlets
from orifice pressure reducer and bellows valve.

particles to the sample stream. The entire sample system should be manufactured in 316L electropolished stainless steel to facilitate cleaning and minimize particle capture or release.

LARGE-SCALE LABORATORY TESTS OF FILTER CARTRIDGES

We have carried out extensive tests of filter cartridges in a large scale laboratory pressurized test loop. Gas entering the test filter is prefiltered, and controlled aerosol challenges may be added to the gas stream as required. A constant output atomizer was used to generate sodium chloride particles ranging from 0.01 μm to 1 μm in a nitrogen stream at pressure. The resulting particle size distributions were characterized using a TSI Differential Mobility Analyzer (DMA) and Condensation Nucleus Counter (CNC). The characterized aerosols were then injected into the Air Products large scale filter cartridge test loop upstream of the membrane test filter. The upstream particle challenge total concentration was determined and particle penetration studies of the test filter were performed. The atomizer was modified for high pressure operation.

A solution of sodium chloride in microelectronics grade water was used to generate particles. The mean particle size of the aerosol is primarily a function of the sodium chloride concentration in the liquid feed. Feed gas to the atomizer at pressure P_I was first passed through a high efficiency 0.02 μm rated filter as shown in Figure 9. The aerosol was diluted with a second filtered nitrogen stream and dried in a silica gel drier. For measurement purposes the dried high pressure aerosol was then expanded to atmospheric pressure through a dual probe critical-orifice type expansion chamber. Aerosol characterization was performed using a TSI Differential Mobility Particle Sizer (DMPS) system. Simultaneous measurements were made using a high pressure PMS LPC-101 upstream of the expansion chamber and low pressure LPC-101 and LAS-X optical particle counters downstream of the chamber. These devices provided direct comparison with the DMPS readings in the particle size range above approximately 0.1 μm. Pressures P_I and P_0 were independently varied during these tests to determine the aerosol size distribution as a function of atomizer flow conditions.

The mean particle diameter of the aerosol increased with sodium chloride concentration in the liquid feed. However, mean particle size also tended to increase as the aerosol generator pressures P_I and P_0 were simultaneously decreased. In addition, the size distribution of the aerosol was found to deviate slightly from log-normal behavior as shown in the log-probability plot of Figure 10.

The aerosol generator was then incorporated into the Air Products large scale filter test loop as shown in Figure 11. The test loop is a large scale recirculating system capable of full size filter cartridge qualification studies. The pipe loop is constructed of 38 and 76 cm diameter electropolished stainless steel tubing with a single stage Sundyne LMC-311P stainless steel centrifugal compressor to provide compression and flow. Isokinetic sample tubes with critical orifice type expansion chambers are located both upstream and downstream of the test filter. OPC's and CNC's are located for simultaneous measurement at each sample location. Each OPC instrument is equipped with a completely automated data acquisition and disk storage system. Each CNC is interfaced to a corresponding dedicated PC for control and data acquisition. A more detailed description of the test loop was provided in 1986[3]. Further information on the aerosol generation techniques was given in ref. 4.

An important observation of this earlier work was the absence of any increase in downstream particulate concentration when an aerosol challenge of 4 x 10⁶ p/sft³ >0.02 μm was applied to the upstream flow. This demonstrated the very high removal efficiency of commercially available filters. These results are a consequence of specialized cleaning, sampling, and measurement techniques.

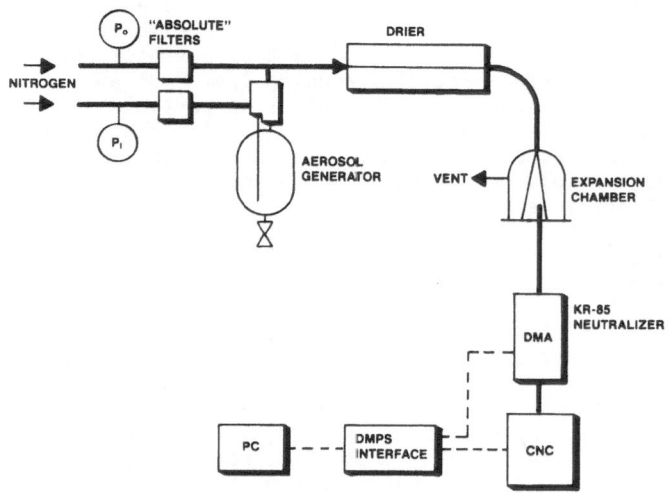

Figure 9. Aerosol characterization test apparatus.

The cleaning method and the maintenance of cleanliness have received substantial effort in our continuing laboratory research and this has been translated into special procedures in the preparation of field installations. Key steps for the laboratory have been the incorporation of the cleaning technology into a cleanroom and the construction of a cleanroom around the filtration test equipment. This has allowed us to maintain all of the test loop surfaces in a clean condition during changeout of test filters in addition to the avoidance of filter element contamination. Rigorous procedures during the assembly of commercial equipment have provided the same benefits.

The benefit of these procedures is seen in a rapid achievement of steady particle concentrations with an absence of spikes and with very low measured particle concentrations. A recent laboratory measurement on a commercial filter cartridge is shown in Figure 12. This demonstrates that filtered nitrogen particle levels can be obtained well below 10 p/sft³ for a measurement sensitivity down to 0.02 μm. The average concentration for this particular instance was 0.6 p/sft³ with an upper 95% confidence limit of 0.83 p/sft³ (Poisson Distribution) at an aerosol challenge concentration of 10⁵ p/sft³. This compared to an average particle concentration of 0.45 p/sft³ >0.02 μm (upper 95% confidence limit 0.7 p/sft³) when no aerosol challenge was present. These low levels of particulate contamination are believed to have originated from surfaces in the piping and sampling system downstream of the filter. There are no significant burst states (spikes) in the data. This result demonstrates the previously reported capability of microporous membranes to essentially completely remove the upstream contamination, the absence of shedding in the filter and measuring system, and the absence of seal leakage.

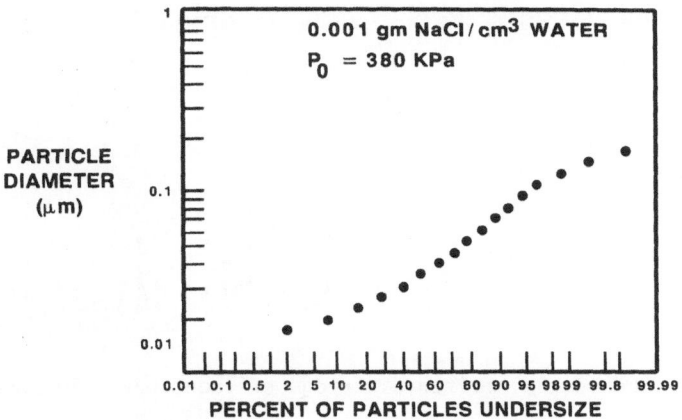

Figure 10.   Measured particle size distribution.

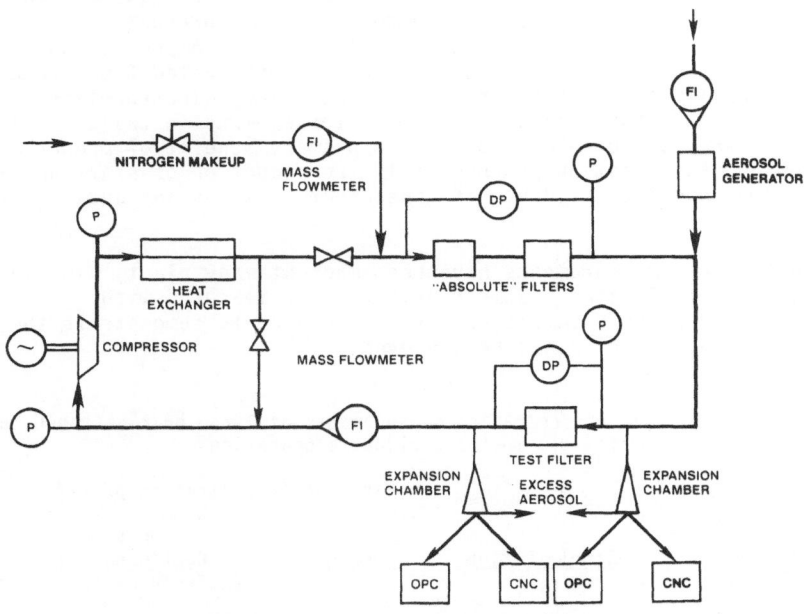

Figure 11.   Air Products large scale filter test loop.

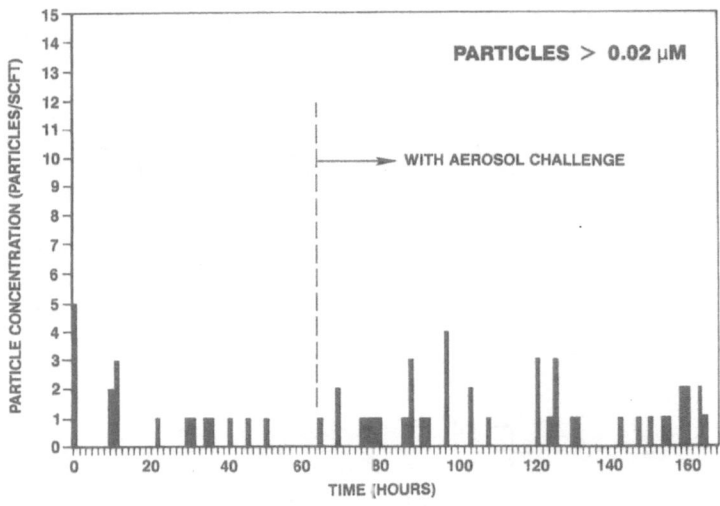

Figure 12. Commercial filter cartridge effluent particle concentration.

FILTER INSTALLATIONS FOR PROCESS NITROGEN GAS

Filter stations in Air Products electronics nitrogen systems have been designed to handle flows from less than 1,000 sft³/hr. to more than 500,000 sft³/hr. Large filter stations with their associated pipelines, flow measurement, and control systems represent a substantial investment and thus need evaluation of component and material choices. New systems are generally constructed from stainless steel to facilitate cleaning and, where necessary, electropolishing is employed. Manufacture, cleaning, and filter element installation procedures are carefully controlled. A primary benefit of proper cleaning and installation procedures is the reduction of start-up times to a minimum and the avoidance of unnecessary gas wastage and cost for purging.

Performance measurements have been made at many plant, pipeline, and filter installations. Some recent results measured with Condensation Nucleus Counters are presented here to demonstrate the cleanliness levels which can be achieved.

Table III. Particle concentration measurements at large bulk nitrogen supply system after 11 months continuous operation.

Sample Point	Sample Period	Average	Particle Concentration p/sft³ >0.02μm Upper Confidence Limit (Poisson Distribution)
Plant Filter Effluent	35 hr	0.95	1.09
Customer Filter Influent	25 hr	1.36	1.55
Customer Filter Effluent	40 hr	1.04	1.48

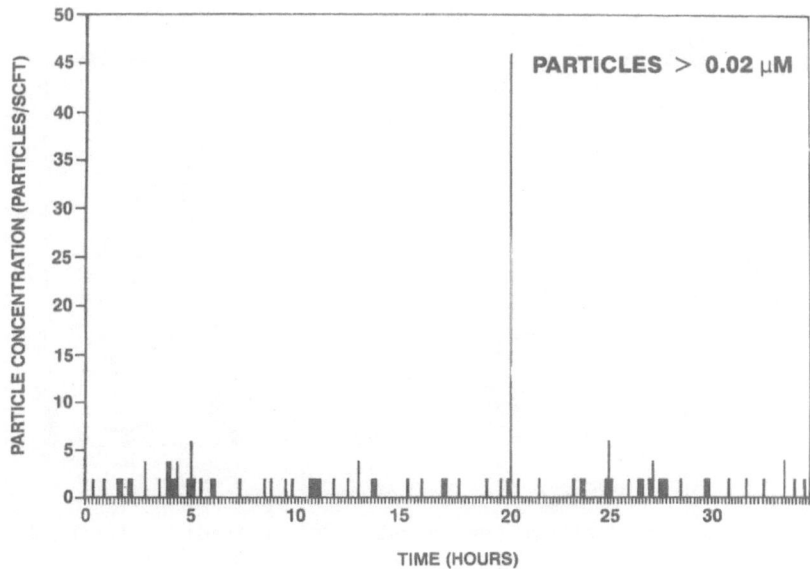

Figure 13. Particle concentration from plant product filler.

These results represent particle concentrations at several points of a nitrogen gas supply to a single large customer through a stainless steel pipeline from a large multipurpose oxygen/nitrogen production plant after eleven months of operation. Data are shown in Figures 13 through 15 and Table III. Special methods of manufacture and cleaning were used to achieve this result. The results measured after 11 months operation show that a bulk gas system can achieve an average particle concentration of 1 p/sft³ larger than 0.02 μm.

CONCLUSION

The production of bulk nitrogen gas with particulate concentration levels below 10 p/sft³ >0.02 μm can be achieved and demonstrated

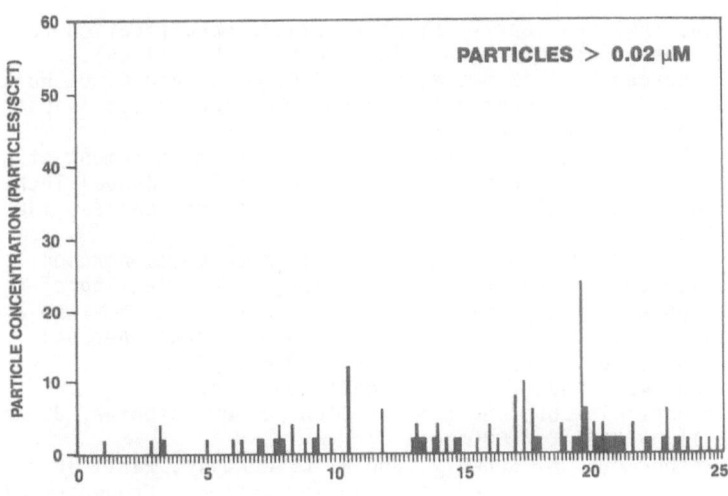

Figure 14. Particle concentration at customer filter inlet. .

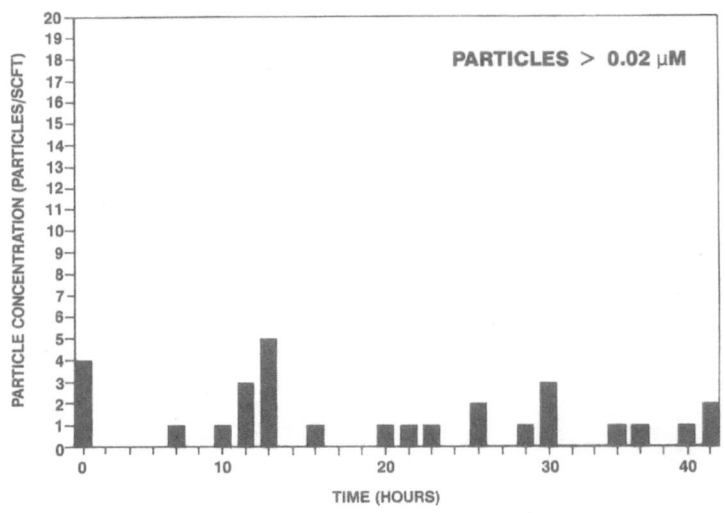

Figure 15. Particle concentration at customer filter exit.

by use of currently existing commercial equipment. Levels as low as 1 p/sft³ can be obtained after reasonable periods of system operation.

The key to success lies in attention to the detail of rigorous equipment selection and manufacturing procedures and especially to cleaning methods. Care must be taken in measurement to ensure that the measured results are a true representation of the process stream and have not been affected by the sampling technique, the method of pressure reduction, and extraneous influences on the particle counter. Success in all of these factors is best achieved by the use of skilled and experienced personnel.

REFERENCES

1. J. King, SEMI develops first particulates specification for VLSI-grade gases. Microcontamination 5(1), 64, (1986).
2. R. M. Thorogood, A. Schwarz, W. T. McDermott, and C. D. Holcomb, Production of ultrapure nitrogen for the electronics industry, Microcontamination 5(8), 28, (1986).
3. R. M. Thorogood and A. Schwarz, Performance measurement of gas ultrafiltration cartridges, Proceedings of 32nd Annual Technical Meeting of Institute of Environmental Sciences, Dallas, 1986, pp. 460-467.
4. A. Schwarz, R. M. Thorogood, and W. T. McDermott, Aerosol formation and characterization for large scale testing of ultrahigh efficiency membrane filters, Aerosols Formation and Reactivity - Proceedings of the 2nd International Aerosol Conference, West Berlin, 1986, pp. 670-673.
5. J. K. Agarwal, and G. J. Sem, Continuous flow, single-particle-counting condensation nucleus counter, J. Aerosol Sci., 11, 343-357, (1980).
6. W. T. McDermott, A. Schwarz, and R. C. Ockovic, Particle deposition in pressurized gas sampling systems, Presented at Institute of Environmental Sciences 33rd Annual Technical Meeting, San Jose, CA, May 1987.

DESIGN AND PRACTICAL CONSIDERATIONS IN USING CASCADE IMPACTORS TO COLLECT PARTICLE SAMPLES FROM PROCESS GASES FOR IDENTIFICATION

W. L. Chiang
California Measurements, Inc.
150 E. Montecito Ave., Sierra Madre, CA 91024

R. L. Chuan
Brunswick Corporation
3333 Harbor Blvd., Costa Mesa, CA 92626

Cascade impactors can collect particles from gases for SEM/EDX analyses. However, additional work is required to verify system design and interface requirements between impactor and gas source. This paper describes an experiment using a 4-stage cascade impactor operating at 0.25 scfm (7 slpm) to collect samples from pressurized nitrogen. The results show that (1) it is feasible to use a simple flow divider as an interconnection between the impactor inlet and the gas source to control system flow, (2) a four-stage impactor is not necessary; two stages would suffice, (3) by using a standard electron microscope specimen mount inside an impactor stage, it could be removed and inserted directly into the electron microscope for analysis. This avoids sample transfers and saves the microscopist's time, and (4) by using a small number of jets in the impactor nozzle, particles can be collected in small concentrated spots, corresponding to the size and number of the jets, making them easier to locate under high magnification.

INTRODUCTION

Information on the morphology and elemental composition of particulate contaminants in process gases helps determine their mechanisms of formation and provides clues as to their origins.[1,2] A two-step procedure is required to get this information. First, samples of the particles must be collected. Subsequently, they must be analyzed, using Scanning Electron Microscopy (SEM) to obtain their morphology, and Energy Dispersive X-ray Spectroscopy (EDX) to identify their elemental compositions.

The feasibility of using an inertial impactor to collect particle samples from gases has been demonstrated by Kasper et. al.[4] However, additional work is required to verify and optimize system design. This paper examines such design and systems considerations. It describes the results of an experiment using a four-stage low-pressure cascade impactor to collect particle samples from ordinary commercial grade nitrogen.[3, 4]

EXPERIMENTAL SETUP

A diagram of the experimental setup and a picture of it are shown in Figures 1 and 2, respectively.

A modified California Measurements MPS-4G1 four-stage cascade impactor system was used. Its basic specifications are in Table I. The main parts of the impactor system, including the nozzles, were made of aluminum with clear anodize finish. The impactor stack was mounted horizontally to conveniently connect to the flow divider, the valves between it, and the gas source.

A vacuum pump, connected to the last stage of the impactor stack, provided the flow through the impactor system and the required pressure drop across it. The pressure drop is necessary to reduce the air density under the nozzles in the lower stages for small particle impaction, and the impactor inlet pressure must be near atmospheric to preserve particle size separation accuracy.

Table I. Cascade Impactor Specifications.

Stage	D_{50}*	No. of Jets
1	1.6	1
2	0.50	2
3	0.20	7
4	0.05	9

*50% particle cut-off size. Particle density 2g/cc.
Flow Rate - 7 std liters/minute.

The particle collection substrate in each impactor stage was the top smooth surface of a standard copper scanning electron microscope specimen mount, with a 0.5 inch diameter stub and a 0.125 inch diameter stem (see Figure 3). Thus the particles impacted directly onto the electrically conductive stub in concentrated dots following the pattern of nozzle jets.

The jet patterns were different in each stage. A single round jet was used in Stage 1. Stage 2 had two jets. Stage 3 had seven jets with one in the center and six equally spaced in a ring around it. Stage 4, the last stage, had nine jets in a 3x3 matrix. Regardless of the number of jets, all were located in the center of the nozzle within a circular area of not more than 0.1 inch diameter. For example, in Stage 4, the particles impacted in nine concentrated spots within the circular region, each spot having a diameter of about 0.012 inch or 0.3 mm. This facilitated particle location under high magnification.

A Nupro Union Tee fitting was connected to the impactor inlet as a flow divider. The 90-degree port was used to bypass air not needed by the impactor. A 1/3 psi check valve was connected to this port to maintain the impactor inlet pressure at slightly above atmospheric and to prevent backflow ambient air from entering the impactor through this port when the cylinder pressure becomes too low.

The nitrogen gas was of commercial grade in a 76-cubic feet cylinder initially pressurized at about 2,200 psi. We chose to use a commercial grade gas rather than a very clean one to increase the probability of particle capture to demonstrate feasibility of the collection methodology.

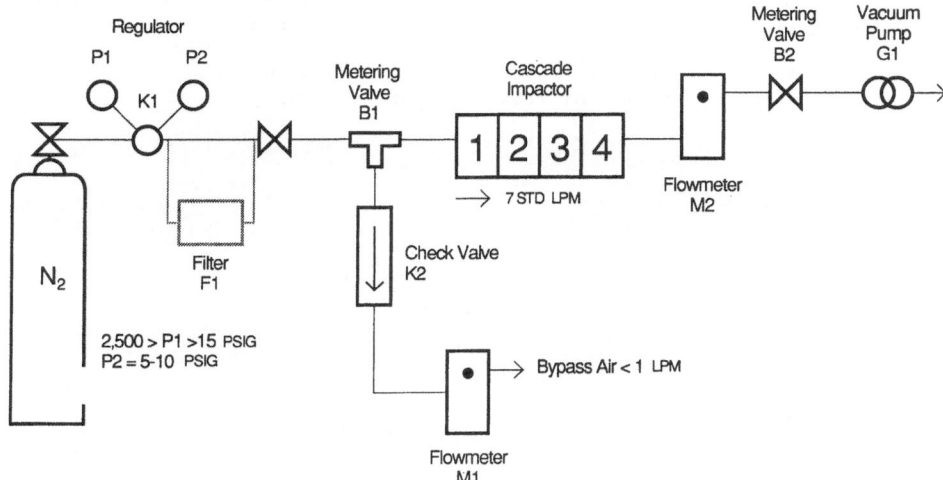

Figure 1.   Schematic of cascade impactor gas sampling setup.

Figure 2.   Photograph of experimental setup.

Metering valve B1 controlled the flow rate of gas delivered to the flow divider. B2 controlled the flow required through the cascade impactor, which was 7 standard liters per minute, or 0.25 scfm, at the impactor inlet. The impactor stack induced a pressure drop of about 400 mmHg, and therefore the M2 indicated flow was about 12 LPM.

PROCEDURES AND RESULTS

The procedure used for each sample run was as follows:

Initially the vacuum pump was left off. Gas was released from the cylinder and the regulator adjusted to an output pressure of about 5 psi. At this time all the flow went through the bypass port of the flow divider and monitored on M1. B1 was used to adjust the flow through M1 to about 8 LPM. The pump was then turned on. The flow through M1 then dropped to about 1 LPM. B2 was used to make vernier adjustments of the impactor flow. To minimize gas waste through the bypass, B1 was further adjusted to make M1 read less than 1 LPM.

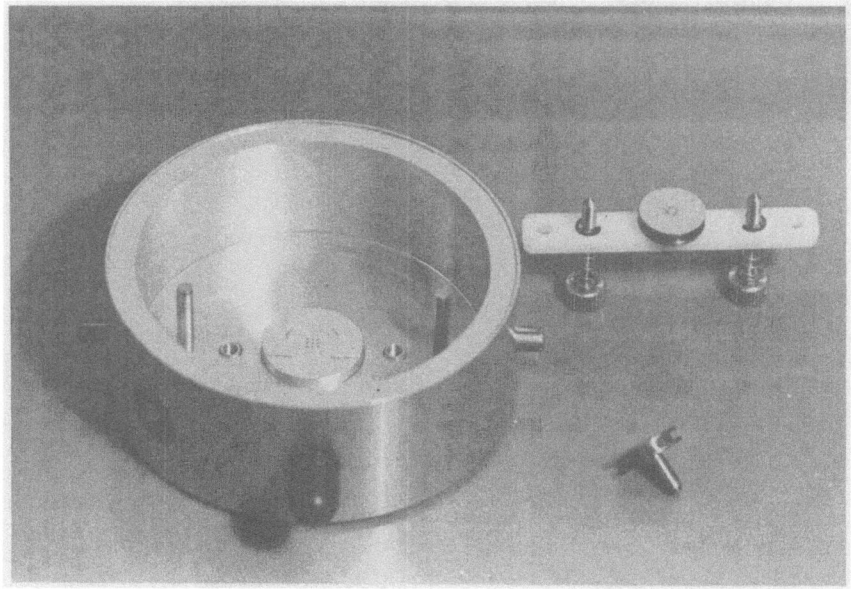

Figure 3.   Collection substrates and impactor stage.

The first sample run was to assess the cleanliness of the plumbing system downstream of the gas supply and regulator. This was a "blanking" run to purge the system of background particulates and to establish the signature of those particles. This was done by adding a Balston Grade AAQ Microfibre Coalescing filter (0.1 micron retention efficiency >99.995%) immediately after the regulator. It was located after the regulator, because this particular filter's casing could not withstand high pressures. The more desirable location would be before the regulator. The duration was about 15 minutes. A thin coating of pure hydrocarbon (Apiezon) high vacuum grease was applied to the particle impaction surface to ensure particle capture.

After sampling, the collection stubs were removed from the impactor stages, coated with a thin layer of carbon and examined in the scanning electron microscope. It was found that only Stage 3, with a cut-off size of 0.2 microns, collected a significant number of particles.

A 2,000X magnification micrograph is shown in Figure 4. It is seen that some of the particles were larger than the size range appropriate for that stage, suggesting there had been some bounce-off of the larger particles from the upper stages.

X-ray spectral analysis showed that these particles contained primarily calcium, as indicated in Figure 5. They were probably calcium oxide or calcium carbonate particles, since no other elements with atomic number larger than 11 (the limit of the X-ray analysis) were evident in the spectrum.

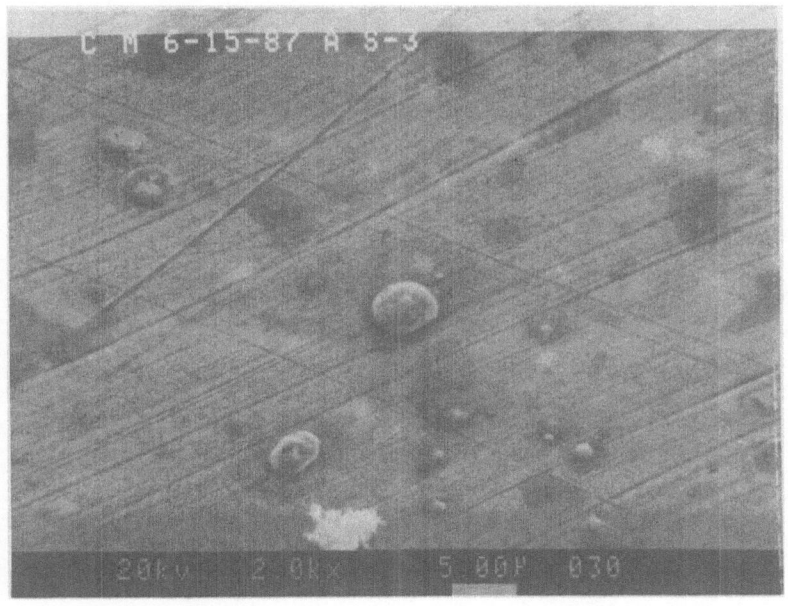

Figure 4.   Micrograph of particles from Run 1.

The bright cluster of material seen in the lower edge of the field in Figure 4 was made up of low atomic number elements, as indicated by the absence of x-ray spectral lines. The morphology and lack of X-ray emission suggest it was probably a soot particle. A count of the particles led to an estimate of 2,300 particles per cubic foot of gas, with particle sizes ranging from 0.5 to 5 microns.

For the next run, we were constrained by the total sampling time of 3 hours to bleed down a full cylinder of gas at a use rate of about 8 liters per minute, 7 for the impactor and 1 for bypass. Since this time was quite short, and the gas might be too clean to provide enough samples, collection stubs in the first three stages were removed. All the particles were allowed to impact on Stage 4 alone. In other words, to expedite the experiment, we did not use the size segregation feature of the cascade impactor.

Thus with the plumbing system cleaned out, the filter was removed, and the unfiltered gas was passed through the cascade impactor for 3 hours. The results verified the high degree of cleanliness of the gas, with a count leading to an estimate of about 5 particles per cubic foot.

Compared to the particles from the plumbing system, the samples showed a variety of composition and morphology. They also had a wide range in size, from about 2 to 50 microns, with very few, if any, smaller particles. Two of the larger ones are shown in Figures 6 and 7.

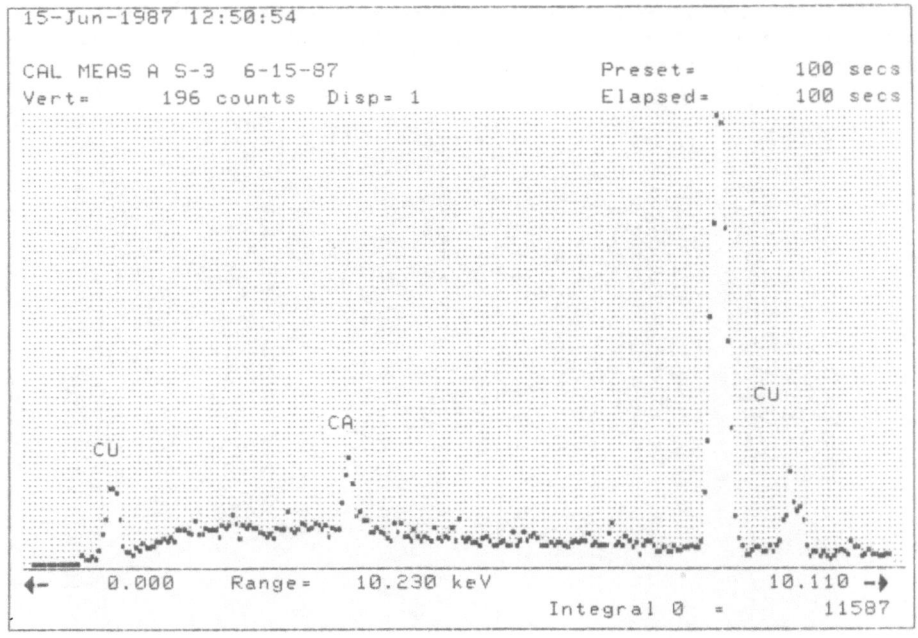

Figure 5.   X-ray spectrum of particles in Figure 4.

Figure 6 shows a 40 micron particle which emitted no X-ray lines, suggesting that it was carbonaceous, possibly a fragment of plastic. Figure 7 shows another quite different 50 micron particle with a complex elemental composition shown in Figure 8. This complex X-ray spectrum, with Al, Si, S, Cl, K and Ca, suggests that it is probably a dust particle of lithic (stone) origin (represented by Al, Si, and Ca) with some agglomeration of salts of S, Cl and K. The Cu line is not from the particle, but from the copper substrate on which the particle impacted.

Other particle types, as represented by their principal elements, were the following:
(Al, S, Ni)
(Cd, Fe)
(Al, S)
A total of at least five different types of particles, each with different elemental compositions, were identified from this run.

162

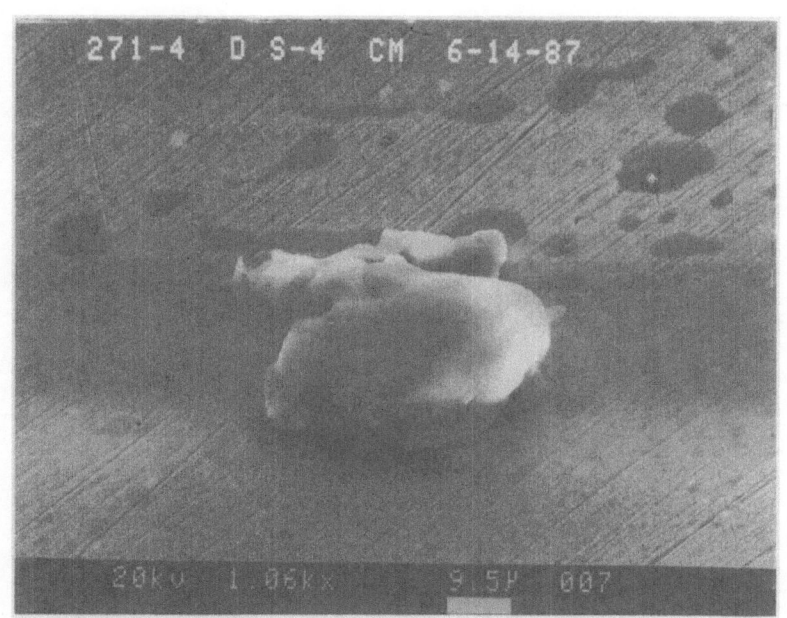

Figure 6. 40 micron particle from Run 2 (without x-ray emissions).

Figure 7. 50 micron particle from Run 2.

Since our main objective was to check the system design aspects of using the cascade impactor to collect samples from a gas source and not to determine the actual cleanliness of the gas sampled, our experiment ended there.

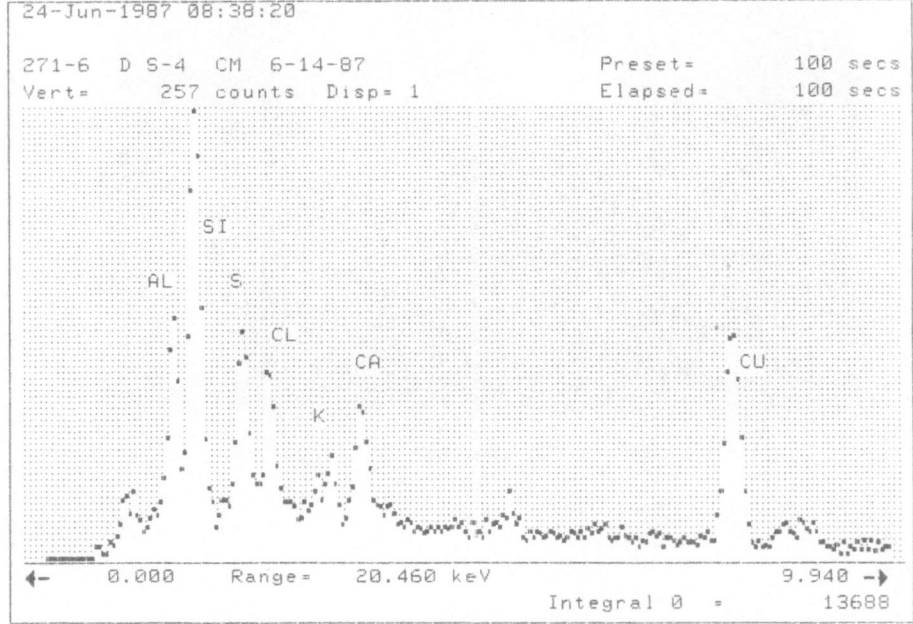

Figure 8.   X-ray spectrum of particle in Fig. 7.

CONCLUSION

The cascade impactor can be simply adapted to sample particles from gas sources. It is a useful tool for gaining a better understanding of the particulate contamination problem in process gases.

It offers the advantage of collecting particles in small concentrated dots so that they can be easily found under high magnification. It also provides flexibility in the choice of material of the particle collection substrates to simplify sample preparation procedures in the scanning electron microscope laboratory, thus saving much of the microscopist's time.

The optimum impactor system flow rate is between 5 and 10 slpm. The sampling time for a particle concentration of 10 particles per cubic foot, with an impactor flow rate of 7 LPM (0.25 cfm), is about 3 hours. This amount of time would be needed to collect about 450 particles for analysis. Based on an impaction pattern of 9 spots, about 50 particles would be collected and available for analysis within each spot. The size of this spot is about 0.3 mm diameter. Doubling this time would provide double the number of samples available for analysis.

At an impactor system flow rate of 7 LPM, plus a bypass (dump) flow of 1 LPM, about 50 cubic feet of gas would have to be sampled in 3 hours, based on a particle concentration of 10 particles per cubic foot. If the gas is 10 times cleaner, at 1 particle per cubic foot, 500 cubic feet of gas would have to be sampled over a period of 24 to 30 hours.

This experiment showed that the flow divider could be a simple "T" without the use of a tapered diffuser or isokinetic inlet. By adjusting the flow entering the "T" to about 8 LPM and the impactor inlet flow at 7 LPM, the potential particle loss is insignificant and isokinetic matching is not necessary.

It also demonstrated that the cascade impactor need not have more than two stages of size separation. Since most gases will be relatively clean, it would be more desirable to concentrate particle collection in the smallest number of stages possible. This will minimize the sampling time. Therefore, a two-stage system, with 50% particle cut-off sizes of 0.5 and 0.05 microns for stages 1 and 2 respectively, would be more optimum. With this arrangement, particles larger than 0.5 micron would be collected in stage 1, and those between 0.5 and 0.05 microns in stage 2.

REFERENCES

1. M. Davidson, Determination of particle origins in high purity gases, Solid State Technol. 63-67, July 1987.
2. W. R. Gerristead Jr., R. Sherman, and J. M. Davidson, Automated SEM/Microprobe finds gas contaminants, determines sources, Research and Development, 93-100, June 1987.
3. R. L. Chuan and W. Chiang, Cascade impactor collects airborne samples for particle identification, Solid State Technol. 133-137, July 1985.
4. G. Kasper, H. Y. Wen and A. Berner, An inertial sampling techniques for identification of submicron particle content in high purity gas streams and clean rooms, in "Proceedings of the 1985 meeting of the Institute of Environmental Sciences", pp. 99-101.

IN-SITU MONITORING OF PARTICULATE CONTAMINATION IN INTEGRATED CIRCUIT PROCESS EQUIPMENT

Steven D. Cheung

Union Carbide Corporation
Linde Division
Tarrytown, New York, 10591

Particulate contamination in wafer processing is recognized as one of the prime contributors to yield loss in advanced integrated circuit (IC) fabrication. Practical approaches for monitoring and controlling particulate contamination during IC manufacture are needed in order for the microelectronics industry to maintain an acceptable level of manufacturing performance. Critical in achieving this goal is the development of methods to identify and eliminate contamination sources. A technique has been developed to provide real-time concentration and size distribution information of particulate contamination at the "true" point-of-use within atmospheric pressure process equipment. Results are presented showing various applications of this technique.

INTRODUCTION

Critical features for ULSI devices are expected to require sub-micron geometries by 1990 (Table I). At this scale unprecedented reductions in defect densities will be required if the industry is to obtain an acceptable level of manufacturing performance. Critical in achieving this goal is the reduction of particulate contamination, a major source of manufacturing defects. Estimates of defects attributable to particles vary from 60 to 70 percent of all defects on a wafer.[1]

Today's IC manufacturing technologies have grown increasingly sensitive to particulate contamination as device geometries have shrunk and die areas and circuit densities have increased. Future generations of microelectronic devices will incorporate geometries requiring an order of magnitude decrease in defect densities over the next five years. To address this issue, new methodologies for monitoring and controlling particulate contamination must be developed. Presented here is one such methodology.

Table I. Predicted Defect Densities and Critical Dimensions
for Advanced IC Manufacture[2],[3]

YEAR	FEATURE SIZE (μm)	DIE SIZE (cm2)	TYPICAL PRODUCT	DEFECTS (D/cm2)	PREDICTED YIELD .
1984	1.6	0.60	256 K (DRAM)	0.18	90%
				0.38	80%
				0.61	70%
1986	1.2	0.96	1 M (DRAM)	0.11	90%
				0.23	80%
				0.37	70%
1988	0.8	1.42	4 M (DRAM)	0.07	90%
				0.16	80%
				0.26	70%
1990	0.5	2.52	16 M (DRAM)	0.04	90%
				0.09	80%
				0.15	70%

IN-SITU PARTICLE MONITORING

A technique for measuring the real-time concentration and size
distribution of particles present in atmospheric pressure process
equipment has been developed. In this technique an isokinetic probe
(described below) is used to deliver a representative sample of the
process ambient to an aerosol monitoring system. The monitoring system
(Figure 1) incorporates both laser and condensation nucleus particle
counters, providing a range of detection of 0.02 μm (optical diameter)
and larger. A part-per-million trace oxygen analyzer is also included
to facilitate the detection of air infiltration. A description of the
instruments is included in Table II.

The sampling system for the instrument package consists of a
quartz isokinetic probe, a custom 3-way flow splitter, and sampling
tubing. All components have been designed to optimize the aerosol flow
path. The preferred material of construction for the sample tubing and
flow splitter is electropolished stainless steel. Quartz construction
is used for the sampling probe, permitting the probe to remain in

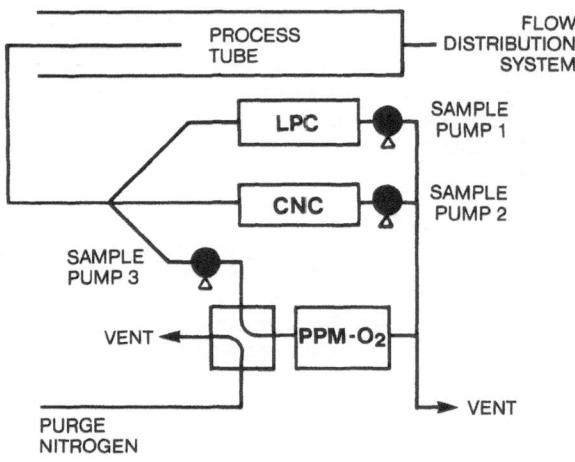

Figure 1. Monitoring System Schematic[4].

continuous high temperature and corrosive service without threat of
ionic or particulate contamination to the process environment. The
design of the probe allows it to be positioned in close proximity to
suspected sources of particulate contamination, thus providing a means
by which particle generators may be isolated and identified.

Table II.   Aerosol Monitoring System

ANALYZER	SENSITIVITY	SAMPLE FLOW	DESCRIPTION.
LASER PARTICLE COUNTER (LPC)*	0.10 μm	0.01 CFM	5 channel resolution ch1: 0.1 - 0.2 μm ch2: 0.2 - 0.3 μm ch3: 0.3 - 0.5 μm ch4: 0.5 - 1.0 μm ch5: > 1.0 μm
CONDENSATION NUCLEUS COUNTER (CNC)*	0.02 μm	0.01 CFM	Dual mode operation single particle: <1000 part/cc photometric mode: >1000 part/cc
TRACE OXYGEN MONITOR	0.1 ppm	0.01 CFM	Detection Limit: 0.2 ppm

* Background noise levels for the CNC and LPC have been determined
  experimentally to be 0.98 and 10.6 particles per cubic foot,
  respectively.

EXPERIMENTAL RESULTS

To illustrate the utility of the in-situ aerosol monitoring system
described above, the following case studies are presented:

CASE 1: Particle generation due to a flow
        distribution system.

CASE 2: Particle shedding due to diffusion
        tube heat-up.

CASE 3: Particle shedding due to
        temperature cycling of a diffusion
        furnace.

In each case, particle measurements were made inside a clean,
newly installed, 8-inch diameter, atmospheric pressure diffusion
furnace housed in a Class 100 cleanroom. Samples of the furnace
ambient were withdrawn isokinetically along the axis of flow with an
isokinetic quartz sample probe. Furnace-inlet samples were withdrawn
using a 1/4 inch electropolished stainless steel tube placed between
the flow distribution system and the process tube. Process flow during
experiments was maintained at 55 standard liters per minute of filtered
nitrogen (0.02 μm absolute). No wafers were present in the process
tube during sampling.

In case 1, the particle-generation rate for a typical flow distribution system containing high purity gas regulators, mass flow controllers, high efficiency gas filters, air actuated bellows valves, and electropolished stainless steel tubing, was determined. This was accomplished by (1) placing a 0.02 μm rated filter at the inlet to the flow distribution system; (2) certifying that the gases supplied to the system were particle free (using both CNC and LPC instruments); and, (3) monitoring the particle content of the effluent from the distribution system as it entered the process tube. Results of the evaluation indicate that components of the flow distribution system serve as a major source of contamination of the process gases (Figure 2). The majority of the particles generated by the system were of size 0.02 - 0.1 μm, as determined by the CNC.

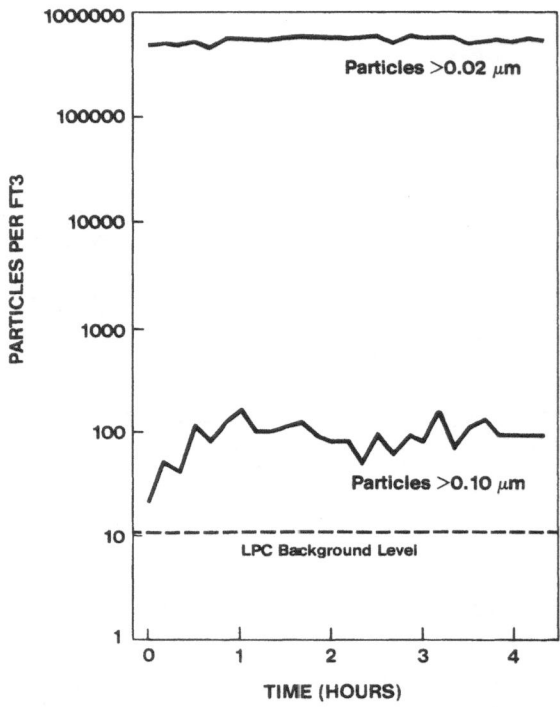

Figure 2. Particle concentrations due to flow distribution system without point-of-use filtration[4].

Based on the knowledge that the flow distribution system contributes particles to the process gases, a point-of-use filter was added to the distribution system as the last flow component contacting the gas prior to its entering the process tube. Following this addition, particle measurements were performed, and it was determined that point-of-use filtration was effective in removing particles from the delivered process gas (Figure 3). The point-of-use filter remained in service for the remainder of the experiments, serving as a means to provide particle free gas to the process environment.

In case 2, the effect of temperature ramp-up upon particulate contamination levels within a clean diffusion tube was investigated. For the experiment the temperature within the tube was maintained at 700°C for several days and certified particle free using both the CNC

and LPC. Following certification a temperature ramp-up from 700 to
1000°C was performed, using a ramp rate of 30°C/minute, and the
response of the system was monitored. For the system under
investigation a large number of particles of size 0.02-0.1 μm were
generated as the process tube was heated above 700°C (Figure 4). Once
the tube reached its setpoint of 1000°C, cleanup back to instrument
background levels took several hours. The source of these particles is
believed to be the quartz walls of the diffusion furnace, with particle
generation resulting from thermally induced stress. Experiments to
substantiate the composition of these particles have been planned.
Results of these experiments will be reported at a future date.

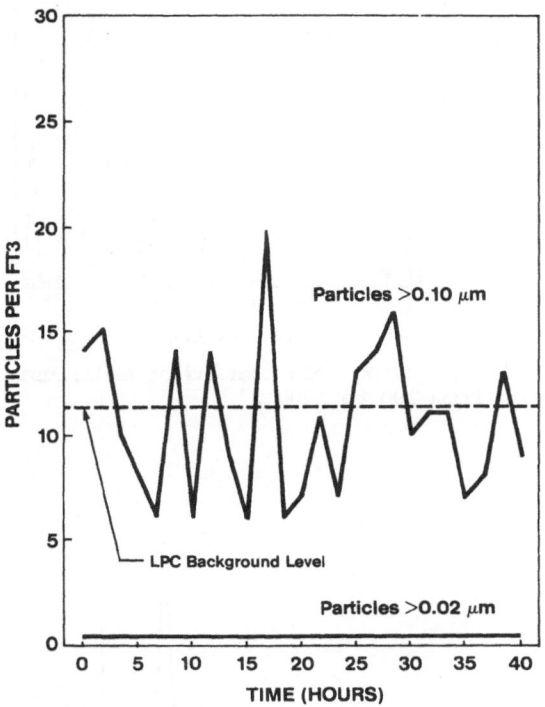

Figure 3. Particle concentrations due to flow distribution system with
point-of-use filtration[4].

In case 3, the effect of temperature cycling on the rate of
particle shedding within a quartz diffusion tube was investigated. For
the experiment the quartz diffusion tube underwent temperature ramp-up
cycles from 500 to 1000°C at a rate of 30°C/minute followed by
temperature ramp-down cycles from 1000 to 500°C at a rate of
10°C/minute. Results of the experiment indicate that there is a
critical threshold temperature at which particle generation is most
prevalent (Figure 5). For the system under investigation this
temperature fell between 600 to 800°C. It is believed that the same
particle generation mechanism as describe in Case 2 is responsible for
the effects seen here.

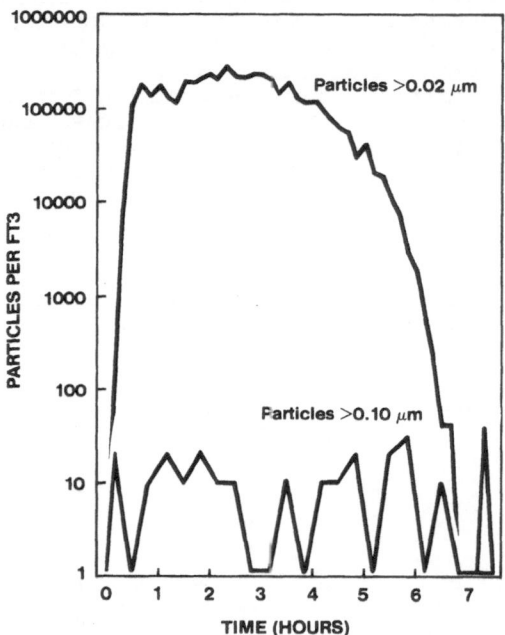

Figure 4. Particle shedding upon heating of a diffusion furnace (Step change from 700 to 1000°C.[4]

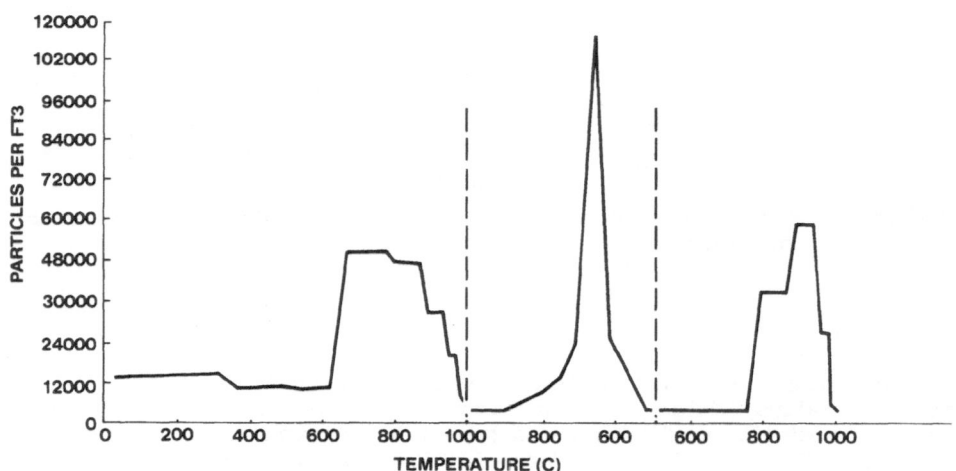

Figure 5. Particle shedding due to temperature cycling (particles >0.02μm).[4]

CONCLUSIONS

Advances in the semiconductor industry require improvements in our ability to detect and eliminate particulate contamination generated within process equipment. Continuous in-situ monitoring of particulate contaminants is one such improvement.[4,5] Unlike other techniques, continuous in-situ monitoring provides the unique capability of correlating particle upsets within the process environment with processing events. Based upon this information contamination sources can be identified and measures can be taken to reduce the levels of contamination at the "true" point-of-use -- where the wafers are.

REFERENCES

1. S. Gunawardena, K. Ulrich, B. Tullis, and J. Vietor; SMIF and its impact on clean room automation, Microcontamination, pp. 55-62 (September 1985).
2. B. T. Murphy; Proc. IEEE, 52, 1537 (1964).
3. W. J. McClean, Editor, "Status 1987: A Report on the Integrated Circuit Industry", Integrated Circuit Engineering Corporation, 1987.
4. S. D. Cheung and R. P. Roberge, An inside look at particles in process equipment, Microcontamination, pp. 44-50. (May 1987)
5. P. G. Borden, Y. Baron, and B. McGinley, In-situ studies of particle events in vacuum processing equipment and gas streams , 1987, Microcontamination Conference Proceedings, Canon Communications, Santa Monica, CA 1987.

A REAL-TIME FALLOUT MONITOR FOR 5-250 MICROMETER PARTICLES

Peter G. Borden, Jon Munson

High Yield Technology
800 Maude Avenue
Mountain View, CA 94043

Donald W. Bartelson*

Lockheed Space Operations Company
K6-1200, Rm. 1011, LSO-246
Kennedy Space Center, FL 32899

In many assembly and clean room applications it is desirable to continuously monitor fallout or sedimentation of large particles, typically greater than 5 microns in diameter. The accepted technique for monitoring sedimentation today is to collect particles on a witness plate and count them at periodic intervals. In many cases, however, it is important to observe sedimentation in real-time. For example, in a location where machinery operates, real-time monitoring can provide immediate warning of a problem.

The paper provides a description of a probe that measures sedimentation by detecting scattering from particles as they fall through a net of light. In this way it measures the current of particles moving through space toward critical surfaces or in critical locations. The sensor is passive and has a volume under 6 cubic inches, allowing installation in small areas without perturbing the local environment.

The sedimentation probe is currently undergoing evaluation in clean assembly areas. The paper presents data on its ability to provide useful data, and how this data correlates to that obtained by witness plate methods.

INTRODUCTION

The Space Transportation System delivers highly sophisticated and specialized payloads into Earth Orbit with projected life spans of 10 to 15+ years. To accomplish mission objectives, payloads have extremely contamination-sensitive components and structures integrated into their total system design configuration. To successfully attain and maintain system performance on-orbit, painstaking efforts must be practiced during ground operations to minimize contamination of critical hardware surfaces.

In many payload assembly and launch facility applications, it is desirable to continuously monitor fallout or sedimentation rates of large particles, typically greater than five microns in diameter. An accepted technique for monitoring sedimentation today is to collect particles on a gridded witness plate over a specified period of time. The particles are counted and sized according to a pre-determined distribution scale. The results are then expressed as cleanliness levels per MIL-STD-1246A.[1]

Automated, real-time acquisition of such data can offer a number of real advantages over the current procedure. With this capability, change-in-rate information can be ascertained at any time during payload processing. For example, in a facility where overhead cranes or telescoping access platforms operate, real-time monitoring can provide immediate warning of a problem. Also, the data are immediately available to a central host computer as part of a full environmental data set, without intermediate analysis steps.

Following a brief discussion of existing techniques for measuring fallout, this paper provides a description of a real-time probe, called a Particle Flux Monitor, that measures sedimentation by detecting scattering from particles as they fall through a net of light. In this way, it measures the current of particles falling toward critical surfaces or in critical environments. The sensor is passive and has a volume under 6 cubic inches, allowing installation in small areas with a minimum perturbation to the local activity. The results of initial experiments comparing the existing techniques and the Particle Flux Monitor are then presented and discussed.

EXISTING TECHNIQUES FOR MONITORING FALLOUT

Small particles fall at a speed approximately equal to the terminal velocity. For spheres, this is given by

$$V_T = \rho d^2 g / 18 \eta,$$

where ρ is the particle density, d the diameter, g the gravitation constant, and η is the viscosity of air.[2] Particles with diameter smaller than about 5 microns fall so slowly that they can be considered airborne. Tools such as airborne particle counters, which use a draw of air to pull particles through an optical sensor, have been designed to monitor such particles, and have been available for some time.[2]

Larger particles have a much higher terminal velocity, and are commonly called fallout. The principle on which the airborne particle counter is based is less effective for these particles, and other techniques are required to monitor them. Historically, the most common technique has been the Particle Fallout Witness Plate, or PFWP. This is a 37 mm diameter membrane filter with a pore size of 0.8 microns and an imprinted grid on 3.08 mm centers. The filter is mounted in a plastic holder having a center section between the top and the bottom parts of the case. This center section serves as a retaining ring to hold the filter in place when the top section is removed for "open" fallout sampling.

The PFWPs are placed at standardized locations within the facility and "activated" for a specified period of time. When the test is complete, the top of the PFWP is reinstalled and the PFWP is transported back to the laboratory for particle size distribution analysis. Once in the laboratory, the filter is removed from the holder and placed onto a petri dish for microscopic analysis. Using a stage micrometer, the particles are sized and counted. If the particles are randomly distributed and it is estimated that there are more than 100 particles in a given range, a statistical count is performed.

The data are then processed through a series of arithmetic exercises to present a final expression per MIL-STD-1246A. This standard is used to specify surface particulate cleanliness levels based on particle size distribution and counts.

Fallout data taken from several aerospace contractor clean room work areas have been consolidated and correlated to the MIL-STD-1246A size distribution criteria by Hamberg and Shon.[3] Their work has shown that the direct adaptation and applicability of MIL-STD-1246A to express cleanroom particle fallout is not recommended, especially in large volume facilities used in the build-up and integration of aerospace payloads. Nevertheless, the standard does show considerably better correlation to solvent cleaned surfaces. Applicability issues of MIL-STD-1246A is outside the scope of this paper and will not be discussed further. Even so, recognizing the drawbacks and limitations of MIL-STD-1246A, the standard is adaptable to present relative particle size distribution information. Further investigations to describe surface cleanliness levels and obscuration ratios (fraction of surface area covered by the cross-sectional area of the particles) are subjects for future discussion.

The PFWP method has been used exclusively at Kennedy Space Center to monitor particle fallout rates for the Space Transportation System (STS) program and an extensive data base is available. Past experience in performing this technique has revealed several drawbacks, and has shown that it is unsatisfactory in providing the payload community with real-time "status-of-health" information on their payload's critical surfaces.

Drawbacks of the method are accentuated in several categories. First, handling and transportation from the field site to the laboratory introduce particle loss or gain errors. Second, human manipulation and interpretation in determining the particle count and size can very consider-

ably. Third, the passive "single event" character of the
technique does not provide real-time information showing
cause-and-effect relationships or alarm conditions.

The first two problems are significant and introduce
error parameters into the final presentation of surface
cleanliness levels. The third problem probably presents the
most significant disadvantage in that the technique does not
provide real-time, early warning information.

Aerospace cleanrooms, by the result of their payload
processing functions, are characterized by massive ground
support equipment, overhead cranes, telescoping access platforms, and personnel activity interactions. In this cleanroom work area environment where the build-up, processing and
integration of payloads and the orbiter payload bay is performed, immediate indication of an abnormal increase in particle generation is highly desirable.

A REAL-TIME FALLOUT MONITOR

A real-time fallout monitor has been developed that has
the potential to overcome the drawbacks inherent in the witness plate technique. This monitor is based on a passive optical sensor that counts large particles as they fall through
a net of light.

The sensor optics are shown in figure 1. An AlGaAs
laser diode emits a 15mW, 780 nm beam through collimating
optics to generate a beam that reflects several times back
and forth between two mirrors. This creates a net of light
with dimensions of 0.5" x 1.0". Particles falling though
this net scatter light to photocells mounted above each
mirror. This is similar to a sensor designed to detect small
particles in vacuum process equipment, which has been described in detail elsewhere.[4] In the present case, the beam
is much wider, reducing the intensity so that the minimum
detectable particle size is 5 microns, but providing a high
probability of detecting a particle passing through the net.

Sizing information is obtained by measuring the amount
of light scattered by the particle. Particles of any size
up to the dimensions of the net will be detected, but particles larger than 250 microns will scatter enough light to
saturate the detector.

In the present configuration, a controller monitors up
to four sensors in real time at a count rate of up to 100
particles per second. This is well in excess of rates that
are encountered in typical environments. The controller
couples to a host computer through an RS-232 link, allowing
the fallout monitor to pass real-time information directly
to a central environmental monitoring and control system.

Experimental Evaluation of the Real-time Fallout Monitor

Initial experiments were carried out at Kennedy Space-flight Center in the Orbiter Processing Facility (OPF), on a platform adjacent to the orbiter Atlantis.

Results from three experiments are reported. In each case, the monitor was mounted near ten PFWP plates. The monitor output was collected using a printer. The PFWP plates were read using the standard method. The conditions differentiating the three experiments were as follows:

Experiment 1: Results from two monitor sensors are reported. The first sensor was set to record any particles moving through it with a nominal diameter greater than 5 microns. The second was set to record particles greater in diameter than 100 microns. Data were collected over a weekend, for a period of 71 hours.

Experiment 2: Results are reported from one sensor set to monitor particles greater in diameter than 10 microns. Data were collected overnight, for a period of 18 hours.

Experiment 3: Conditions were the same as in experiment 1, but the data were collected during a weekday afternoon for a period of 4 hours and 40 minutes.

Two types of units are used in the presentation of the data. Fallout counts as a function of time obtained with the real-time monitor are reported directly. Summary data, in which the real-time monitor counts are compared to the PFWP and square foot plate counts are reported in units of particles per square foot in 24 hours. This is a standard unit for reporting fallout data in the aerospace industry. A real-time monitor sensing area of 0.5 square inches is assumed.

Correlation of Real-time Data to Local Activity

The value of real-time fallout data collection is shown in figures 2 and 3, which are plots of the hourly counts in experiments 1 and 3, respectively. Consider first the 5 micron and larger data from experiment 1, shown in figure 2. This appears consistent with activity near and orbiter. Note how the counts drop at midnight on Friday, rise again during the day shift on Saturday, stay low through most of Sunday, and rise again during the day shift on Monday. Periods of very high counts are seen Saturday afternoon, Sunday around 5 p.m., and Monday before noon. The Saturday afternoon peak corresponds to crane activity in the vicinity of the monitor. Other peaks may correspond to activity close to or above the monitoring site. Since the data are collected in real time, they can serve as the basis for issuing a

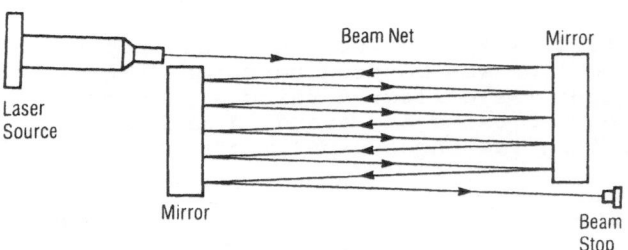

Figure 1. The optical configuration of the particle flux monitor.

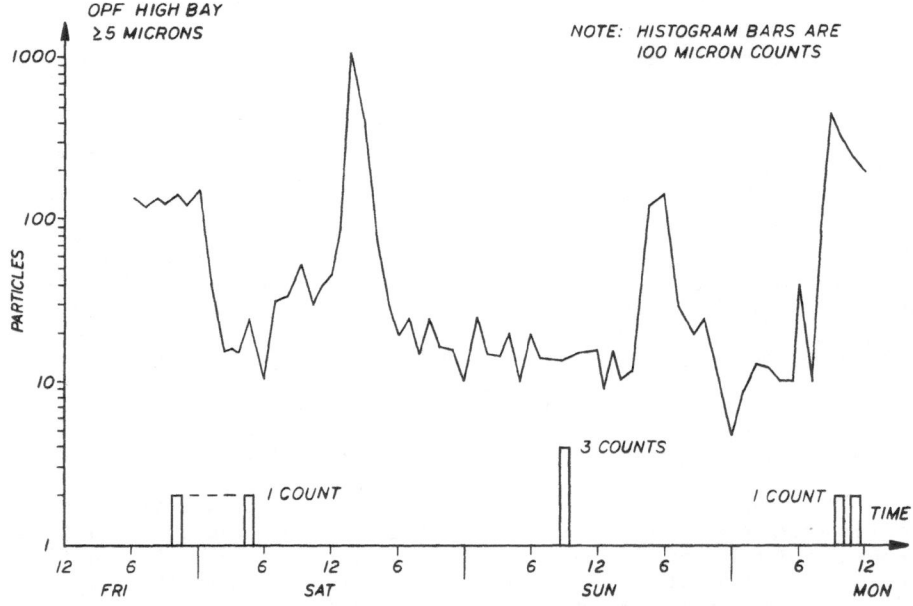

Figure 2. Fallout Counts Over 3 Days. Fallout counts obtained with the real-time fallout monitor mounted next to the Orbiter Atlantis in the Orbiter Processing Facility High Bay 2 over a period of 71 hours. The connected dots are hourly counts for the sensor monitoring particle fallout with diameters greater than 5 microns. The vertical bars are the hourly counts for the sensor monitoring particle fallout with diameters greater than 100 microns.

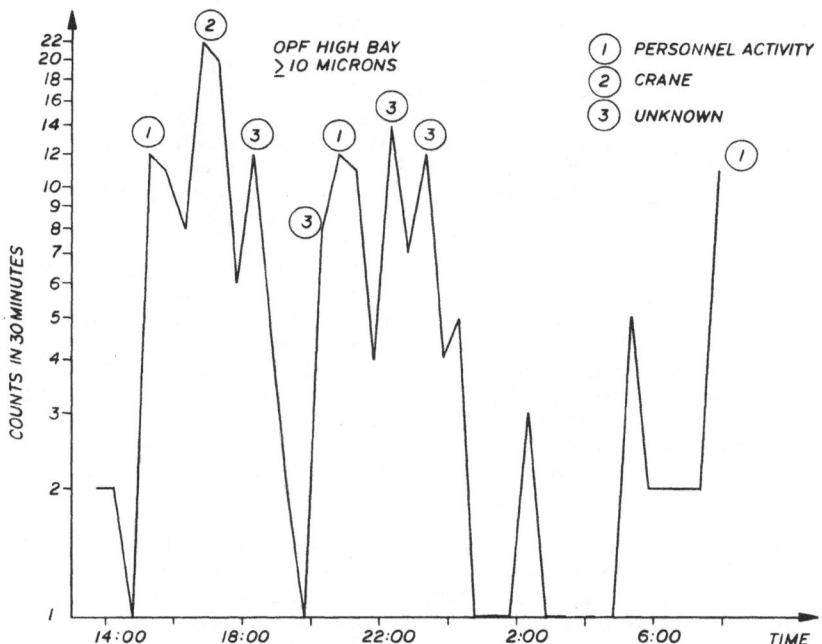

Figure 3. Fallout Counts. Fallout counts obtained in the same location as used in figure 2, over a period of 4 hours and 40 minutes. The sensor size threshold was increased to 10 microns. Major peaks are identified as to the type of activity that was occurring near the sensor.

warning that too much fallout is being generated near the monitoring location.

The 100 micron and larger data are also plotted as vertical bars in figure 2. The data rate is much lower. With the exception of the three counts at 9 a.m. on Sunday, events correlate with increased count rate on the 5 micron and larger sensor.

The 10 micron and larger data obtained in experiment 3 and shown in figure 3 are similarly consistent with local activity. Known activity of a crane or personnel corresponds to count peaks. In some cases, peaks occurred during times when it was not known whether personnel were nearby. These peaks often bracket periods of known activity, and may correlate to personnel activities.

In summary, it appears that to the extent that activity in the area of the experiment was observed, peaks in real-time monitor data correlate to local activity. This indicates that the real-time monitor may be useful to warn of high local fallout levels.

Correlation of Real-time and PFWP Data

Comparison of the data from the real-time and collection plate techniques is shown in Table 1. The data are presented in units of particles per square foot per 24 hours. Because of the limited amount of trials, and because the data have been obtained only recently, it is difficult to state the likely statistical variation. For this reason, mean values are presented.

As seen in Table 1, the real-time monitor consistently counts more particles than the collection plates. The best hypothesis to explain this is that a certain number of the particles counted by the real-time monitor are airborne, and flow or circulate through the sensor without being collected on the plates. Indeed, while the real-time sensor in each case was mounted above a PFWP plate, no attempt was made in these experiments to seal the real-time sensor to the PFWP plate, to ensure that air currents could not carry particles away.

This hypothesis is fortified by noting that as the particles get larger, the results obtained by the two techniques converge. This is shown in figure 4. The 10 micron and larger results are about a factor of 10 apart. The 100 micron and larger results are 6 to 10 times greater using the real-time approach.

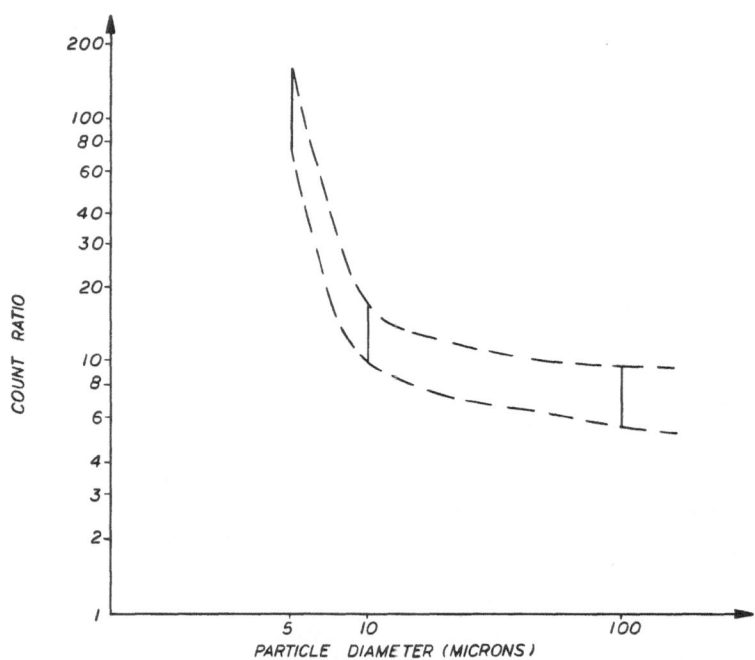

Figure 4.   Count Ratios, Real Time vs. Collection Plates. Ratio of real-time monitor counts to collection plate counts. Data come from Table 1. Curves represent upper and lower limits of error bars.

182

This suggests two things. First, care has to be taken to reduce the effect of air currents in counting, especially when smaller, somewhat airborne particles are to be observed. Second, work is needed to define the cutoff between airborne and falling particles as it pertains to this measurement technique. Indeed, the settling velocity of a 1 gram per cubic centimeter particle with a diameter of 10 microns is very slow, about .3 cm/sec, while a 100 micron particle has a settling velocity of 30 cm/sec. Furthermore, while these numbers are true for spheres, particles with high surface area to density ratios will fall more slowly.

Table 1. Comparison of counts from the three fallout monitoring techniques. Units are particles per square foot in 24 hours. Only mean values are presented; statistical variations can be large due to both sample size and counting technique.

	Real-time	PFWP	Sq. Ft. Plate
Experiment 1			
5 microns	482,866	3,045	5,956
10 microns	681	70	121
Experiment 2			
10 microns	83,368	5,339	
Experiment 3			
10 microns	136,264	14,324	

CONCLUSION

A sensor has been described that has the ability to monitor fallout in real-time. Evaluation is proceeding in the clean assembly areas in the Orbiter Processing Facility at Kennedy Space Center. The results at this point are preliminary. The evaluations so far have been aimed at obtaining an understanding of the ability to provide a real-time indication of fallout generating activity in an area. Further work will be performed to statistically validate results, and to understand the effects of air currents.

The results obtained to date have been encouraging, and indicate that high count rates obtained with the monitor may correlate to increased fallout generating activity in the local area. Count rates are somewhat higher than those observed with PFWP methods. This may be due to the motion of smaller and low density particles induced by air currents.

The second phase will be to better establish the adaptability of the real-time monitor to cause-and-effect relationships. To accomplish this, the monitor will be installed at locations of increased activity. Data to date were taken in the Orbiter Processing Facility High Bay 2 where the Orbiter Atlantis is currently undergoing ground process operations. Activity in this High Bay is currently low. Experiments will be shifted to an adjacent OPF High Bay, where the Orbiter Discovery is being readied to support the first STS mission following the Challenger tragedy. This will allow cause-and-effect evaluation in a high activity environment.

Future applications are visualized where multi-sensor locations are networked to a central data processing center for real-time information feedback. The sensor locations will be selected based on established high particle generation probabilities and proximity to critical surfaces. The real-time sensor system will be used to monitor both baseline facility environmental health conditions and localized payload-unique critical surfaces.

ACKNOWLEDGEMENTS

The authors would like to thank Martin McClellan of the Lockheed Space Operations Company, and Brian McGinley and Yuda Baron of High Yield Technology, for their assistance in developing the sensor and taking the data presented here.

REFERENCES

1. MIL-STD-1246A, "Product Cleanliness Levels and Contamination Control Program".
2. W. C. Hinds, editor, "Aerosol Technology," Chapter 3, Wiley Interscience, New York, 1982.
3. O. Hamberg and E. M. Shon, Particle size determination on surfaces in clean rooms, Proceedings of the Institute of Environmental Sciences Conference, May 1-3, 1984.
4. P. Borden, Y. Baron, and B. McGinley, Monitoring particles in vacuum-process equipment, Microcontamination, 30-34, October 1987.

MEASUREMENT AND CONTROL OF PARTICLE-BEARING AIR CURRENTS

IN A VERTICAL LAMINAR FLOW CLEAN ROOM

G. S. Settles
Center for Electronic Materials and Processing, and
Mechanical Engineering Dept., Pennsylvania State
University, University Park, PA 16802, and

G. G. Via
IBM Corp., Federal Systems Div., Manassas, VA 22110

Contamination is a crucial issue in the economics of present-day semiconductor manufacturing technology. Such contamination, which may spring from many sources, is controlled in part by the use of special facilities such as the VLF (Vertical Laminar Flow) clean room. However, the current emphasis on high-performance semiconductors requires submicron lithography, where even 0.1 micrometer and smaller-sized particles result in significant product yield losses. Such particles lack the mass to be affected significantly by gravity, but may be convected by even the weakest air currents. In essence, these particles behave as aerosols. Thus, the issue of particle transport by clean room air currents is of great significance. Unfortunately, this issue is poorly understood and is probably not given sufficient attention in clean room design. Despite the VLF designation, the airflows in these clean rooms may not be casually assumed to be either vertical or laminar. In the present study, flow visualization techniques have been applied to reveal such airflow patterns. Among several available techniques, the schlieren optical technique is non-intrusive, non-contaminating, and is itself a detector of the weak thermal gradients in air currents. The study was carried out in a full-scale, state-of-the-art VLF clean room (the IBM Manassas Submicrometer Lithographic Facility). Attention was centered on the airflow about a multistation silicon wafer processing tool. The results emphasize the need for redistribution and rebalancing of the airflow after tool installation. The consideration of secondary flows self-induced by the tool and the compromise between airflow distribution and maintenance access are also discussed.

INTRODUCTION

In modern microelectronics manufacturing the most serious current problem is the low yield of usable products, stemming from serious defects and losses during the manufacturing process. This problem is responsible for many millions of dollars lost annually in the industry. It arises, in particular, due to the extremely small scale of microelectronic circuitry, thus imposing harsh constraints on the cleanliness of the manufacturing environment.

Recent trends in components per microchip and the corresponding minimum size of features on the chip reveal that the smallest features are already below one micrometer (μm) in size. This is illustrated in Figure 1, which is an electron micrograph showing 0.5 μm lines etched in photoresist. This example was produced in IBM FSD facilities.

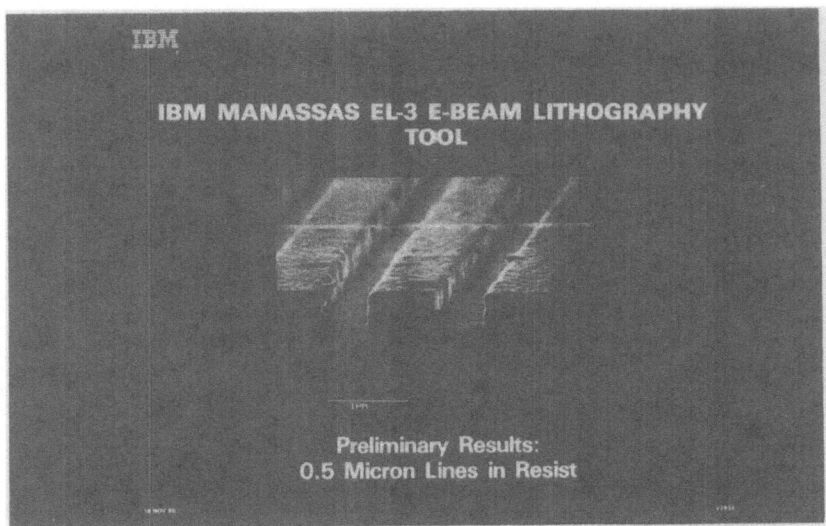

Figure 1. Example of 0.5 μm photoresist lithography.

This trend toward sub-μm feature sizes in microelectronics carries with it the problem of contamination due to sub-μm particles. Such small contaminant particles were not serious threats to product yield in the early years of microelectronics manufacturing, when feature sizes were much larger. Now, however, they are of great concern. Aside from chemical contamination, the threat is that sub-μm particles can cause short or open circuits in the extremely small linewidths which are important now and in the future.

The frequency of occurrence and behavior of sub-μm particles differs significantly from past experience in contamination control. For example, Cooper[1] has estimated the surface flux of contaminant particles as a function of particle diameter. His work reveals that many more sub-μm particles may be expected than those 1 μm and above in clean room environments. Further, Cooper[1] has shown results indicating that, While gravitational settling is a key deposition mechanism for large particles, it ceases to be important in the sub-μm range. Instead, Brownian motion and motion due to thermal and electrical gradients take precedence. Such small particles are especially susceptible to convective transport by air currents.

Contamination resulting from airborne particulates is one of the largest contributors to yield loss in microelectronics manufacturing.[2] In order to reduce such contamination and thus promote reasonable yields, VLF (Vertical Laminar Flow) clean rooms are used. In such installations, filtered air enters across the entire ceiling and is then exhausted through a perforated, raised floor, whence it is recirculated through the ceiling HEPA filters. The downdraft speed of the filtered air is of the order of 0.5 m/s, corresponding to 650 "changes" of the room air per hour. With proper design, the particle count in the air entering the room from the ceiling can be kept well below 1 μm-range particle/cubic meter/minute. Sub-μm particle counts and their control in clean room facilities are not well understood, however. These are issues for current and future study.

Unfortunately, little effort has been made to understand what happens after the air enters the clean room from the ceiling filters, and how the particles generated by the tooling and human operators behave within the clean room airflow. Once constructed and qualified, VLF clean rooms are densely populated with tooling. The presence and structure of such tooling will definitely modify the airflow within a clean room, possibly allowing the "trapping" of contamination generated by the tools themselves or by human operators working in the clean room. Further, the presence within certain tools of heat sources (e.g., silicon wafer drying plates) may result in powerful vertical updrafts which may interact with the main VLF downflow to create localized, unsteady turbulent vortices. These vortices can trap and transport particulate contamination. This trapped contamination may then be redistributed on the product undergoing processing. All these potential problems are exacerbated when sub-μm particles are considered.

This paper examines such airflow problems in VLF clean rooms, with special emphasis on the ready transportability of sub-μm particles. Methods for visualizing such airflows and the constraints placed upon these methods are then discussed. Finally, an example is given of a portable schlieren optical instrument used for visualizing airflow patterns in a clean environment. Initial results are shown from the IBM Sub-μm Lithographic Clean Room Facility. These results emphasize the importance of rebalancing the airflow in a clean room after the installation of tooling.

AIRFLOW PROBLEMS IN VLF CLEAN ROOMS

Including other work by the current authors[3-5], only a few literature citations have been found[6-13] on the subject of airflow problems in VLF clean rooms. Model studies such as that of Akabayashi, Murakami, Kato, and Chiriful[3] have been useful in this regard. They observed a variety of flows about clean room equipment and operators, and arrived at some helpful guidelines. Several full-scale clean room studies have also been done, primarily using smoke for flow visualization. The work of Sodec and Veldboer[11] was especially productive. They observed flow disturbances from ceiling lights, heat sources, hot walls, perforated and unperforated work surfaces, and various floor venting arrangements. Some suggested solutions to these problems were presented.

At the outset there is a semantic problem to be dealt with. The "VLF" designation applied to a clean room does not mean that the flow is laminar in any strict fluid dynamical sense. Laminar flow is flow in which momentum transport occurs only on a molecular basis, while turbulent flow denotes transport by gross eddy motion. All laminar flows eventually succumb to natural oscillations (transition) leading to turbulence, given sufficiently high values of the Reynolds number:

$$Re = VL/\nu \qquad (1)$$

where V is the airspeed, ν is the kinematic viscosity of the air, and L is a length dimension characteristic to the problem at hand. For typical VLF clean room conditions V = 0.5 m/s, ν = 1.6×10^{-5} m^2/s, and L (the distance from floor to ceiling) is about 2.5 m, yielding a Reynolds number of about 80,000. Transition to turbulence is such a complex phenomenon that no exact Reynolds number criterion can be stated, but it appears likely from the above that clean room flows will be transitional except under ideal conditions. Ceiling supports, light fixtures, and any equipment in the clean room act to trigger early transition, so that true laminar flow is doubtful in a real clean room. Indeed, our observations thus far have consistently shown turbulent eddies in clean rooms airflows.

Accepting the airflow to be turbulent rather than laminar, the immediate consequence is that much higher particle transport rates may be expected due to turbulent eddy convection. Additional consequences arise from the interaction of the downflow with tooling and personnel. Some typical resulting airflow problems are categorized in Figure 5.

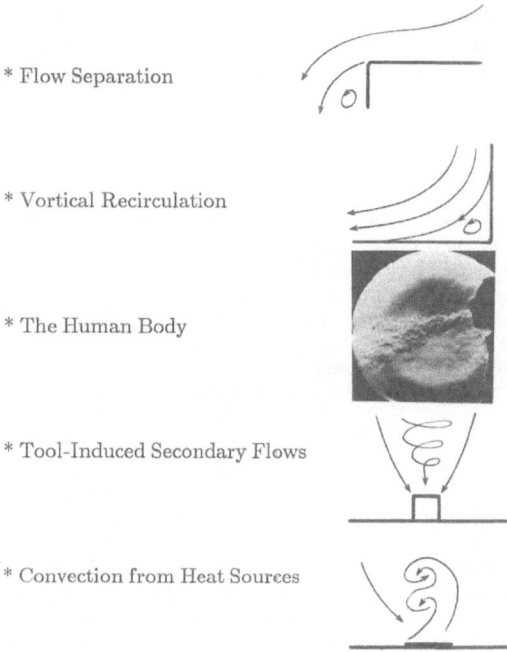

* Flow Separation

* Vortical Recirculation

* The Human Body

* Tool-Induced Secondary Flows

* Convection from Heat Sources

Figure 2. Some typical VLF clean room airflow problems.

Viscous boundary layers naturally form on all solid surfaces in a clean room, including the walls of the room itself. Freestream turbulence (introduced by filter grids), roughness, and protuberances on these surfaces usually dictate that these boundary layers are transitional or turbulent. It is well known that such a boundary layer cannot sustain the pressure gradients produced by either an interior or exterior 90 degree corner without separating from the surface along which it flows. When boundary layer separation occurs, a vortical recirculation region (separation bubble) always results.

Vortical recirculation is a serious contamination control problem. The air exchange rate between such a recirculation zone and the main downflow is far less than the 650/hr exchange rate of the main flow itself. Thus, contaminant particles become trapped within such recirculation zones. Any contamination sources within such zones are effectively isolated from the main flow, allowing contamination to build up. Contamination-sensitive manufacturing processes in the vicinity of air recirculation zones are thus likely to experience far greater rates of surface particle flux than expected from the overall clean room specification. In this regard the laminar or turbulent nature of the main downflow is a secondary issue; flow separation and recirculation are primarily related to the interaction of the downflow with tooling installed after the clean room is qualified. An example study by the present authors of such a recirculation zone in a VLF workstation is given in References 3 and 4.

The human body is a critical issue in clean room work, since it is a prime contamination source. Special garments attempt to contain this contamination with varying degrees of success. For example, it has been observed[10,14] that the breath plume from a human cough has sufficient inertia to penetrate the surrounding airflow up to 1 or 2 meters distance (see Figure 2). Standard clean room face coverings prevent this, but do not prevent local airborne contamination.[10]

Tool-induced secondary flows may cause additional problems of VLF clean room airflow distribution. Some tools contain internal blowers, ducting, and ventilated work surfaces. A particular multipurpose wafer processing bench to be discussed later in this paper moves some 200 m^3/min. of the main downflow for local ventilation and cooling purposes. The proper integration of such strong secondary flows with the overall airflow balance of the room does not happen automatically upon tool installation.

Finally, locally strong thermal sources may be present in particular tools, leading to possible thermal updrafts. Wafer drying stations are one example. It was established in previous work[4] that thermal drafts of sufficient strength can flow against the main downflow of the room, thus exposing wide areas to possible contamination.

FLOW VISUALIZATION

It is a guiding principle of experimental fluid dynamics that one should first study and understand the phenomenology of an unknown flow before attempting detailed pointwise measurements. Since airflows are normally invisible, flow visualization is required for this purpose. The principles of a variety of flow visualization techniques including surface indicators, tracers, and optical methods are well known.[15,16]

However, in applying flow visualization in clean room facilities one must be aware of certain constraints. Primarily, contamination from the flow visualization method itself is generally forbidden. This eliminates one of the most powerful visualizations methods: the addition of particle tracers such as smoke.[8,10,11] A visible fog produced by steam and liquid nitrogen vapor has been tried as a non-contaminating flow tracer. Unfortunately, our experience with this technique has shown it to be too coarse and uncontrollable to be useful. Similarly, "streamers" made from audio microcassette tape proved to have too much inertia to faithfully trace the flow streamlines.

The remaining category of flow visualization methods involves optical instruments such as the shadowgraph, schlieren, and interferometer.[15,16] These methods render temperature differences or foreign gases visible in air and do not inherently generate particulate contamination. The shadowgraph is simple but lacks the sensitivity required for clean room studies. Interferometry shows future promise, but is currently precluded because it requires a laser beam of a potentially hazardous power level. The schlieren instrument, the intermediate of the three, appears quite suitable for clean room studies, however.

The present authors have had considerable success in applying schlieren visualization to clean room-type flows in the laboratory.[3,4,14] An example of schlieren applied to the study of downflow in an open clean room workstation is shown in Figure 3. Here, turbulent flow separation is observed at the operator's forehead, extending down to the work surface. A separation bubble is thus formed which exposes the item being handled to recirculated contamination from the operator's body.

Figure 3. Schlieren photograph of downflow in an open clean room workstation and its interaction with a human operator.

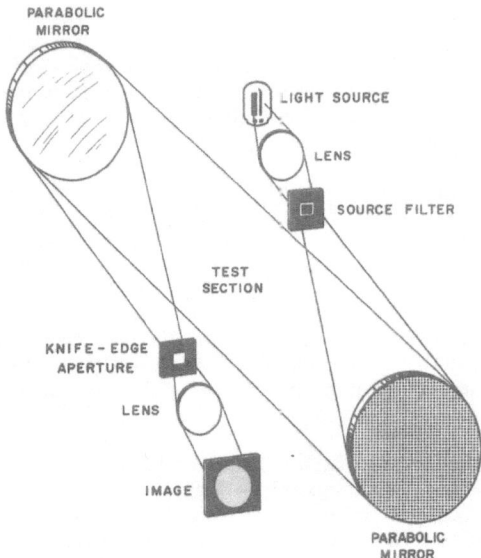

Figure 4. Diagram of "Z-type" schlieren flow visualization instrument.

A prototype, portable schlieren instrument for use inside actual VLF clean room facilities has also been fabricated and tested. Its defining equations and operating principles have been described in a previous paper.[4] Briefly, the components of this instrument are illustrated in Figure 4. A small white-light source is collimated by a parabolic reflector through a colored source filter, producing a beam of 32 cm diameter. This beam is refocused by a second parabola onto a knife-edge aperture or mask. The mask blocks most of the light, but allows rays refracted by disturbances in the test section to illuminate the schlieren image in color. The components up

190

through the first mirror comprise the sending assembly, while the remaining components comprise the receiving assembly. Each assembly is mounted on a separate, portable dolly. 45 degree planar mirrors are also provided to fold the 32 cm light beam twice as necessary for access and compactness in clean room studies.

SCHLIEREN STUDY OF VLF CLEAN ROOM AIRFLOW

The prototype schlieren instrument described above has been used to observe airflow patterns in the IBM FSD Sub-μm Lithographic Clean Room Facility at Manassas, Va. In particular, the airflow in the vicinity of a multipurpose wafer processing bench (shown photographically in Figure 5) was the subject of the study. This apparatus contains wafer transfer tracks, coating/development units equipped with air exhaust systems, heat sources, and other devices. It is obviously an area which is highly contamination-sensitive. The schlieren system was positioned so as to align the parallel test beam with the long axis of the bench at several locations just above and in front of the bench top. Because other tooling blocked one end of the bench, the schlieren transmitting assembly was fitted with a 45 degree folding mirror as shown in Figure 6. The schlieren receiving assembly is in the foreground in this photograph.

Figure 5. Photo of multipurpose silicon wafer processing bench.

Schlieren images of the airflow patterns were recorded on videotape. To augment the visibility of flow streamlines, a "thermal wand" was used. This consisted of a small resistive heating element at the end of a long thin probe. The probe could be hand-held or tripod-mounted, with automatic extension and retraction via a servo motor. The heating element produced a filament of warm air which served to reveal the local flow direction in a non-contaminating manner. This filament was observed to roll up into a characteristic Karman vortex street within a few centimeters, which convected away along the local streamline direction, In highly-turbulent regions, the filament was observed to break up almost immediately. Experience has shown that this thermal wand technique does not produce significant interference with the flow, in that the overall features of the flow do not change when it is inserted.

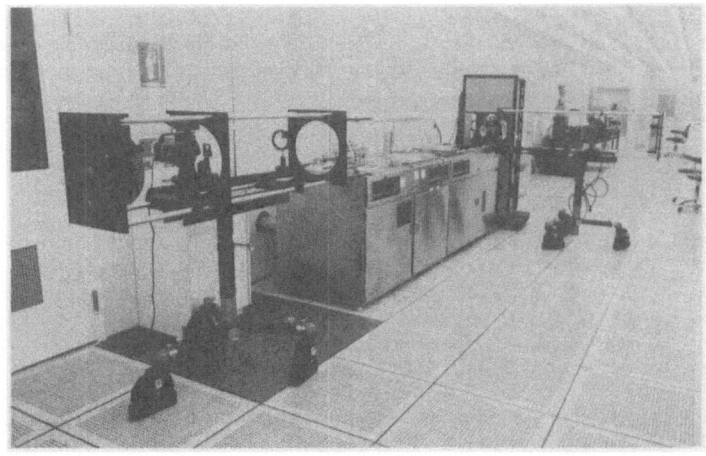

Figure 6. Photo of prototype schlieren instrument in place in the IBM Sub-
μm Lithographic Clean Room Facility.

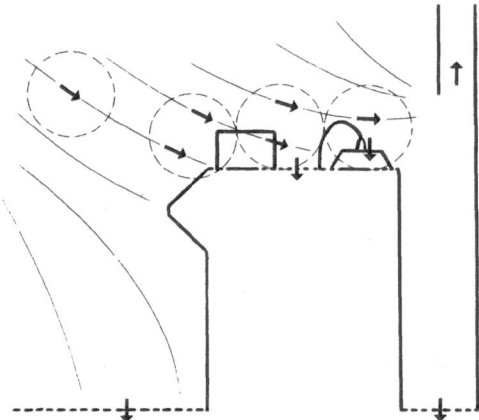

Figure 7. Diagram of airflow about multistation wafer processing bench
(before modification).

The schlieren observations revealed a strong crossflow from the center
aisle of the clean room toward the back of the multipurpose wafer processing
bench and the wall. The observed streamlines are sketched in Figure 7, where
the dashed circles reveal the actual positions of the schlieren beam in four
consecutive tests. The view is along the long axis of the bench and parallel
to the wall.

The strong crossflow was puzzling at first. It certainly was not
present when the clean room was qualified before the tooling was installed.
It was also highly undesirable in that airborne contamination from human
operators was swept directly across wafers undergoing processing. With the
aid of the schlieren observations, the problem was traced to a modification
made during the tool installation. The lower wall panels behind the bench
had been removed, allowing air from the clean room to be drawn directly into
the return plenum rather than through the floor grating. The existing
pressure differential caused a considerable air removal behind the bench,

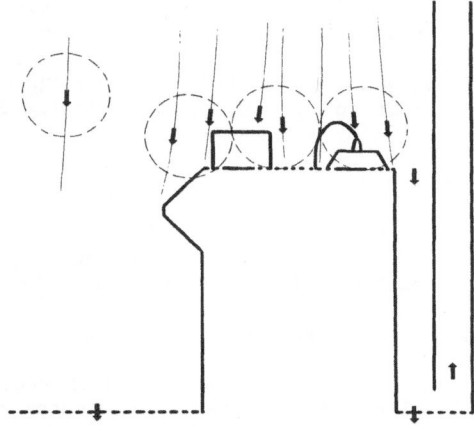

Figure 8. Diagram of airflow about multistation wafer processing bench (after modification).

inducing the observed crossflow. This problem had not been detected in routine clean room operations prior to the schlieren study. Of course, it would have been detected soon enough by other means, but it serves as a good example of how useful it is to be able to see the airflow by way of schlieren flow visualization.

The solution to this problem is illustrated in Figure 8. By replacing most of the wall panel behind the bench, the crossflow was eliminated. Some suction was still retained there to avoid the formation of a recirculation zone over the benchtop. The criterion for a proper airflow balance was established by schlieren observation at the front edge of the benchtop: a streamline from the ceiling HEPA filters was forced to meet this edge, preventing contamination due to operators in the aisle from reaching the sensitive benchtop area. Note that the airflow balance varied along the length of the bench due to the suction and exhaust systems of the bench itself. Note also that this balance could not have been achieved before the tooling was installed, or without the aid of flow visualization.

SUMMARY AND CONCLUSIONS

A flow visualization study of particle-bearing air currents in a VLF clean room has been carried out. The following observations and conclusions have been reached:

1. Sub-μm lithography is important now and in the foreseeable future.

2. Sub-μm particles are expected in large numbers in clean room environments.

3. Such particles are freely transported by the prevailing airflow patterns.

4. Transport and entrapment of particles is exacerbated by several phenomena occurring in clean room airflows.

5. Non-contaminating flow visualization is called for to identify and solve these flow problems.

6. A portable schlieren optical tool has been used successfully for this purpose in a VLF clean room.

7. Results emphasize the need for airflow redistribution and rebalancing *after* the installation of tooling in a clean room.

8. It is recommended that a flow-visualization mock-up or scale-model study of a clean room be carried out before final design and construction of the actual room.

REFERENCES

1. D. W. Cooper, Particulate contamination and microelectronics manufacturing: An introduction, *Aerosol Sci. Technol.*, 5, 287-299 (1986).
2. IBM scales designs to 0.5 micron, *Semiconductor International*, 8, 22 (June 1985).
3. G. S. Settles, B. C. Huitema, S. S. McIntyre, and G. G. Via, Visualization of clean room flows for contamination control in microelectronics manufacturing, in "*Flow Visualization IV*," C. Veret, editor, Hemisphere Press, 833-838 (1987).
4. G. S. Settles and G. G. Via, A portable schlieren optical system for clean room applications, in "*Proceedings of the 8th International Symposium on Contamination Control*," Milan, Italy, 381-392, Sept. 1986.
5. G. S. Settles, Indoor environments, Chapter IV-14, "*Handbook of Flow Visualization*" W.-J. Yang, editor, Hemisphere Press, to be published in 1988.
6. D. M. Deaves and D. Malam, Advanced analysis techniches for the optimum design of clean rooms, *J. Environmental Sci.*, 17-20 (September/October 1985).
7. N. Shi, The water model experimental equipment for research on airflow field of laminar clean room, in "*Proc. 8th Intl. Symposium on Contamination Control*," Milan, Italy, 859-868 (1986).
8. T. Hayashi, T. Okonogi, and M. Takemura, Proposal of air supply method for clean tunnel system, in "*Proc. 8th Intl. Symposium on Contamination Control*," Milan, Italy, 332-339 (1986).
9. S. Kato, S. Murakami, and S. Chirifu, Study of airflow in conventional type clean rooms by means of numerical simulation and model test, in "*Proc. 8th Intl. Symposium on Contamination Control*," Milan, Italy, 781-791 (1986).
10. G. Sullivan and J. Trimble, Evaluation of face coverings, *Microcontamination*, 4, No. 5, 64-70 (1986).
11. I. F. Sodec and W. Veldboer, Influences on the stability of airflow in clean rooms, in "*Proc. 8th Intl. Symposium on Contamination Control*," Milan, Italy, 869-885 (1986).
12. K. Toshigami, H. Kanayama, and S. Yashima, Finite-element analysis of airflow and advection-diffusion of particles in clean rooms, in "*Proc. 8th Intl. Symposium on Contamination Control*," Milan, Italy, 278-285 (1986).
13. S. Akabayashi, S. Murakama, S. Kato, and S. Chirifu, Visualization of airflow around obstacles in laminar-flow type clean rooms with laser light sheet, in "*Proc. 8th Intl. Symposium on Contamination Control*," Milan, Italy, 691-697 (1986).
14. G. S. Settles and J. W. Kuhns, Visualization of airflow and convection phenomena about the human body, *Bull. Amer. Phys. Soc.*, 29, No. 9, 1515 (1984).
15. W. Merzkirch, "Flow Visualization," 2nd ed., Academic Press, Orlando, Florida (1987).
16. G. S. Settles, Modern developments in flow visualization, *AIAA J.*, 24, No. 8, 1313-1323, (1986).

PARTICLE DEPOSITION VELOCITY STUDIES IN SILICON TECHNOLOGY

D. S. Ensor, A. C. Clayton, T. Yamamoto, and R. P. Donovan

Research Triangle Institute
P.O. Box 12194
Research Triangle Park, North Carolina 27709

Mechanisms of particle deposition on silicon wafers are briefly reviewed by reference to and citation of results from several recent theoretical and experimental papers. Two mechanisms--one originating from electrical forces and a second from thermal forces--are shown to strongly affect particle deposition velocity based on measurements carried out primarily in a specially designed particle deposition chamber but also on supplementing measurements of particle deposition made in the semiconductor clean room at the Microelectronics Center of North Carolina.

INTRODUCTION

The accumulation of particles on the surface of a silicon wafer as a result of exposure to contaminated environments during manufacturing represents an important limitation in the yield of contemporary and future silicon device chips. The traditional methods for controlling surface particle accumulation are to isolate production wafers from all unnecessary sources of particles and to eliminate/minimize particle generation from those sources which cannot be isolated from the wafer.

Isolation from the many particle sources that make up the typical urban air ambient environment is usually achieved by conducting wafer manufacturing in a clean room--all outside ambient air supplied to the manufacturing area enters through at least one stage of HEPA (High Efficiency Particulate Air) or ULPA (Ultra Low Penetration Air) filtration. The quality of the air entering the manufacturing area is thus very high, typically being less than 100 particles/ft^3 as measured by a condensation nucleus counter (detects particles >0.01 μm) and with essentially no detectable particles greater than 0.5 μm. When, in addition, this high quality air is continuously recirculated through the HEPA filter banks so that any given air volume makes multiple passes through the filters, the contribution to the aerosol concentration within a clean room attributable to the outside ambient air can be expected to be negligibly small. However, it is not always negligible under all conditions as predictions of concentrations and size distributions of aerosol particles in clean rooms at rest show (Figure 1). The four calculated curves of Figure 1 come from operating on two typical ambient air size distributions, labelled urban and rural, with two typical filter

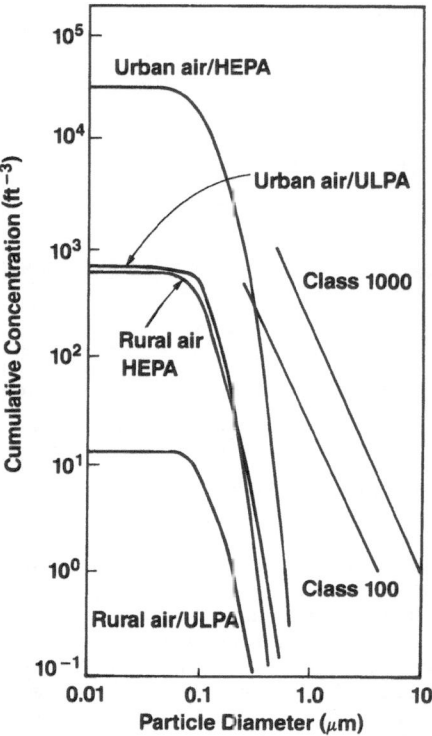

Figure 1. Computed cumulative size distribution curves for clean
rooms.[1]

efficiency curves, labelled HEPA and ULPA. These calculations assume
that 20% of the air passing through the filter is outside air, the
balance being recirculated clean room air. If this 20% make-up air is
itself HEPA filtered prior to its introduction into the recirculating air
loop as is common practice, its contribution to the steady state particle
concentration within the clean room is, of course, further reduced.

Operating clean rooms seldom achieve the air qualities depicted in
Figure 1, being dominated by emission sources within the clean room
itself--sources such as people, equipment, and processes. Thus, the
second major traditional particle control strategy is to minimize
particle emissions from these sources: people are dressed in bunny suits
and urged to avoid cosmetics; equipment designs use materials and
configurations that reduce particle emissions in the vicinity of wafers;
and processes use those chemicals, gases and reactions that introduce or
produce minimal contaminating particles.

In spite of these control measures, particulate contamination still
exists, and chips still fail because of it. The next level of protection
from such particle-induced failures is to learn to carry out manufacturing in the presence of particles by understanding and controlling the
particle transfer process--it is not the particles in the air or the
processing chemicals that cause chip failure but rather those particles
that deposit on the chip surface at the wrong place and the wrong time.
This paper addresses particle deposition and the factors that affect it.

Appreciation of this aspect of particle control already reflects itself in the widespread use of the vertical laminar flow design of clean rooms. High quality air entering the clean room through ceiling-mounted HEPA filters sweeps particles generated within the room into the return ducts located beneath the raised floor. Thus, particles introduced by sources within the clean room are kept away from wafers by the forced air flow as long as the particle source is downwind of the wafers. Similar considerations apply to equipment design--particle generating motions or parts are deliberately excluded from the regions upstream of the wafers.

Again, while these design rules help, they do not guarantee particle-free air flow past the wafers. The capture of aerosol particles by the wafer surface remains a threat. Not all particles flowing past a wafer will be captured, of course. The coefficient relating the accumulation of surface particles to the aerosol particle concentration in the adjacent air is called the deposition velocity:

$$\text{deposition velocity (cm/s)} = \frac{\text{surface particle flux (cm}^{-2}\text{s}^{-1})}{\text{aerosol concentration (cm}^{-3})}$$

Deposition velocity has been calculated for wafers oriented differently with respect to the direction of air flow[2] and under the assumption that diffusion and sedimentation are the only important deposition mechanisms. Figure 2 reproduces a typical family of

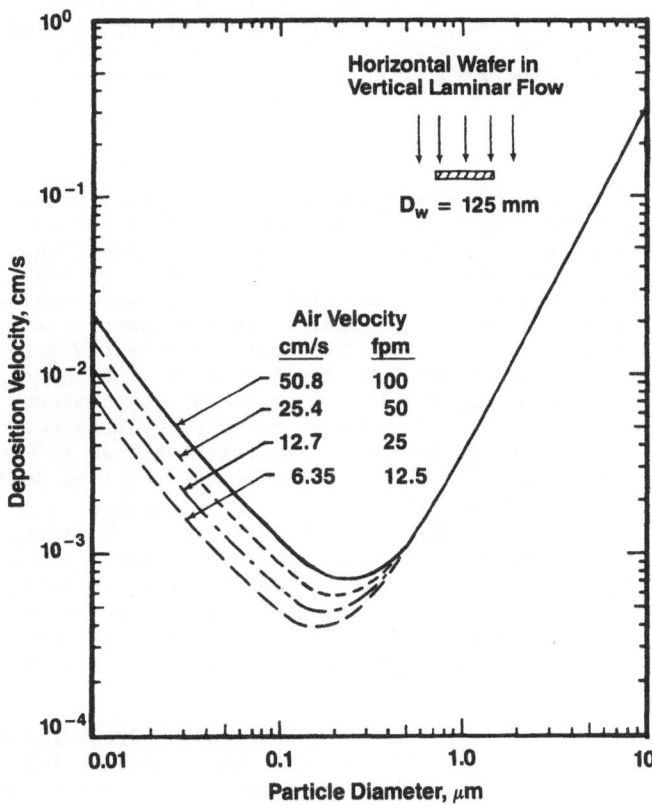

Figure 2. Calculated deposition velocity on a horizontal wafer in vertical laminar flow.[2]

calculated curves from this analysis. Deposition velocity based on these two deposition mechanisms is predicted to have its minimum values for particles in the 0.1 to 0.3 μm size range. While this result is favorable for device fabrication when the killer particle size is greater than 0.1 μm, the sharp rise in deposition velocity for sub 0.1 μm particles bodes ill for the fabrication of future device chips whose even smaller dimensions increase the chip vulnerability to these smaller particle sizes. Deposition by diffusion means these particles will be captured more efficiently by the wafer surface.

Other predictions contained in the Liu and Ahn paper are:

1.   Larger wafer sizes (8" vs 5' wafers) always result in a slightly reduced mean deposition velocity. This conclusion holds for horizontally oriented wafers placed in either a vertical or horizontal air flow and for vertically oriented wafers in a vertical air flow.

2.   The location of a wafer on a flow barrier such as a solid table top makes a difference, a wafer centered on a table top having a lower mean deposition velocity than the same wafer placed at the edge of the same table.

OTHER VARIABLES AFFECTING DEPOSITION VELOCITY

The work to be reported here describes the influence of two other mechanisms upon the deposition of aerosol particles onto silicon wafers: electrical forces and thermal gradients. Both of these forces have been analyzed with respect to particle deposition on silicon wafers. This paper now provides experimental verification of the general effects predicted.

Electrical Forces

Electrical forces have been predicted to be important in the capture of submicron particles by wafers[3], and some experimental evidence in support of such predictions now exists.[4,5] The calculations of Liu, Fardi, and Ahn assume a surface electric field value of 100 V/cm. Their calculated values of deposition velocity are very sensitive to particle charge over the submicron size regime. Figure 3 summarizes a set of their typical calculations for the horizontally oriented wafer of Figure 2--the neutral curve in Figure 3 is similar to the 100 fpm curve of Figure 2. The addition of just Boltzmann charge to submicron particles is predicted to increase the deposition velocity by a factor of 5 to 10. For large particles, the influence of electrical charge is much smaller until, under the conditions of this calculation, settling dominates the deposition for 5 μm and larger particles. Note, however, that the assumed electric field of 100 V/cm used in these calculations is quite modest and that the experimentally determined voltages of 5 to 10 kV measured at various stations in a semiconductor manufacturing area by Blitshteyn and Martinez[5] could easily create electric fields well in excess of that value. Higher electric fields would, of course, increase the calculated deposition velocities of the charged particles even more than shown in Figure 3.

The experimental data reported by Donovan, Clayton, and Ensor[4] varies in form from that used in Figure 3, thus complicating comparisons. The particle charge in most of the measurements was that associated with the Boltzmann distribution and achieved by passing monodisperse polystyrene latex (PSL) spheres through a Kr[85] neutralizer. What was

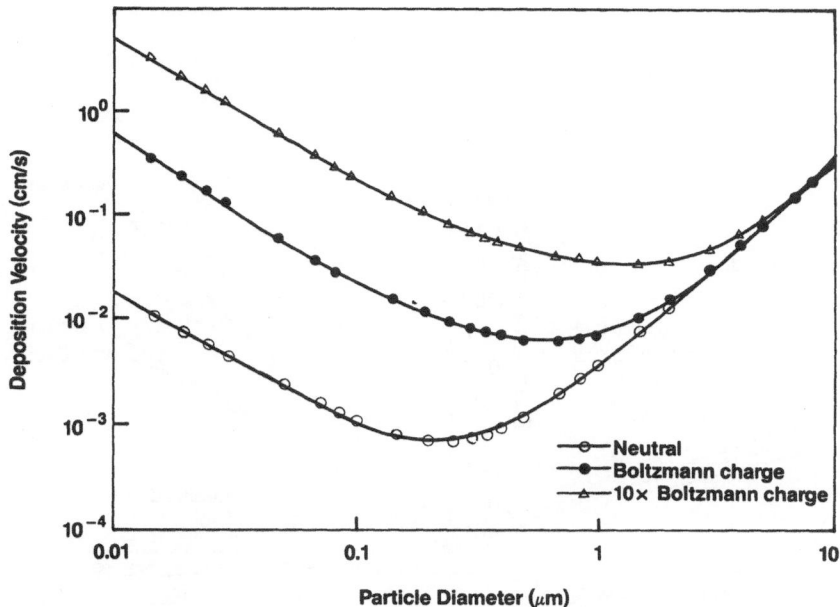

Figure 3. Calculated deposition velocity including effects of
diffusion, settling and electric field (E = 100 V/cm).[3]

varied was the magnitude of wafer surface potential. Figure 4 typifies
the results of these experiments.

The Figure 4 results are consistent with Couloumb enhanced
deposition, in that deposition velocity increases with increasing dc
field and is independent of ac field. The asymmetry in dc polarity may
mean that a true Boltzmann charge equilibrium does not exist--that there
is an asymmetry in the charge distribution of the PSL spheres by the time
they come under the influence of the wafer electric field. The larger
deposition velocities associated with the largest spheres of Figure 4 are
not predicted by the Figure 3 calculations. This dependence may be a
property of the deposition chamber in which the measurements were carried
out. To achieve a uniform aerosol concentration within the chamber
during the deposition, a stirring paddle was continuously rotated at 36
rpm. The mixing created by this stirring action does indeed achieve a
uniform aerosol concentration throughout the deposition chamber, but it
also alters the air flow past the wafer from that depicted in Figure 2.
Thus, the experiment departs from the conditions modeled, and departures
from predicted performance are to be expected.

Increasing the particle charge to some indeterminate value also
causes increases in the deposition velocity as shown in Figure 5. The
increase in particle charge illustrated in Figure 5 was brought about by
simply having the feed aerosol bypass the Kr[85] neutralizer. No measure
of the actual particle charge was made, but it was obviously much higher
as evidenced by the increased values of deposition velocity,
qualitatively in agreement with the Figure 3 calculations and by the
significant reduction in chamber aerosol concentration (not shown),
reflecting the increased tube losses of the more highly charged aerosol.
Even at zero wafer potential, the deposition velocity showed a
significant increase, suggesting either that the wafer surface electric
field was not really zero in spite of electrically grounding the plate on
which the wafer rested or that the charged particles created an image
force electric field which enhanced their deposition velocity.

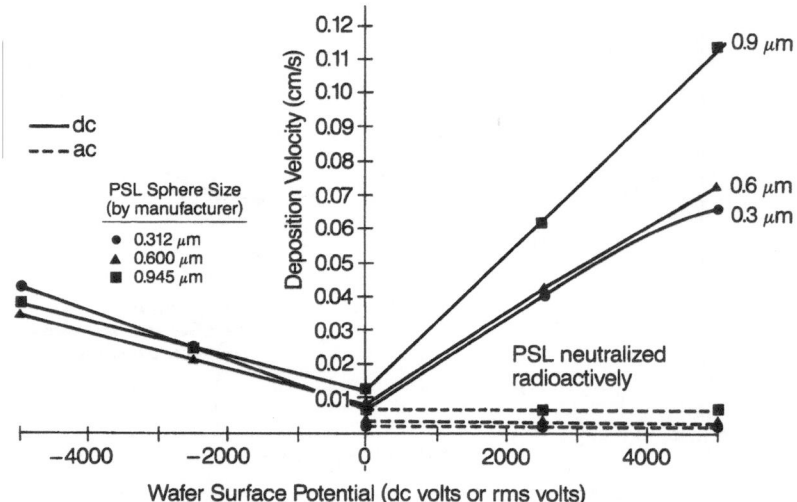

Figure 4. Mean values of deposition velocity as a function of wafer surface potential.

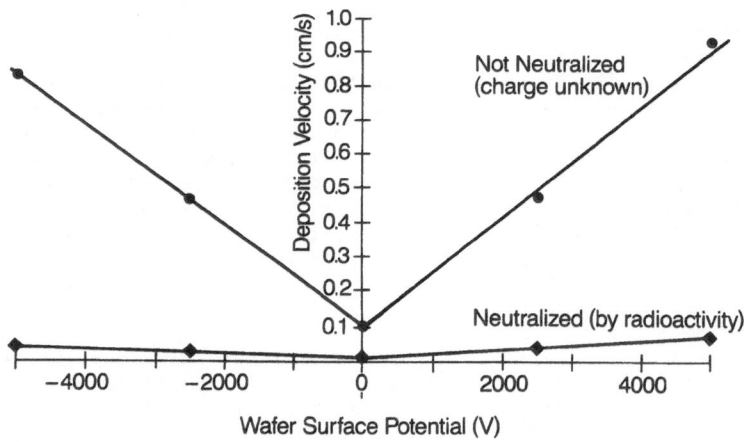

Figure 5. Increase in deposition velocity brought about by increased electrical charge on 0.6 μm polystyrene latex spheres.[4]

In summary, Coulomb forces, depending on both wafer charge and particle charge, have been shown both theoretically and experimentally to significant increase particle deposition velocity under a variety of conditions. Prudent practice would thus dictate that these charges be minimized throughout the silicon chip manufacturing sequence.

Thermal Forces

Recognition of the particle-shielding properties of holding a body above the ambient temperature has been in the literature for years. More recently, this phenomenon has been proposed as a mechanism for reducing particle deposition velocity on silicon wafers.[7] Calculations of the width of the dust-free space attributable to thermophoresis have been made by Stratman, Friedlander, Fissan, and Papperger,[8] and measurements of the reduction in particle deposition velocity brought about by heating silicon wafers have been made by Donovan, Clayton, Yamamoto, and Ensor[9] hereafter abbreviated to DCYE. This section reviews the basic experiment of the last cited publication and summarizes the key conclusions.

The experimental appartus used by DCYE[9] was identical to that used in earlier measurements of surface potential effects on deposition velocity[4] except that the aluminum disks on which the silicon wafers were mounted were increased in thickness (to 3/8") to accommodate the insertion of an electrical rod heater into a drilled side hole. The side of each aluminum disk was thermally insulated to help maintain a uniform temperature distribution. A thermocouple well also drilled into the aluminum disk provided a set point for proportional control of the disk temperature. This modified fixture was then placed in the deposition chamber as before and exposed to controlled and measured well mixed aerosol. Simultaneous exposure of a group of silicon wafers, held at various temperatures, led to the type of relationship illustrated in Figure 6. As before, the surface particle flux was determined by measuring the surface particle areal density before and after a known exposure time to an aerosol of known particle concentration. These surface particle measurements were made with the Aeronca Wafer Inspection Station 150, a laser surface scanner. The scanner and the deposition chamber are located next to each other in one bay of the clean room at the Microelectronics Center of North Carolina.

The data of Figure 6 were collected from wafers positioned horizontally in the deposition chamber during exposure. They show impressive reductions in deposition velocity brought about by just modest elevation of wafer surface temperature above ambient. The surface temperatures plotted in Figure 6 are the average surface temperatures of the aluminum disks. The set point temperature, measured in the interior of the aluminum disk, was higher because of heat losses through the various disk surfaces.

While Figure 6 is based on just one size of PSL sphere, the data of Figure 7 show that over a narrow submicron range sphere size is not important. Figure 7 also includes a datum collected by placing heated wafers in an open clean room environment. Here the air flow differs significantly from what it is in the deposition chamber, and the aerosol is of uncontrolled particle size, shape, and composition and of much lower concentration. Even with all these differences, the reduced deposition velocity accompanying wafer heating remains obvious so that under certain conditions slightly elevated temperature should be considered a practical method of preventing particulate contamination of silicon wafers.

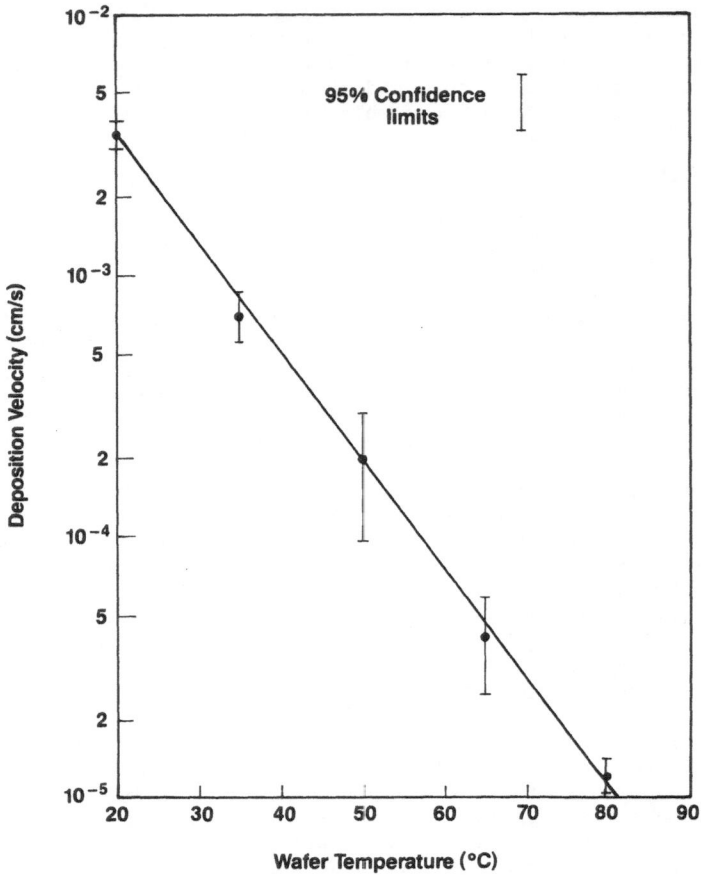

Figure 6.   The effect of wafer temperature on the particle deposition
velocity of neutralized 0.6 μm polystyrene latex spheres.

PROCESSING LIQUIDS

The preceding sections have dwelt on particle deposition from an
aerosol. Silicon wafers are exposed to many liquids during processing,
all of which have higher particle concentrations than the ambient clean
room air. Many of these liquids are used in cleaning steps although it
often turns out that wafers have more surface particles following such a
cleaning procedure than before.[10] A next study in particle deposition
velocity is of particles from liquid sources using much the same
technique as described here for the aerosol studies. Such research is
now underway at the Research Triangle Institute as another step in a
broad, long-term program of understanding and controlling particulate
contamination is silicon device manufacturing.

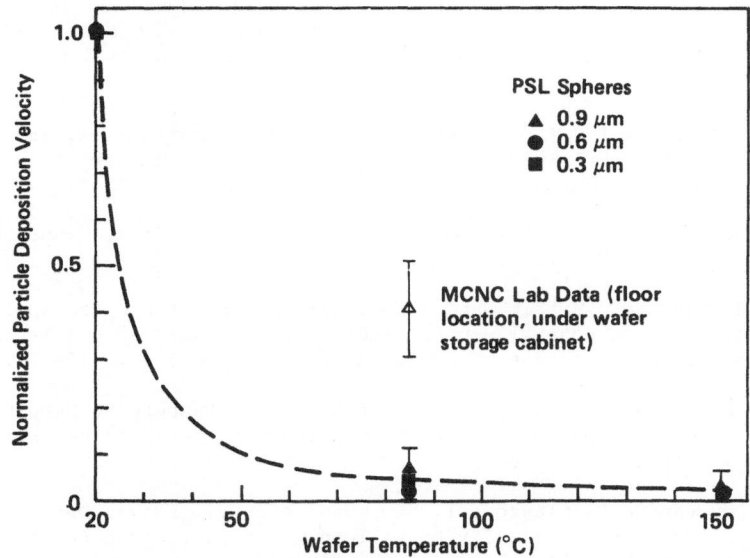

Figure 7.   The effect of wafer temperature on the deposition velocity
of differently sized polystyrene latex (PSL) spheres.

CONCLUSIONS

  Particulate contamination of silicon wafers depends on particle
transport to the wafer surface in addition to the presence of particle
generators. Two actions that have been shown to minimize particle
deposition velocity are: 1) the elimination/minimization of particle and
wafer electrical charge, and 2) the elevation of wafer temperature above
that of its surroundings.

ACKNOWLEDGEMENT

  This research was sponsored by the Semiconductor Research
Corporation through a Manufacturing Science Contract with the
Microelectronics Center of North Carolina for whom RTI is a
subcontractor.

REFERENCES

1. D.S. Ensor, R.P. Donovan, and B.R. Locke, Particle size distributions in
  clean rooms, J. Environ. Sci. <u>30</u> (6), 44-49 (1987).

2. B.Y.H. Liu and K.H. Ahn, Particle deposition on semiconductor wafers,
  Aerosol Sci. Technol., <u>6</u> (3), 215-224 (1987).

3. B.Y.H. Liu, B. Fardi, and K.H. Ahn, Deposition of charged and uncharged
  aerosol particles on semiconductor wafers in Proceedings of the 33rd
  Annual Technical Meeting of the Institute of Environmental Sciences, pp.
  461-465, 1987.

4. R.P. Donovan, A.C. Clayton, and D.S. Ensor, The dependence of particle deposition velocity on surface potential, in Proceedings of the 33rd Annual Technical Meeting of the Institute of Environmental Sciences, pp. 473-478, 1987.

5. M. Blitshteyn and A.M. Martinez, Electrostatic charge generation on wafer surfaces and its effect on particulate deposition, Microcontamination, $\underline{4}$ (11), 54-60, 132 (November 1986).

6. H.H. Watson, The dust free space surrounding hot bodies, Trans. Faraday Soc., $\underline{32}$. 1073-1084 (1936).

7. S.A. Hoenig, S.R. Martin, and I.J.M. Slinski, Electrostatic applications in contamination control, in Proceedings of the 33rd Annual Technical Meeting of the Institute of Environmental Sciences, pp. 479-482, 1987.

8. F. Stratman, S. Friedlander, H. Fissan, and A. Papperger, Suppression of particle deposition to surfaces by the thermophoretic force, Aerosol Sci. Technol., to be published.

9. R.P. Donovan, A.C. Clayton, T. Yamamoto, and D.S. Ensor, Dependence of particle deposition velocity on silicon wafer surface temperature, paper presented at the Fall Meeting of the Electrochemical Society, Honolulu, Hawaii, October 1987.

10. I. Bansal, Particle contamination during chemical cleaning and photoresist stripping of silicon wafers, Microcontamination, $\underline{2}$ (4), 35-39, 90 (August/September 1984).

INFLUENCE OF PARTICLE CHARGE ON THE COLLECTION EFFICIENCY OF ELECTRIFIED FILTER MATS

M. K. Mazumder and A. M. Siag

University of Arkansas at Little Rock
Graduate Institute of Technology
2801 South University, ETAS Room 575
Little Rock, Arkansas 72204

A quantitative determination of the electrostatic effects on aerosol filtration is difficult since currently available methods do not provide direct measurements of coulombic and polarization parameters. This paper presents an experimental approach to determine the effects of both coulombic and polarization forces on the particle collection efficiency of filter mats. Two filter mats were tested against submicron aerosols to determine the effect of particle charge on the filtration efficiency with and without a polarizing electric field at two face velocities: 5 and 15 cm/s. The challenge aerosols consisted of either solid particles or liquid droplets in the size range 0.4 to 1.0 μm in aerodynamic diameter either in electrostatically charged or in Boltzmann charge distribution conditions. An Electrical Single Particle Aerodynamic Relaxation Time (E-SPART) analyzer and a Differential Mobility Analyzer (DMA) were used for measuring size and charge distributions of aerosol particles at the upstream and downstream of the filters. Experimental data show that measured single fiber efficiency due to electrical effects increased by factors ranging from 5 to 12 depending upon the electrostatic charge and electrical field involved. Particulate penetration in the submicron range decreased exponentially as a function of the electrostatic filtration parameter K_{EO} defined by the ratio of electrical migration velocity to the face velocity.

INTRODUCTION

Electrostatic enhancement of filtration efficiency of filters for removing particles from fluid media has a long history. During the last fifty years, the following three types of collectors have been studied for removing particles from fluids by applying electrostatic forces: (a) electrostatic precipitators, (b) electrified fibrous filters, and (c) electrets. A comprehensive study performed by Bergman et al.[1-4] presents an extensive review of the earlier works and demonstrates the advantages of applying an electrical field to the fibrous filters. Electrostatic forces increase the filter efficiency and decrease the pressure drop across the filter. The filtration mechanisms of the electrostatically enhanced air filters are complex because of the interaction between the fluid mechanical and electrostatic forces acting on the particles.

In spite of this complexity, several theoretical models have been developed[5-12] that account for the following collection mechanisms: diffusion, interception, inertial impaction, polarization and Coulombic and image attractions involving particles. A large number of experimental studies on the electrostatically enhanced fabric filters (ESFF) demonstrated conclusively that both the external electric field and the particle charge play important roles in collection efficiency and pressure drop characteristics. The effect of an external electric field on filtration characteristics was studied in detail since the applied field can be measured and controlled. Experimental difficulty arises in the study of the effect of electrostatic charge on the filtration mechanism due to problems in measuring and controlling the electrostatic charge. Experimental data are lacking on the macroscopic electric field developed inside the filter media as a result of external field and space charge distributions.

For submicron particles, the single fiber collection efficiencies of fibrous filters due to mechanical collection parameters are generally small. Therefore, the electrostatic enhancement of collection efficiency is of great importance in this size range. The effect of electrostatic charge on the collection efficiency of the fibrous filter for submicron aerosol particles is investigated in this study. This investigation utilizes a new instrument, namely the electrical single particle aerodynamic relaxation time (E-SPART) analyzer[13], which is capable of measuring both aerodynamic diameter and electrostatic charge (both magnitude and polarity) on a single particle basis and in real-time. This technique enables filter efficiencies to be measured in real-time over a wide range of electrostatic charge distribution. The results of the filter performance are presented in terms of penetration and single fiber efficiency for different (a) particle sizes, (b) filtration velocities, and (c) applied electrical fields. This experimental method allows the determination of the correlation between the particulate penetration and the electrical filtration parameter defined by the ratio of the electrical migration velocity to the face velocity.

MATERIALS AND METHODS

The Experimental Set-Up

A schematic diagram of the experimental set-up is shown in Figure 1. The experimental set-up consists of (1) several aerosol generators to produce aerosols consisting of solid particles or liquid droplets of known size distribution, (2) a dryer, (3) a neutralizer using a Krypton-85 radioactive source, (4) an aerosol charger, (5) a precipitator, (6) a filter test column, and (7) instrumentation for size and charge analyses. The filter mats were tested for penetration against the aerosol particles for the following conditions of electrostatic charge:

A. naturally produced bipolar charge distribution of the particles acquired during the generation process,
B. particles neutralized to acquire the Boltzmann charge distribution, and
C. particles charged by using a charger where the neutralized aerosol particles acquired a desired charge distribution.

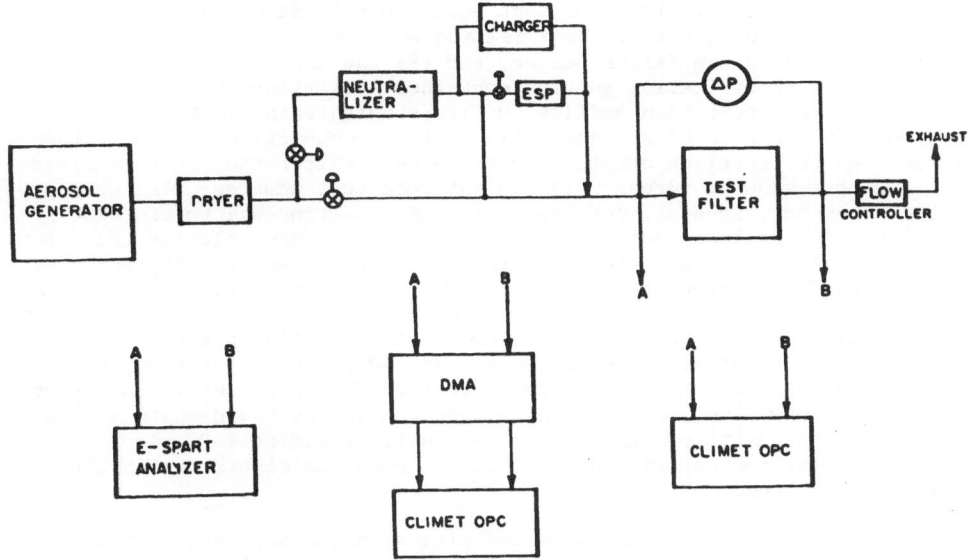

Figure 1. Experimental set up for testing penetration of aerosol particles through filter mats as a function of particle charge.

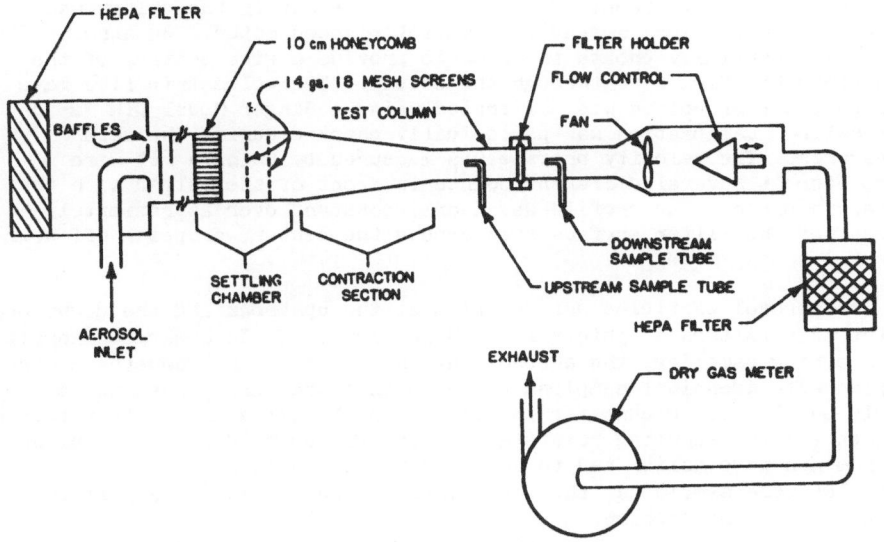

Figure 2. Filter test column.

Filter Test Column

A filter test column (Figure 2) was constructed to test the filters in a nonuniform flow field. The test column consisted of three components: (1) an air flow stabilization section, (2) the filter test section containing the filter holder, and (3) the air flow control section. The stabilization section contained a settling chamber, an aerosol delivery port with baffles, a flow straightening section with a honeycomb, and a pair of screens followed by a contraction section. Room air entered the settling chamber through a HEPA filter and was then mixed with the test aerosol before entering into the test section. The purpose of the honeycomb, screens, and the contraction section was to minimize turbulence in the test section and to develop a uniform velocity profile across the filter surface area. Baffles, honeycomb, and the screens also provided uniform mixing of aerosol particles in the test column.

The filter holder was made of Delrin^R, an insulating plastic material. Two steel screen electrodes, with 90 percent open area, were used for applying an external electric field. The screen at the upstream was maintained at the ground potential and the one at the downstream was connected to a positive high voltage terminal. A variable DC power supply was used for applying high voltage across the electrodes in the range 0 to 5 kV/cm.

The filter holder was designed and placed inside the test section to avoid leakage and minimize flow perturbations. The sampling of aerosols was performed using two copper tubes with 7 mm internal diameter, one at the upstream of the filter and the other downstream with the sampling tube facing the direction of flow. The pressure drop across the filter was measured by using a micromanometer capable of measuring the pressure drop with a resolution of 0.01 mm of butanol.

The air flow control section was made of 15.2 cm-diameter PVC pipe and was located downstream of the filter. The air in the column was driven by a propeller powered by a variable speed motor. An annular flow controller partially choked the flow to provide a fine control of the volumetric air flow rate through the column. The volumetric flow rate through the test column was calibrated using a Singer Model DTM 325 dry test meter. Calibration was periodically checked during the test procedures. The velocity profile was measured by using a hot wire anemometer at several different points in front of the filter with the filter in place. The profile was nearly constant over approximately 50 percent of the filter surface area around the center, dropping off near the wall.

The aerosol particles were sampled at the upstream and the downstream of the test filters at points A and B in Figure 1. In order to minimize error during sampling, the aerosols at the upstream and downstream were sampled with identical sampling tubes so that the line losses would be nearly equal. To account for the losses in the walls of the test section between the two sampling points and in the filter holder, the sampling instruments were calibrated to read 100 percent penetration when particles were sampled at the two points (A and B) without any filter in place in the test section.

Particles were sampled using the following three different instruments: (1) a TSI differential mobility analyzer (DMA) and a Climet optical particle counter (OPC) for classifying aerosol particles with a

given charge and then measuring the concentration by the OPC (Figure 3);
(2) an E-SPART analyzer to determine the electrostatic charge and
aerodynamic size distributions at both sampling points (Figure 4); (3) a
Climet OPC to measure the concentration at the upstream and downstream of
the filter (Figure 1).

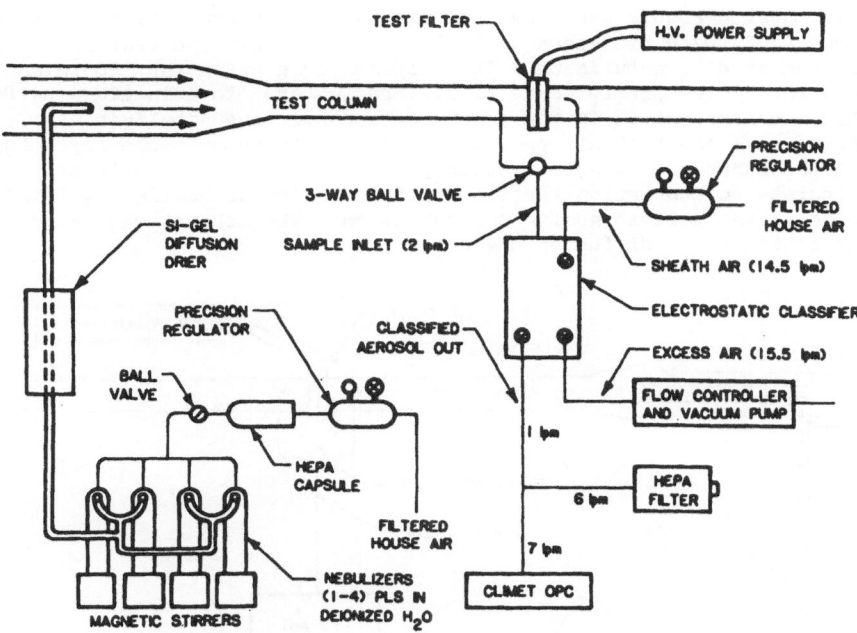

Figure 3.   Experimental arrangements for measuring penetration of
monodisperse aerosol particles as a function of particle
charge using a DMA and an OPC.

Filter Media

Two types of fiber glass filter media were used in this study, the
AF-11 and AF-18, manufactured by Johns Manville Corporation. These
filter mats were studied extensively[1,2] for their filtration performance
with a superimposed electric field. The characteristics of the fiber
mats are:

Fiber diameter: AF-11: 2.7 μm; (number-median diameter)
                      AF-18: 3.9 μm; (number-median diameter)

Solidity: AF-11: 0.0038
(Packing fraction)   AF-18: 0.0086

Thickness: AF-11: 0.6 cm
                      AF-18: 0.6 cm

```
Efficiency (Dust Spot Test):       35 percent
Dielectric constant $\varepsilon_f$:       3.81
Diameter of the filter mat tested: 11.4 cm
Face velocities:                   5 cm/s; 15 cm/s
Pressure drop:                     5 Pa  at 3 cm/s
```

Test Aerosols

Two test aerosols were used containing monodisperse polystyrene latex spheres (PLS) with diameters of 0.460 and 0.913 μm. The test aerosols were generated by nebulizing a PLS suspension in water that contained approximately 10^9 particles/cc of filtered water. To maintain a uniform particle concentration in the generated aerosol, the nebulizer was operated at a constant air pressure using precision pressure regulators while the suspension was continuously stirred by magnetic stirrers. The particulate concentration throughout the experiments remained within ± 3 percent of the desired number concentration. The aerosol was dried by using a silica gel diffusion dryer.

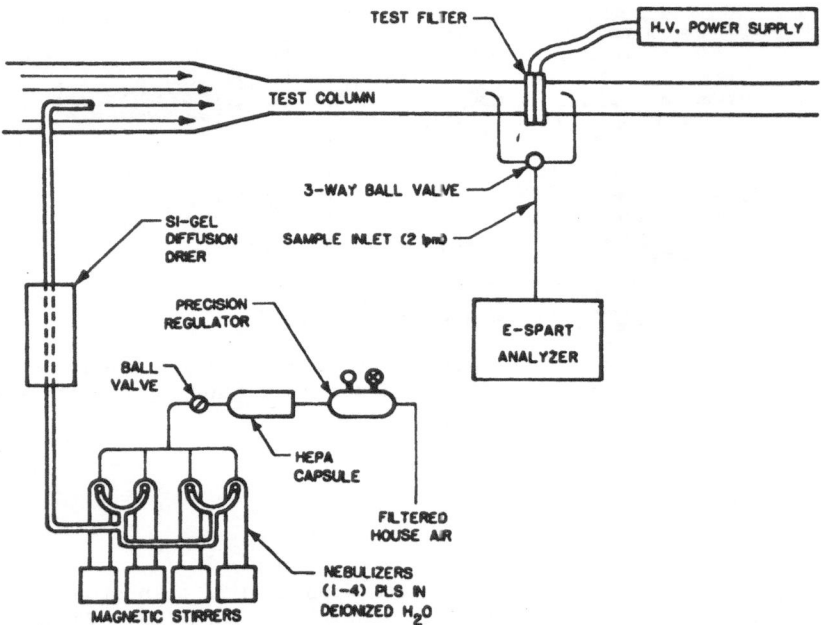

Figure 4. Experimental arrangements for measuring penetration of monodisperse aerosol particles as a function of particle charge using an E-SPART analyzer.

Measurement of Aerosol Penetration

Aerosol particle penetration through the filter mats was determined by measuring the particulate concentration at the upstream and downstream of the filter. Penetration (P) is given by

$$P = n_o(q)/n_i(q)$$

where $n_o(q)$ and $n_i(q)$ are the concentrations of aerosol particles at the output and input of the test filter, respectively, with a given average charge q. Experimental studies reported here relate to test filters under clean condition, i. e., the particulate loading effects were minimal.

Concentrations of particles with a Boltzmann charge distribution were measured using a Climet Optical Particle Counter (OPC). No further charge distribution measurement was necessary. For measuring filter penetration for particles with discrete charge levels (q = ± 5e, ± 10e, and ± 21e) the filter was challenged with bipolarly charged particles with the charge distribution that was present after the aerosols were generated. The particle concentrations $n_o(q)$ and $n_i(q)$ were measured by sampling the particles at the points A and B and classifying them for a predetermined charge level q and then measuring the concentration using the OPC. The arrangement is shown in Figure 3. Since the sampling rate of the DMA was less than that of the OPC, the DMA output was appropriately diluted using filtered air at a controlled volumetric flow rate.

When particles were charged with a corona charger for obtaining a wide charge distribution, the charge distributions at the upstream and downstream were measured by using two E-SPART analyzers (Figure 4). Aerosol particles were sampled simultaneously at both points A and B and the ratio of $n_o(q)/n_i(q)$ was determined for q in the range of 0 to ± 100e/particle.

Determination of Single Fiber Efficiency

The efficiency of the filter is calculated by the following equation:

$$E = 1 - P, \text{ where} \tag{1}$$

$$P = EXP-[4\alpha L n_T/\pi D_f(1 - \alpha)] \tag{2}$$

where n_T is the total single fiber efficiency,
    E = filter efficiency,
D_f = fiber diameter,

    L = filter thickness, and
    α = solidity defined by the ratio of the fiber volume to the total volume of the filter. (Also called packing fraction.)

The total single fiber efficiency (n_T) incorporates both mechanical and electrostatic capture mechanisms. The total single fiber efficiency was computed from Equation (2),

$$n_T = -\pi(1-\alpha)D_f \ln P/4\alpha L \tag{3}$$

for the measured value of P.

Estimation of n_T from Theoretical Models

Theoretical models[1,5,6,8,10,11,14-17] are generally based on the assumptions that the fibers have uniform diameter, that the solidity of the medium is uniform, that the particles are monodisperse and that the loading effects are negligible. With these and other simplifying assumptions, some of the individual parameters can be estimated as follows.

Mechanical Capture Mechanisms

(A) Interception. The single fiber efficiency due to interception depends on the interception parameter[7,15] defined by

$$R = D_p/D_f \tag{4}$$

where

R = interception parameter
D_p = particle diameter.

Yeh and Liu[15] derived the single fiber efficiency due to the interception mechanism based on Kuwabara flow field and found:

$$\eta_{in} = (1/2Ku)[2(1+R)\ln(1+R)-(1+R)+1/(1+R)] \tag{5}$$

$$Ku = -(\ln\alpha/2)-3/4+\alpha-\alpha^2/4 \tag{6}$$

where

η_{in} = single fiber efficiency due to interception
Ku = Kuwabara constant.

(B) Inertial Impaction.

$$\eta_{im} = J(Stk)/2(Ku)^2 \tag{7}$$

where

Stk = Stokes number = $\rho D_p^2 V C/18\mu D_f$
V = face velocity
C = slip correction factor
μ = viscosity.
ρ = particle density

$J = (29.6 - 28\alpha^{0.62})R^2 - 27.5R^{2.8}$

(C) Diffusion. The η_D due to diffusion is given by Davies[7] in the following equation:

$$\eta_D = 2P^{-2/3} \tag{8}$$

where

η_D = single fiber efficiency due to diffusion

$$P = D_fV/D \tag{9}$$
= Péclet number.
D = diffusion coefficient

Electrical Capture Mechanisms

(A) Polarization. The polarization mechanism refers to the attractive force between polarized fibers and polarized particles. The single fiber efficiency due to polarization was derived by Zebel[17] as

$$\eta_p = [(\varepsilon_p - 1)/(\varepsilon_p + 2)][(\varepsilon_f - 1)/(\varepsilon_f + 1)][D_p^2E^2C/12\pi\mu D_fV] \tag{10}$$

where

η_p = single fiber efficiency due to polarization

ε_p = particle dielectric constant

ε_f = fiber dielectric constant

E = electric field.

(B) _Coulombic Attraction_. The Coulombic mechanism refers to the attractive forces between polarized fibers and charged particles. The single fiber efficiency due to Coulombic attraction was derived by Zebel[17] as

$$\eta_c = (neEC/3\pi\mu D_p V)[\{1 + (\varepsilon_f - 1)/(\varepsilon_f + 1)\}/\{1 + neEC/3\pi\mu D_p V\}] \quad (11)$$

where

η_c = single fiber efficiency due to Coulombic attraction
n = number of elementary charge units/particle
e = unit of charge.

In the above equation, the inertial and image forces for charged particles were neglected.

Billings and Gussman[5] derived the single fiber efficiency due to the attractive forces between charged fibers and charged particles, and between charged fibers and neutralized particles.

$$\eta_{pq} = (4/3)[(\varepsilon_p - 1)/(\varepsilon_p + 2)][Q^2 D_p C/D_f^3 \mu V] \quad (12)$$

where

$$\eta_{pn} = (4/3)(neQC/\mu D_p D_f V) \quad (13)$$

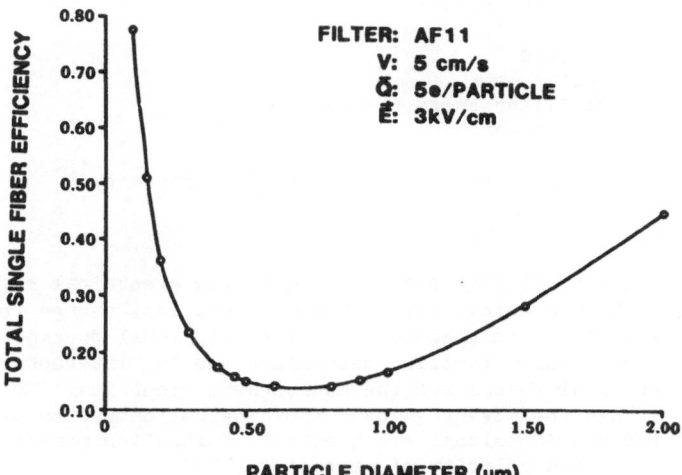

Figure 5. Calculated values of total single fiber efficiency as a function of particle diameter.

where

η_{pq} = single fiber efficiency for charged particles

η_{pn} = single fiber efficiency for neutral particles

Q = charge per unit length on the fiber

The total single fiber efficiency η_T was estimated by summing the individual efficiencies. The summation neglects the mutual interaction between the individual collection mechanisms and provides only a general guideline of the complex filtration process. Figure 5 shows a plot of η_T as a function of particle diameter for the following filtration parameters

Filter: AF-11
Velocity: 5 cm/s
Charge: 5 e/particle
Field: 3 kV/cm

Electrical Migration Velocity

Another approach to model aerosol penetration for ESFF was developed on the basis of the electrical migration velocity of the individual particles in the filter media, similar to the calculation of P from the electrostatic precipitator. This model was proposed by Nelson, et al.[18] for fibrous filter, and by Shapiro et al.[19] for granular filter beds. The penetration can be expressed as

$$P = EXP-(V_E/V)(S/A) \tag{14}$$

where

V_E = electrical migration velocity, which is equal to BqE,

and

B = mobility of the particle
q = charge
E = electric field
V = face velocity
S = effective filter surface area
A = filter face area

The electrostatic filtration parameter (K_{EO}) is given by[18-20]

$$K_{EO} = V_E/V. \tag{15}$$

In this model it is possible to estimate the electrical migration velocity from the mechanical mobility (B), electrical charge (q), and the external applied electric field (E). While this model does not take into account the mechanical filtration mechanisms and the interaction between different electrical forces and the macroscopic electrical field inside the medium, it has the advantage that the penetration can be plotted as a function of the nondimensional electrostatic filtration parameter for a given filter and face velocity of aerosol.

RESULTS AND DISCUSSION

Experimental data for two particle sizes (0.460 and 0.913 μm in diameter), four particle charges (5, 10, 15, and 21 electronic charges per particle), and two face velocities (5 and 15 cm/s) were obtained for each filter media, AF-11 and AF-18. Experiments were also conducted with particles having Boltzmann equilibrium charge conditions. The applied electric field varied from 0 to 3 kV/cm in steps of one kV/cm.

Experimental data are presented in Tables I through VII. The filter medium was changed from one set of experiments to another so that the filters were relatively clean and the effects of loading were minimal. However, for each section of the same type of filter (AF-11 or AF-18), it was observed that the initial penetration was different from one filter sample to another.

In each table experimentally determined penetration is presented. From these experimental values of penetration, the total single fiber efficiency was calculated. Experimental values of penetration of charged and equilibrium charged aerosol particles in the presence of an external electric field show that penetration decreases as the applied electric field increases or as the electrostatic charge for a given particle size increases. When both particle charge and external fields are present, the filtration is enhanced due to the Coulombic attraction forces and polarization effects.

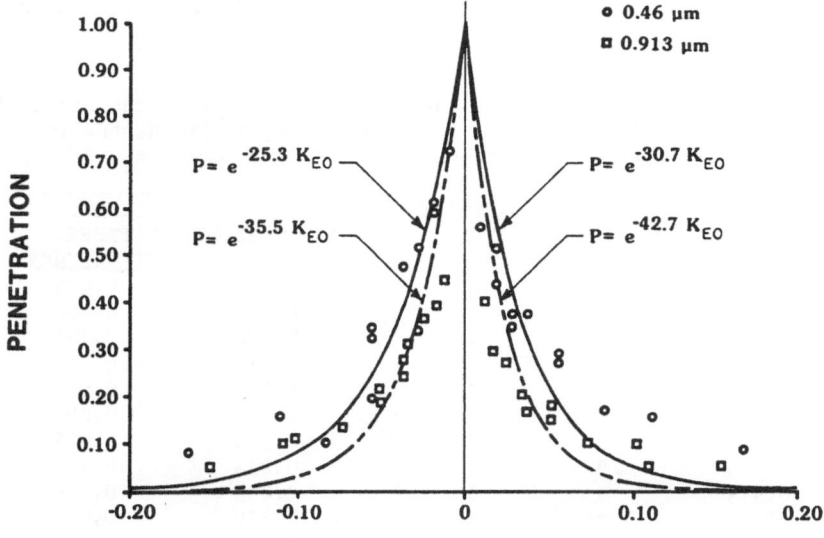

Figure 6. Experimental data on particle penetration through AF-11 and AF-18 filter mats as a function of electrostatic filtration paramenter K_{EO}.

Table I. Experimentally Determined Filter Penetration and Single Fiber Efficiency for AF-11 Filter Challenged with Charged PLS Particles and at a Face Velocity of 5 cm/s.

PARTICLE SIZE (μm)	FACE VELOCITY (cm/s)	ELEMENTARY CHARGE UNIT	ELECTRIC FIELD (kV/cm)	PENETRATION	SINGLE FIBER EFFICIENCY
0.460	5	+5	1	0.210	0.131
"	"	"	2	0.113	0.183
"	"	"	3	0.061	0.240
"	"	+10	1	0.152	0.158
"	"	"	2	0.064	0.231
"	"	"	3	0.029	0.297
0.913	"	+15	1	0.043	0.264
"	"	"	2	0.016	0.350
"	"	"	3	0.007	0.420
"	"	+21	1	0.039	0.273
"	"	"	2	0.013	0.362
"	"	"	3	0.006	0.430
0.460	"	-5	1	0.274	0.110
"	"	"	2	0.122	0.177
"	"	"	3	0.073	0.220
"	"	-10	1	0.175	0.146
"	"	"	2	0.073	0.220
"	"	"	3	0.028	0.300
0.913	"	-15	1	0.088	0.204
"	"	"	2	0.044	0.262
"	"	"	3	0.024	0.312
"	"	-21	1	0.082	0.210
"	"	"	2	0.033	0.290
"	"	"	3	0.017	0.342

Table II. Experimentally Determined Filter Penetration and Single Fiber Efficiency for AF-11 Filter Challenged with Neutralized PLS Particles and at Face Velocities of 5 cm/s, and 15 cm/s.

PARTICLE SIZE (μm)	FACE VELOCITY (cm/s)	ELECTRIC FIELD (KV/CM)	PENETRATION	SINGLE FIBER EFFICIENCY
0.460	5	0	0.692	0.031
"	"	1	0.353	0.088
"	"	2	0.196	0.137
"	"	3	0.113	0.183
"	15	0	0.690	0.031
"	"	1	0.490	0.060
"	"	2	0.330	0.093
"	"	3	0.234	0.122
0.913	5	0	0.653	0.036
"	"	1	0.310	0.098
"	"	2	0.159	0.155
"	"	3	0.088	0.204
"	15	0	0.610	0.042
"	"	1	0.490	0.060
"	"	2	0.340	0.091
"	"	3	0.220	0.130

*For Boltzmann Distribution

$\overline{q} = 1.5$ e/particle for $d_p = 0.460$ μm

$\overline{q} = 2.23$ e/particle for $d_p = 0.913$ μm

Table III. Experimentally Determined Filter Penetration and Single Fiber Efficiency for AF-11 Filter Challenged with Charged PLS Particles and at a Face Velocity of 15 cm/s.

PARTICLE SIZE (µm)	FACE VELOCITY (cm/s)	ELEMENTARY CHARGE UNIT	ELECTRIC FIELD (kV/cm)	PENETRATION	SINGLE FIBER EFFICIENCY
0.460	15	+5	1	0.412	0.075
"	"	"	2	0.263	0.112
"	"	"	3	0.171	0.150
"	"	+10	1	0.329	0.095
"	"	"	2	0.200	0.136
"	"	"	3	0.119	0.174
0.913	"	+15	1	0.156	0.160
"	"	"	2	0.088	0.204
"	"	"	3	0.053	0.250
"	"	+21	1	0.130	0.171
"	"	"	2	0.064	0.231
"	"	"	3	0.038	0.275
0.460	"	-5	1	0.431	0.071
"	"	"	2	0.300	0.100
"	"	"	3	0.230	0.124
"	"	-10	1	0.420	0.073
"	"	"	2	0.298	0.110
"	"	"	3	0.229	0.124
0.913	"	-15	1	0.195	0.137
"	"	"	2	0.125	0.175
"	"	"	3	0.082	0.210
"	"	-21	1	0.160	0.154
"	"	"	2	0.091	0.200
"	"	"	3	0.058	0.240

Table IV. Experimentally Determined Filter Penetration and Single Fiber Efficiency for an AF-18 Filter Challenged with Charged PLS Particles and at a Face Velocity of 5 cm/s.

PARTICLE SIZE (µm)	FACE VELOCITY (cm/s)	ELEMENTARY CHARGE UNIT	ELECTRIC FIELD (kV/cm)	PENETRATION	SINGLE FIBER EFFICIENCY
0.460	5	+5	1	0.371	0.049
"	"	"	2	0.270	0.065
"	"	"	3	0.168	0.089
"	"	+10	1	0.287	0.062
"	"	"	2	0.154	0.093
"	"	"	3	0.084	0.124
0.913	"	+15	1	0.164	0.090
"	"	"	2	0.099	0.115
"	"	"	3	0.049	0.150
"	"	+21	1	0.179	0.086
"	"	"	2	0.097	0.117
"	"	"	3	0.050	0.150
0.460	"	-5	1	0.337	0.054
"	"	"	2	0.195	0.082
"	"	"	3	0.100	0.115
"	"	-10	1	0.325	0.060
"	"	"	2	0.156	0.093
"	"	"	3	0.080	0.130
0.913	"	-15	1	0.241	0.071
"	"	"	2	0.133	0.100
"	"	"	3	0.100	0.115
"	"	-21	1	0.190	0.083
"	"	"	2	0.110	0.110
"	"	"	3	0.050	0.150

Table V. Experimentally Determined Filter Penetration and Single
Fiber Efficiency for an AF-18 Filter Challenged with
Neutralized PLS Particles and at Face Velocities of 5 cm/s
and 15 cm/s

PARTICLE SIZE (μm)	FACE VELOCITY (cm/s)	ELECTRIC FIELD (kV/cm)	PENETRATION	SINGLE FIBER EFFICIENCY
0.460	5	0	0.758	0.015
"	"	1	0.444	0.045
"	"	2	0.263	0.074
"	"	3	0.156	0.103
"	15	0	0.759	0.015
"	"	1	0.600	0.030
"	"	2	0.440	0.045
"	"	3	0.340	0.060
0.913	5	0	0.810	0.012
"	"	1	0.483	0.040
"	"	2	0.310	0.065
"	"	3	0.197	0.090
"	15	0	0.790	0.013
"	"	1	0.620	0.030
"	"	2	0.470	0.042
"	"	3	0.370	0.055

Table VI. Experimentally Determined Filter Penetration and Single Fiber
Efficiency for an AF-18 Filter Challenged with Charged PLS
Particles and at Face Velocity of 15 cm/s.

PARTICLE SIZE (μm)	FACE VELOCITY (cm/s)	CHARGE UNIT	ELECTRIC FIELD (kV/cm)	PENETRATION	SINGLE FIBER EFFICIENCY
0.460	15	+5	1	0.5570	0.0292
"	"	"	2	0.4370	0.0413
"	"	"	3	0.3440	0.0532
"	"	+10	1	0.5110	0.0340
"	"	"	2	0.3720	0.0500
"	"	"	3	0.2690	0.0660
0.913	"	+15	1	0.4000	0.0460
"	"	"	2	0.2700	0.0653
"	"	"	3	0.1650	0.0900
"	"	+21	1	0.2930	0.0613
"	"	"	2	0.2020	0.0800
"	"	"	3	0.1483	0.0952
0.460	"	-5	1	0.7220	0.0163
"	"	"	2	0.6110	0.0250
"	"	"	3	0.5130	0.0333
"	"	-10	1	0.5880	0.0265
"	"	"	2	0.4730	0.0374
"	"	"	3	0.3420	0.0540
0.913	"	-15	1	0.4460	0.0403
"	"	"	2	0.3630	0.0510
"	"	"	3	0.2760	0.0642
"	"	-21	1	0.3910	0.0470
"	"	"	2	0.3100	0.0600
"	"	"	3	0.2140	0.0800

Table VII. Experimentally Determined Filter Penetration and Single Fiber Efficiency for an AF-11 Filter Mat Challenged with Charged PLS Particles (data taken by the E-SPART Analyzer).

PARTICLE SIZE (μm)	FACE VELOCITY (cm/s)	ELEMENTARY CHARGE UNIT	ELECTRIC FIELD (kV/cm)	PENETRATION	SINGLE FIBER EFFICIENCY
0.460	5	+4	2	0.2922	0.1040
"	"	+12	"	0.1836	0.1424
"	"	+20	"	0.2045	0.1330
"	"	+28	"	0.2184	0.1280
"	"	+36	"	0.1725	0.1480
"	"	+44	"	0.2120	0.1304
"	"	+52	"	0.1565	0.1558
"	"	+60	"	0.1611	0.1534
"	"	+68	"	0.0839	0.2100
"	"	+76	"	0.0611	0.2350
"	"	+84	"	0.0156	0.3500
"	"	+92	"	0.0452	0.2600
"	"	-4	"	0.2383	0.1210
"	"	-12	"	0.2514	0.1160
"	"	-20	"	0.2810	0.1100
"	"	-28	"	0.2240	0.1260
"	"	-36	"	0.3720	0.1830
"	"	-44	"	0.2220	0.1270
"	"	-52	"	0.1242	0.1753
"	"	-60	"	0.0691	0.2250
"	"	-68	"	0.1006	0.1930
"	"	-76	"	0.1003	0.1932
"	"	-84	"	0.0301	0.2943
"	"	-92	"	0.0533	0.2500
"	"	-100	"	0.0301	0.2944

In order to examine the overall effect of electrostatic enhancement on the filtration efficiency, the penetrations for the two particle sizes are plotted in Figure 6 as a function of the electrostatic filtration parameter K_{EO} defined by Equation (15). The variation of P versus K_{EO} is plotted for both positive and negative charges. It was found that penetration for the same values of K_{EO} was consistently less for positively charged particles than for the negatively charged particles. While there is a wide scatter in the experimental data for the values of K_{EO}, it is instructive to note that the overall pattern shows the exponential decay of penetration with respect to the electrostatic filtration parameter as expected from the simplified theoretical model. The scatter in the experimental data was primarily due to the uncertainty in the filtration parameters such as fiber diameter, solidity, filter thickness and the actual macroscopic electrical field present inside the filter medium. Because of these uncertainties, there were discrepencies between the measured single fiber efficiency and the calculated values of single fiber efficiency from equation (10). Figure 5 is a plot of the variation of single fiber efficiency as a function of particle size calculated from equation (10) for filter AF-11, with an external electric field of 3 kV/cm, a particle charge of 5 e/particle, and a face velocity of 5 cm/s. These calculated values underestimate the filtration efficiency since the models do not take into account the influence of neighboring fibers and the interaction between different collection mechanisms. In Figure 5 there is a sharp rise in the single fiber efficiency because of the hypothetical assumption of a constant charge level, i.e., 5 e/particle in the entire size range.

Electrostatic enhancement is most significant when an external electric field is applied against the filter medium and particles are charged. For example, for particles with a Boltzmann charge distribution, the single fiber efficiency for 0.460μm-diameter particles is 0.031 (Table II). When these particles were charged with 10 e/particle and an electric field of 3 kV/cm was applied, the single fiber efficiency increased to 0.297 (Table I), a factor of almost 10. With comparable enhancement at a velocity of 15 cm/s, the single fiber efficiency increased by a factor of 5.6. Similar enhancement was also achieved for 0.913μm-diameter particles where the single fiber efficiency increased from 0.036 (for particles with a Boltzmann charge distribution) to 0.342, a 9.5 fold increase, when the particles were charged to 21 e/particle and a field of 3 kV/centimeter was applied. The corresponding increase for the AF-18 filter for 0.46μm-diameter particles was from 0.015 to 0.124, an enhancement by a factor of 8.3. For 0.913μm-diameter particles, the enhancement was by a factor of 6.7. The efficiencies for AF-18, which is a coarser filter, were consistently lower than for the AF-11 filter.

CONCLUSION

Filtration efficiency of fibrous filters can be increased significantly by electrostatic enhancement if the filters are operated at a low face velocity. For submicron aerosol particles (e.g. 0.46μm-dia. PLS), filtration efficiency increased from 31 to 97 percent with electric field polarizing filter media and by electrostatic charging of the particles. Experimental determination of the relationship between particulate penetration (P) and the electrostatic filtration parameter (K_{EO}) can provide data on electrostatic enhancements for practical applications.

ACKNOWLEDGEMENTS

The authors are grateful to many colleagues who gave valuable assistance during the entire course of the work. They are particularly grateful to K. Tennal and P. Sinha for technical assistance, to S. Turner, P. Archer, and D. Belk for their help in the preparation of the manuscript, to D. Watson and B. Nelson for editing the manuscript.

REFERENCES

1. W. Bergman, A. Biermann, W. Kuhl, A. Bogdanoff, H. Hehard, M. Hall, D. Banks, M. Mazumder, and J. Johnson, J., "Electric Air Filtration," Lawrence Livermore National Laboratory (LLNL) Report, UCID-19952, September, 1983.
2. A. Biermann, and W. Bergman, Measurement of aerosol concentration as a function of size and charge, Aerosol Sci. Technol. 3:293-304 (1984).
3. W. Bergman, W. D. Kuhl, W. L. Russell, R. D. Taylor, H. D. Hebard, A. H. Biermann, N. J. Alvares, D. G. Beason, and B. Y. Lum, "Electrofibrous Prefilters for Use in Nuclear Ventilation System," LLNL Report, February, 1981.
4. W. Bergman, A. H. Biermann, H. D. Hebard, B. Y. Lum, and W. D. Kuhl, "Electrostatic Air Filters Generated by Electric Fields," LLNL Report, January, 1981.

5. C. E. Billings, and R. A. Gussman, Dynamic behavior of aerosol, in "Handbook on Aerosols," R. Dennis, Ed., Chapter 3, National Technical Information Service (1976).

6. C. N. Davies, "Air Filtration," Academic Press, London, New York, (1973).

7. C. N. Davies, Filtration of aerosols, J. Aerosol Sci, 14(22), 147-161 (1983).

8. F. Henry and T. Ariman, Cell model of aerosol collection by fibrous filters in an electrostatic field, J. Aerosol Sci, 12(2), 91-103 (1981).

9. William C. Hinds, "Aerosol Technology," John Wiley and Sons Inc, NY, New York (1982).

10. H. F. Kraemer and H. F. Johnstone, Collection of aerosol particles in presence of electrostatic fields, Engineering, Design and Equipment, 47(12), 2426-2434 (December 1955).

11. K. W. Lee, and B. Y. Liu, Experimental study of aerosol filtration by fibrous filters, Aerosol Sci. Technol., 1, 35-46 (1982).

12. M. K. Mazumder, and K. T. Thomas, Improvement of the efficiency of particulate filters by superimposed electrostatic forces, Filtration and Separation, 4(1), 25-30,66 (January/February 1967).

13. R. G. Renninger, M. K. Mazumder, and M. K. Testerman, Particle sizing by electrical single particle aerodynamic relaxation time analyzer, Rev. Sci. Instrum. 52(2), 242-246 (1981).

14. A. T. Rossano Jr, and L. Silverman, Electrostatic effect in fiber filters for aerosols, Heating and Ventilating Reference Section, (1954).

15. H. C. Yeh, and B. Y. H. Liu, Aerosol filtration by fibrous filters-1, theoretical, J. Aerosol Sci, 15, 191-204 (1974).

16. H. C. Yeh, and B. Y. H. Liu, Aerosol filtration by fibrous filters-2, experimental, J. Aerosol Sci, 5, 205-217 (1974).

17. G. Zebel, Deposition of aerosol flowing past a cylindrical fiber in a uniform electric field, J. Colloid Sci., 20, 522-543 (1965).

18. G. O. Nelson, W. Bergman, H. H. Milla, R. D. Taylor, C. P. Richards, and A. H. Biermann, Am. Ind. Hyg. Assoc. J., 39, 472 (1978).

19. M. Shapiro, G. Laufer, and C. Gutfinger, Experimental study on electrostatically enhanced granular filters, Aerosol Sci. Technology, 5, 435-445 (1986).

20. G. I. Tardos, E. Yu, R. Pfeffer and A. M. Squires, Experiments on aerosol filtration in granular sand beds, J. Colloid Interface Sci., 71, 616-621 (1979).

PARTICLE RETENTION AND DOWNSTREAM CLEANLINESS OF POINT-OF-USE

FILTERS FOR SEMICONDUCTOR PROCESS GASES

Mauro A. Accomazzo and Donald C. Grant

Millipore Corporation
80 Ashby Road
Bedford, Massachusetts 01730

Particles in gas streams are captured within a filter matrix by several mechanisms and their removal efficiency is dependent on particle size. Hence, for a given type of filter matrix and typical process flow conditions, there exists a most penetrating particle size.

The most penetrating particle size for fibrous filter media and microporous membrane filter media have been determined to be typically 0.15 μm and 0.05μm, respectively. The retention ratings at the most penetrating particle size for fiberglass filters have been determined to be ≥99.99% and for membrane filters ≥99.9999999%.

The downstream cleanliness of filters during typical operating conditions is also of importance. The shedding of particles from the downstream side of various types of Point-of-Use filter devices has been evaluated, using laser and condensation nucleus type particle counters. Differences were seen between various POU filter designs. Fiberglass filters can shed particles initially at concentrations in the hundreds to thousands of particles per cubic foot (>0.2μm) during steady and pulsing flow operation. Pleated membrane filters intially released particles in the tens to hundreds of particles per cubic foot range. POU filters, using stacked disc or molded support structures, achieve very low levels of particle shedding (<1 particle/ft^3) during both steady and pulsed flow conditions for particles >0.2μm.

INTRODUCTION

The need for the removal of particles in the fluids that contact wafers during the manufacturing process of semiconductors is well established. Today's VLSI (very large scale integration) and ULSI (ultra large scale integration) devices require even greater contamination control as the dimensions of the circuit components shrink in both the horizontal and vertical directions.[1] Class 10 clean rooms (<10 particle /cubic foot >0.5μm) and localized Class 1 areas are becoming commonplace for semiconductor manufacturers. More and more of the processing steps for these new devices require dry processing and use a variety of specialty gases.[2,3]

The need for particle free gases used in the production of semiconductor devices is readily accepted.[4,5] Particles in gases can come from many sources, such as, the gas manufacturing process, walls of the gas container, gas purifiers, the gas delivery system (pressure regulators, valves, flow controllers, piping, fittings, welds, and filters), and the processing equipment.[6,7]

Many different gases and gas mixtures are used in semiconductor processing as shown in Table I. Gases like nitrogen, argon, oxygen, and hydrogen are often supplied from central storage facilities or even pipelines, whereas the other gases are supplied from cylinders. SEMI (Semiconductor Equipment and Materials Institute) standards have been established for the bulk gases[8] and presently cylinder gases are under consideration. The preferred materials of construction for the gas delivery system is stainless steel. However, there are still some systems of carbon steel and aluminum for nitrogen and argon, and copper for oxygen.[9,10]

Table I. Typical Process Gases used by the Semiconductor Industry.

Process	Gases Used
Chemical Vapor Deposition	AsH_3, CO_2, SiH_4, N_2, N_2O, H_2, SiH_2Cl_2, HCl, O_2, B_2H_6
Diffusion	Ar, H_2, N_2, O_2, AsH_3, BCl_3, B_2H_6, PH_3
Oxidation	Ar, N_2, Cl_2, H_2, HCl, O_2
Etching	Ar, BCl_3, Cl_2, ClF_3, CF_4, F_2, He, H_2, HCl, N_2, O_2, OF_2, C_3F_8, SiF_4
Photolithography	Ar, N_2
Ion Implantation	Ar, AsH_3, BCl_3, BF_3, Cl_2, He, H_2, N_2, PH_3, SiH_4, PF_5, SiF_4

Filters are used at various points in the gas distribution system. The size and type of filter depends on the flow rate and specific gas being filtered. This paper will discuss the performance of so-called Point-of-Use (POU) filters. A POU filter is typically the last filter in the gas distribution system and is very near the processing equipment (oxidation/diffusion furnaces, epitaxial reactors, chemical vapor deposition reactors, plasma etchers/strippers, ion implanter, etc.).

Figure 1 shows schematically a typical gas distribution system for a semiconductor manufacturing facility. As previously mentioned, the bulk gases supplied usually exceed the SEMI specifications of <20 particles/cubic foot >0.2μm. Most manufacturers install filtration just downstream of the suppliers line to protect their distribution system. Additional filtration is then installed in the chase areas of the clean rooms with POU filters capable of handling flow rates up to 300 standard liters per minute (SLPM). Smaller POU filters are also utilized in the clean room and in the process equipment, downstream of valves and mass flow controllers and as close to the wafer as possible. These mini type filters can handle flow rates up to 30 SLPM.

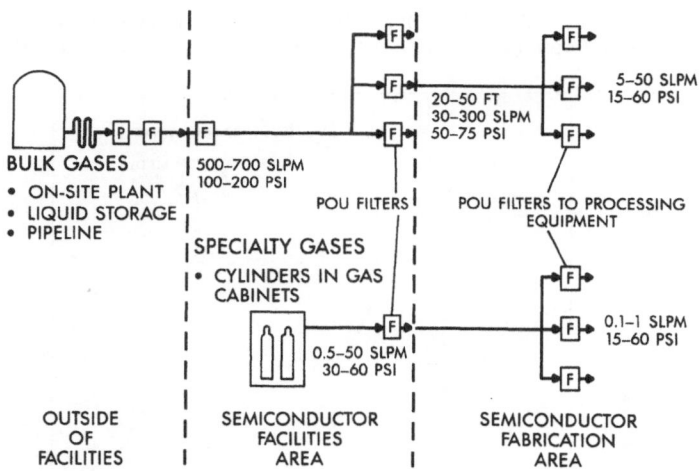

Figure 1. Schematic of typical process gas distribution system.

Today's POU filters are sealed in stainless steel housings with welds and end fittings that are inboard helium leak rated typically in the 10^{-6} to 10^{-8}cc/sec range. The design of the filter, the materials of construction, the surface finish of the housing, and the assembly process are all critical to filter performance. The filter media are made from glass fibers or fluoropolymers in the form of a microporous membrane. The POU membrane media used are either PTFE (polytetra-fluoroethylene) or PVDF (polyvinylidenefluoride). Support materials for the filter media range from porous stainless steel and nonwoven polypropylene fabrics to injection molded components of PFA (poly-fluoroalkoxy), ECTFE (ethylene chlorotrifluoroethylene), and polysulfone.

This paper will cover two areas of POU filter performance: particle retention (the ability of the filter to remove particles from the process gas stream) and downstream cleanliness (the contribution of particles to the process gas stream by the filter device).

PARTICLE RETENTION

Particles in gas streams are retained in filter media by five differing particle deposition mechanisms: gravitational settling, electrostatic deposition, impaction, interception, and diffusion.[11] Since the microelectronics industry is primarily concerned about the removal of small particles (below one micron), the interception and diffusion mechanisms can provide effective particle capture for typical process gas applications.

The particle removal efficiency of filter media is dependent on the properties of the particle and the gas in which it is flowing. The theory of aerosol filtration by both fibrous and membrane media is well established and predicts a most penetrating particle size (MPPS) for a given set of operating conditions.[12,13]

Recent studies at the University of Minnesota have been conducted with filter media typically utilized in POU filters supplied to the semiconductor industry. The results of these studies are shown in Figures 2 and 3 for fiberglass[14] and membrane filter[15] media, respec-tively. As shown in Figure 2, the most penetrating particle size (MPPS) for ULPA (ultralow penetration aerosol) fiberglass media is in the 0.1-0.2μm particle size range for face velocities ranging from

7—1cm/sec, respectively. The MPPS for membrane filter media (Figure 3) is in the 0.04-0.06µm size range and will be discussed further below.

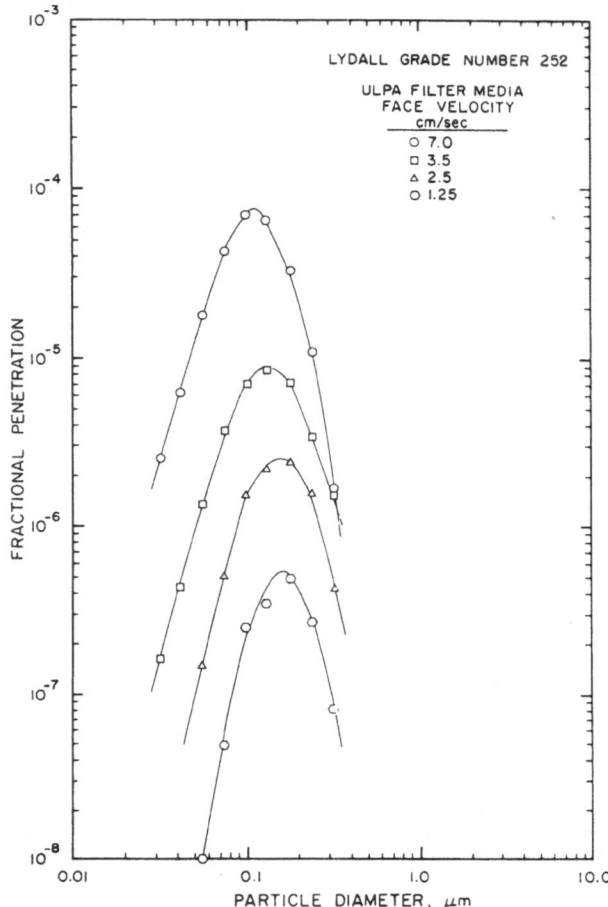

Figure 2. Fractional penetration of DOP aerosol through ULPA filter media.[14]

Initial particle retention studies on membrane filter media (PVDF) with sodium chloride aerosols showed less than 10^{-9} penetration over the face velocity range of 5-50cm/sec.[16] Based on Rubow's filter model, the most penetrating particle size was predicted to lie in the 0.04-0.06µm range, as shown in Figure 4. The model also predicted a penetration in the 10^{-25} to 10^{-30} range. Since theory also predicted that particle penetration is a function of membrane thickness, a thinner membrane was prepared that had essentially the same pore size (based on bubble point and porosity measurements). The penetration of sodium chloride particles as a function of particle size and face velocity was measured and the results are those previously shown in Figure 3. By taking into account the thickness of typical PVDF membranes used in semiconductor applications, one would predict a particle penetration of 6×10^{-32} and 4×10^{-23} at the MPPS for face velocities of 5-50cm/sec, respectively.

Retention studies have also been carried out on POU filters using sodium chloride particles over the MPPS range. Rubow and Liu[17] reported that stacked disc filters had less than 10^{-9} penetration at flow rates up to 700 SLPM. Recent studies conducted on mini POU filters (Millipore

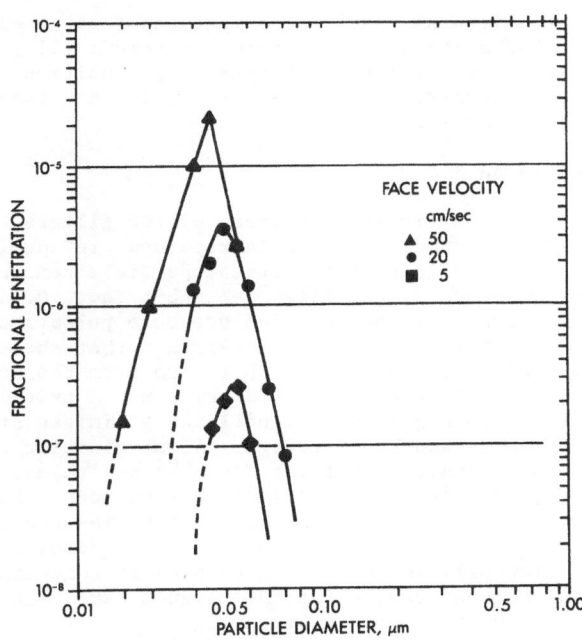

Figure 3. Fractional penetration of sodium chloride aerosol through
ultrathin PVDF filter media.[15]

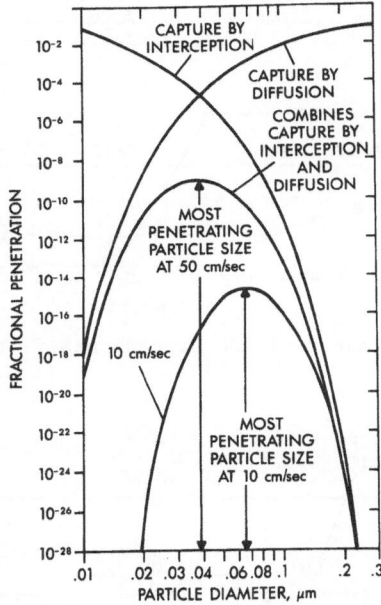

Figure 4. Theoretical plot showing the effect of velocity on the most
penetrating particle size.(Rubow Model[13])

Mini In-Line WGFG01HR1) have verified less than 10^{-9} penetration with
sodium chloride particles over the MPPS range (0.03-0.09 μm) at flow
rates up to 200 SLPM. In addition, these POU filters have been
challenged with ultrafine sodium chloride particles down to 0.003μm in
size and found to have unmeasurable penetration ($<10^{-9}$).[15]

The above experimental results for membrane filters clearly demonstrate that POU membrane gas filters can remove all particles in the inlet gas stream over all typical operating conditions. However, particles have been observed downstream of filters and their source is the subject of the next section.

DOWNSTREAM CLEANLINESS

The detection of particles downstream of POU filters, sometimes referred to as filter shedding, has been reported previously.[18-20] The earlier studies reported initial downstream particle concentrations in the hundreds to thousands of particles per cubic foot >0.1μm, when the filters were subjected to mechanical and pressure pulsations. Recent studies conducted at Millipore[21] have determined that the particle sources were not only from the filters but also from the downstream fittings and tubing that make up the test system. A system was constructed as shown in Figure 5 that utilized stainless steel tubing with a <10μ-inch finish and VCO fittings. After cleaning the components in an ultrasonic bath containing first Freon(R) and then deionized water, the system was assembled in a laminar flow hood. Filtered air was then passed through the system at high velocities (near sonic) to further clean the system. Once this procedure was completed, air was kept flowing continuously in the system to prevent contamination from the room. The background counts for this system have been reported previously and are shown in Figure 6.

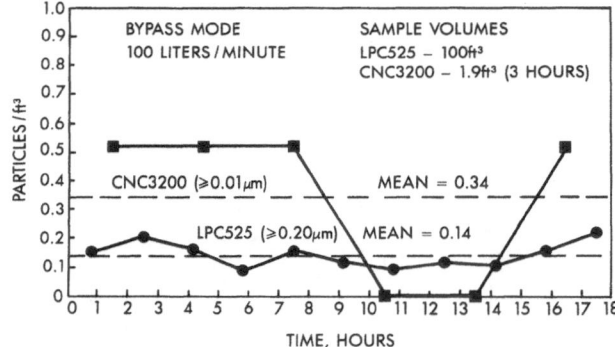

Figure 5. Schematic of POU process gas filter test system.

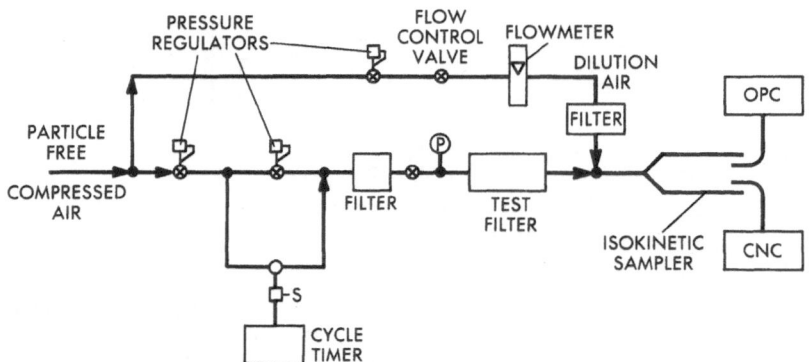

Figure 6. Typical test system background particle counts.[21]

We have recently completed downstream cleanliness studies of five
different types of POU filters that are commercially available today for
semiconductor applications. Figure 7 is a photograph showing the five
filters. The two filters on the left can be used with gas flow rates up
to 300 SLPM whereas the three filters on the right are limited to 30
SLPM.

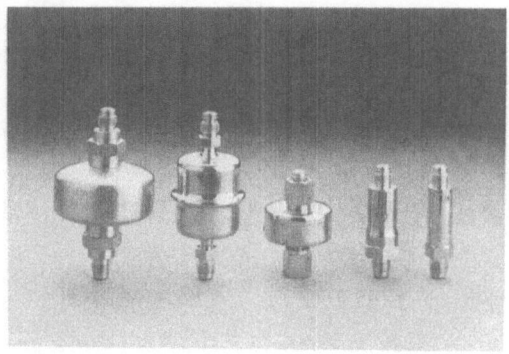

Figure 7. POU process gas filters.

The first filter on the left, subsequently referred to as Filter A,
utilizes a PTFE membrane attached to a molded disc of ECTFE. Six discs
are stacked on each other to provide a filter with 300cm^2 of membrane
area. The second filter from the left (Filter B) also uses PTFE membrane in a pleated configuration. The PTFE membrane is sandwiched
between two layers of nonwoven polypropylene fabric and the pleated
structure is supported by a molded polypropylene central core and end
caps.

The center filter (Filter C) uses fiberglass filter media in disc
form supported on porous stainless steel. The fourth and fifth filters
from the left (Filters D and E) both utilize PTFE membrane filter media
wrapped around a molded Teflon^R core. These latter filters have been
designed to fit into process equipment as shown in Figure 8.

Filter Testing Results

Four to seven of each type of the POU filters described above were
evaluated in the test system shown in Figure 5. The typical background
counts for the system shown in Figure 6 are essentially the current
limits of the laser optical particle counter (OPC) and the condensation
nucleus counter (CNC).

The test protocol consisted of installing the filter under a
laminar flow hood and then initiating steady flow through the filter.
Particle counting was also begun when flow commenced. Three constant
flow regimes were evaluated for particle shedding followed by a regime
of pulsed flow (one cycle per second for 15 minutes). The results of
the shedding studies are shown in Figures 9 and 10.

Figure 9 indicates that the stacked disc configuration releases
about an order of magnitude less particles than the pleated filter
during both steady flow and pulsed flow operation. The stacked disc
filter typically reached particle concentrations of <1 particle per
cubic foot within 10 minutes of each flow regime (steady or pulsing
flow). Previously we reported that the downstream particles from
stacked disc POU filters were generated by the fittings during installation.[21] Hence, once a filter is operated for a short period of time,
no further particle shedding will be experienced.

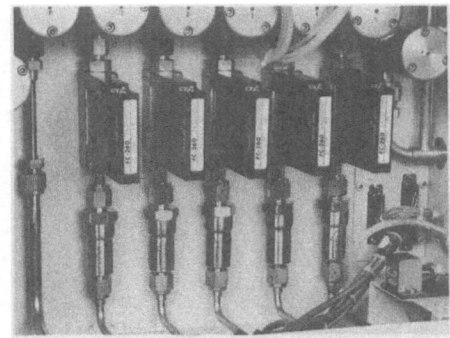

Figure 8. Mini POU process gas filter installation.

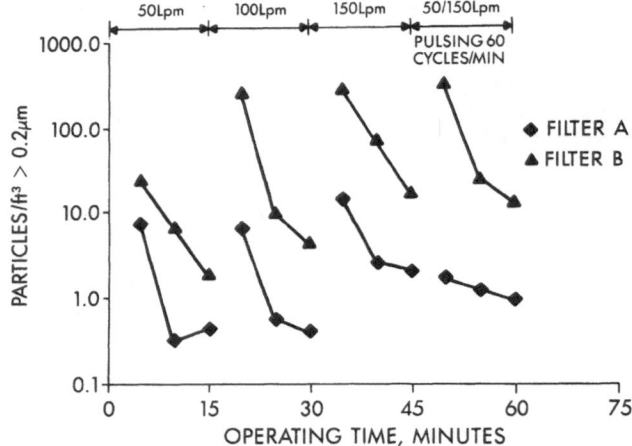

Figure 9. Particle shedding performance of POU process gas filters.

Figure 10 shows the average performance of the mini type POU filters. There is a significant difference in the shedding characteristics between the fiberglass filter and the membrane filters. The initial shedding, during steady flow, from the fiberglass filter was in the 100 to 1000 particles/cubic foot (pt/cf) range. During pulsing the particle concentrations were in excess of 1000 pt/cf. After the pulsing regime, the concentration of particles was reduced to less than 10 per cubic foot at a steady flow rate of 20 SLPM.

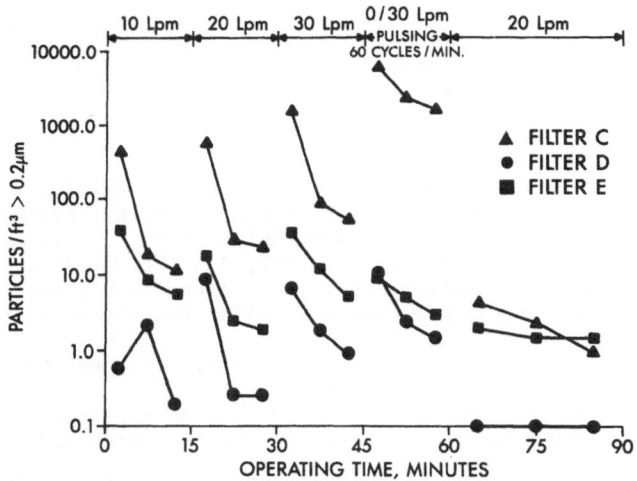

Figure 10. Particle shedding performance of mini POU process gas filters.

On the other hand, the membrane filters were able to achieve <1 particle per cubic foot (>0.2μm) concentrations during each flow condition. Since both filters are of similar design, the differences in downstream shedding performance between filters D and E are probably attributable to the filter manufacturing process.

CONCLUSION

The semiconductor industry uses POU filters to remove particles from the gases utilized in the manufacturing process of wafers. These filters use either fiberglass or microporous membranes as filter media. Studies have shown that the most penetrating particle size (MPPS) for fiberglass and membrane media are 0.15 and 0.05μm, respectively. The fractional penetration for fiberglass media has been determined to be less than 10^{-4} and less than 10^{-9} for membrane media.

The downstream cleanliness of various commercially available POU filters has been determined with a test method that simulates use conditions (steady and pulsed flow). Differences can be seen between the various types of filters.

Using fiberglass media, the POU filter initially shed particles in the hundreds to thousands per cubic foot during steady and pulsing flow. After pulsing for 15 minutes, lower shedding concentrations were observed in the 10 particles per cubic foot range (>0.2μm) during steady flow. However, very high concentrations were still observed during pulsing.

Pleated membrane filters initially shed particles in the tens to hundreds of particles/ft^3 range during steady and pulsing flow. These structures tended to release an order of magnitude more particles than membrane filters supported by molded discs.

Using stacked discs or molded support, POU membrane filters rapidly achieved low levels of particle shedding ($\leq$1 pt/ft^3) during both steady and pulsed flow conditions.

ACKNOWLEDGEMENT

The authors wish to thank John Jaillet for careful experimental work and Frances Carlson and George Witham for manuscript preparation.

REFERENCES

1. D. L. Tolliver, New needs in contamination control: A critical cornerstone in integrated circuit manufacturing, presented at the Fifth Annual Microelectronics Technical Symposium sponsored by Millipore Corporation, May 18, 1987, San Jose, California.
2. J. W. Mitchell, Chemical analysis of electronic gases and volatile reagents for device processing," Solid State Technol., p. 131, March 1985.
3. C. Murry, Improving gas handling safety, Semiconductor International, p. 60, August 1985.
4. L. Faure and H. Thebault, Perspectives on contamination control: Parts I and II, Microcontamination, p. 16, March and p. 10, April 1987.
5. D. W. Cooper, Particulate contamination and microelectronics manufacturing: An introduction, Aerosol Sci.Technol., 5,287-299 (1986).
6. J. M. Davidson and T. P. Ruane, Gas-handling: Considerations for ensuring gas purity, Microcontamination, p. 35, March 1987.
7. D. Jensen, Reducing microcontamination generation and entrapment within high purity gas distribution systems, Microcontamination, p. 52, May 1987.
8. J. King, SEMI develops first particulates specifications for VLSI-grade gases, Microcontamination, p. 64, January 1986.
9. R. Zawierucha, Materials for high purity gas distribution systems, Technical Proceedings of SEMICON/WEST '85, May 21-23, San Mateo, California.
10. R. M. Thorogood, A. Schwartz, W. J. McDermott, Measurement and submicron particle removal factors in high microelectronics, paper presented at the Technical Symposium sponsored by Millipore, May 18, 1987, San Jose, California.
11. W. C. Hinds, "Aerosol Technology," John Wiley & Sons, New York, 1982.
12. K. W. Lee and B. Y. H. Liu, On the minimum efficiency of the most penetrating particle size for fibrous filters, J. Air Poll. Control Assoc., 30,377-381 (1980).
13. K. L. Rubow, "Submicron Aerosol Filtration Characteristics of Membrane Filters," Ph.D. Thesis, University of Minnesota, Mechanical Engineering Department, Minneapolis, MN, 1981.
14. B. Y. H. Liu, K. L. Rubow, and D. Y. H. Pui, Performance of HEPA and ULPA filters, "Proceedings of the 31st Annual Meeting of the Institute for Environmental Sciences," pp. 25-28, 1985.
15. K. L. Rubow, B. Y. H. Liu, and D. C. Grant, Characteristics of ultra-high efficiency membrane filters, "Proceedings of the 33rd Annual Meeting of the Institute of Environmental Sciences," pp. 383-387, 1987.
16. M. A. Accomazzo, K. L. Rubow, and B. Y. H. Liu, Ultrahigh efficiency membrane filters for semiconductor process gases, Solid State Technology, p. 141, March 1984.
17. K. L. Rubow and B. Y. H. Liu, Evaluation of Ultrahigh-Efficiency Membrane Filters, Microcontamination, p.39, March 1985.
18. R. L. Duffin, Process gas filtration in integrated circuit production, Microcontamination, p. 34, December 1983/January 1984.
19. M. A. Accomazzo, Particulate retention and shedding characteristics from Point-of-Use process gas filters, in "Technical Program Proceedings SEMICON/WEST," San Mateo, California, May 21-23, 1985.

20. M. A. Accomazzo and D. C. Grant, Mechanisms and devices for filtration of critical process gases, "Fluid Filtration: Gas," Volume 1, p. 402, ASTM STP 975, R. R. Raber, Ed., American Society for Testing and Materials, Philadelphia, 1986.

21. D. C. Grant, Particle shedding from semiconductor process gas filters, methods for measurement and elimination, in "Proceedings of the 33rd Annual Meeting of IES," San Jose, California, May 5–7, 1987.

A FLUID DYNAMIC STUDY OF A MICROCONTAMINANT PARTICLES REMOVAL PROCESS

Ahmed A. Busnaina, M.A.R Sharif, and Glenn Gale
Mechanical and Industrial Engineering Department
Clarkson University, Potsdam, New York 13676

and

Frederick W. Kern, Jr.
General Technology Division, IBM Corporation
Essex Junction, Vermont 05452

Flow visualization techniques are used to study microcontaminant particles removal using drag and centrifugal forces (wafer rinsing). This removal process is an effective and commonly used method in the semiconductor industry. Understanding the flowfield around the rotating wafers is essential in determining the effectiveness and efficiency of the process. Flow visualization is used to observe the effect of rotational speed, jet flow rate and nozzle type on the flow pattern on the wafers. Optical and novel chemical flow visualization techniques are used. The qualitative and quantitative results show that the rotating wafers were completely covered with a water-film (that provides the drag force) only at certain values of the governing parameters. An empirical correlation between the water-film coverage and the operating parameters is introduced. A three-dimensional turbulent time-dependent prediction procedure for swirling flows is used to predict the flow around the wafers. Predictions of the trajectories of the water droplets issued from a peripheral jet into the flowfield between two rotating disks are also performed and the results presented.

1   INTRODUCTION

Microcontaminant particles removal constitutes a major problem to wafer fabrication processes. Particle removal using drag and centrifugal force (often called wafer rinsing) is proving to be an effective and commonly used method in microelectronics manufac-

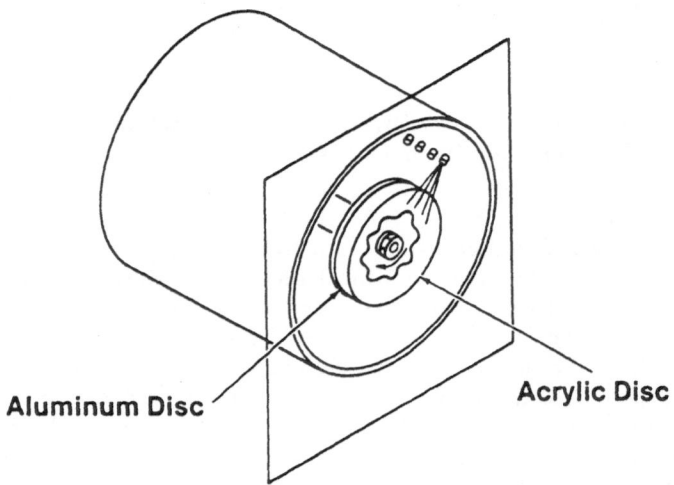

Aluminum Disc

Acrylic Disc

Figure 1 Problem schematic

turing. In this process, deionized water jets are introduced into the flowfield between rotating disks from a peripheral jet. The rinsing and drying process of silicon wafers comprises a very complex fluid dynamics problem. Understanding the flowfield around the rotating wafers is essential in determining the effectiveness and efficiency of the process. The water film on the wafer represents the mechanism by which the contaminant particles are removed. The water-film moving radially outward on the wafer provides the drag force on the particles. This indicates that in order for the rinsing process to be completely effective every wafer has to be completely covered by a water-film. In this study a typical horizontal rinser/dryer configuration (STI) is used. Flow visualization is used to observe the flow pattern for different flow rates and rotational speeds. The second part of this experimental study involves studying the shielding effects of the carrier and the rotor on the flow pattern on the wafers. Numerical predictions of the three-dimensional turbulent flowfield between two rotating disks is presented. Also the trajectories of the water droplets introduced into this flowfield from a peripheral jet is predicted and the results are presented.

2   EXPERIMENTAL PROCEDURE

The first phase of the experimental procedure investigates the effectiveness of the rinsing process under the best (ideal) possible conditions the wafer can be rinsed. The study provides a fundamental functional relationship between the governing parameters in the absence of shielding presented by the carrier and rotor. The second phase of the project investigates the flow patterns on the shielded wafers (using the actual geometry) and the effects of the geometry (i.e., the inclusion of the shielding effects of the carrier, rotor and other wafers).

2.1   Phase I

In this phase flow visualization is used to observe the flow pattern on two free rotating wafers for different flow rates and rotational speeds. A dye is used to color the jet injected between two rotating disks. The first disk is made of a transparent acrylic material (to permit flow visualization) and the second is made of polished aluminum. The disks are five or eight inches in diameter to simulate the actual wafers. Figure 1 shows a schematic of the considered experimental geometry of Phase I. A stroboscope that employs intermittent light permits the visual observation of the flow pattern on the rotating disks (wafers). When the stroboscope flashes are adjusted to the same frequency as the rotational speed, the two disks will appear stationary. This enables us to observe the flow pattern between and on the rotating disk in the absence of rotation. These flow patterns are then recorded using a 35 mm camera equipped with a telephoto lens. The shutter speed is adjusted to expose the film to one flash per exposure only.

2.2   Phase II

In the second phase, the flow pattern on several wafers in a carrier is investigated. The flow can not be visualized or seen through the rotor, carrier, and wafers. Only after the

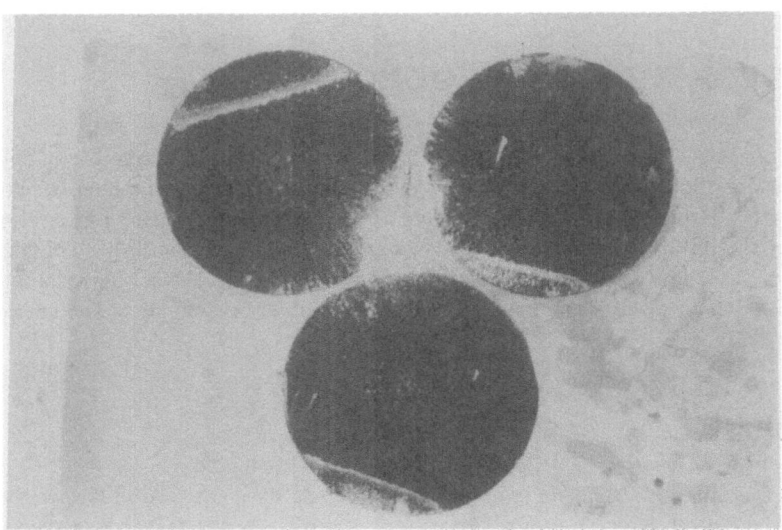

Figure 2   Three wafers which are completely covered with a water-film and completely black in color.

experiment is over, the flow pattern can be visualized with the aid of some chemicals on the surface of the wafer and in the water jet. These chemicals (template lacquer on the wafer and 20 % ammonia in the water jet) react when the water jet impinges on the wafer surface or comes in contact with it. When the reaction takes place, the thin lacquer film cover turns from a golden yellow to black color. This technique is being used for the first time for such an application[1]. The technique has worked very well, despite the initial difficulty associated with spreading and drying the lacquer on the wafers. Photographs of the flow pattern on the wafers were taken for ten wafers

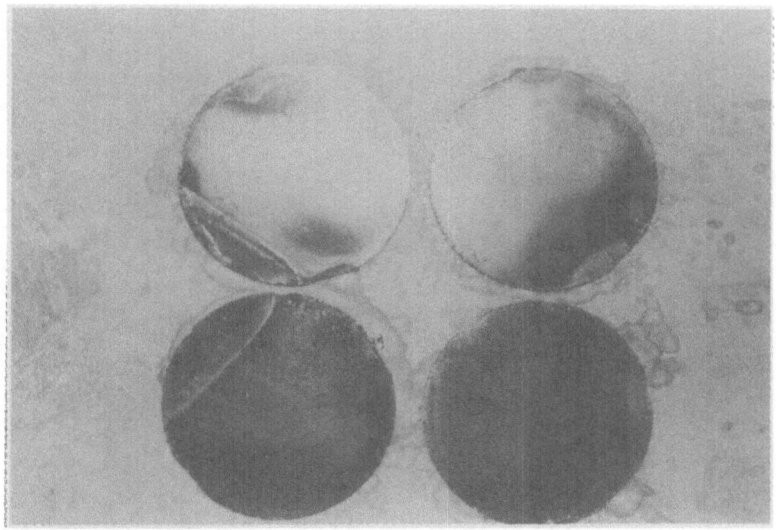

Figure 3 Four wafers show only a small part of the wafers covered with a water-film.

in every run. If a wafer is completely black, that would indicate total coverage with a water-film. If, on the other hand, the wafer had yellow, green and black colors, that would indicate partial coverage only by the water-film. Figure 2 shows three eight inch wafers which are completely covered and completely black in color. Figure 3, on the other hand, shows a very poor coverage (for four five-inch wafers), the photograph shows that only a small area of the wafers was covered with a water-film.

This phase represents the actual geometry of the device used. That implies the shielding effect mentioned earlier and one jet for every four wafers approximately compared to one jet per wafer in Phase I.

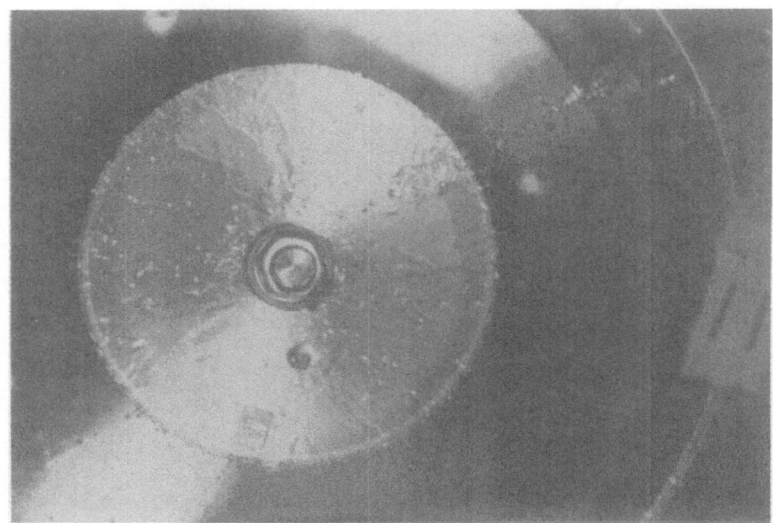

Figure 4   Satisfactory water-film coverage on the aluminum disk.

3 EXPERIMENTAL RESULTS AND DISCUSSION

3.1 Phase I

The photographs show that the fluid on the disks follows different patterns at different rotational speeds and flow rates. The photographs also demonstrate that only in few cases the disk was totally covered with a film of water during the rinsing process (using two free disks supported at the center). An example is shown in Figures 4 and 5. Figure 4 shows a photograph of the water-film on the disk where the coverage is satisfactory. Figure 5 shows a case where the film coverage is poor.

Experiments were conducted at three different locations of the disk with respect to the jet. The distance between the disks was taken as 3/16 inch (the typical distance between wafers in the carrier). The first location situated the jet (or nozzle) axis at the middle of the spacing between the two disks. The second location positioned the jet one quarter of the spacing near one disk. The third location positioned the jet on the disk itself. Every location was examined for five rotational speeds and nine different flow rates.

Two nozzles were used in the study. The first nozzle produces a 120° fan spray (fluid sheet), and the second nozzle produces a 30° hollow cone spray. The second nozzle was found to be superior to the first nozzle.

These experiments were conducted to establish the functional relationship between the flow pattern on the wafer (film coverage) and the governing parameters. Using dimensional analysis yielded the following relationship:

$$\frac{A_c}{D^2} = F[\frac{Q}{\omega D^3}] \tag{1}$$

where
A_c is the covered area of the disk by the water-film, m²,
D is the wafer diameter, m,
Q is the jet flow rate m³/min, and
ω is the rotational speed, rpm.

The data are plotted as a function of the introduced nondimensional coverage parameter, CP,

$$CP = \frac{Q}{\omega D^3} \tag{2}$$

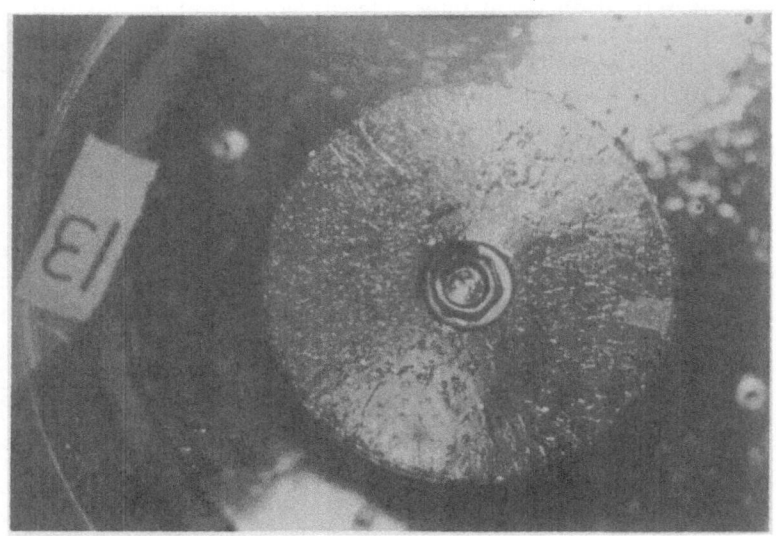

Figure 5   Poor water-film coverage on the aluminum disk.

in Figures 6 and 7. Figure 6 shows the second disk location film coverage. The figure shows that for the second disk location, the data almost collapse to a single curve. This indicates that the data correlate well with the introduced parameter for the second disk location. The data do not correlate as well for the third and first locations. Figure 7 shows the third location film coverage which exhibits weaker correlations with the coverage parameter. The introduced parameter can be used to extend the data to other wafer diameters, flow rates, and rotational speeds.

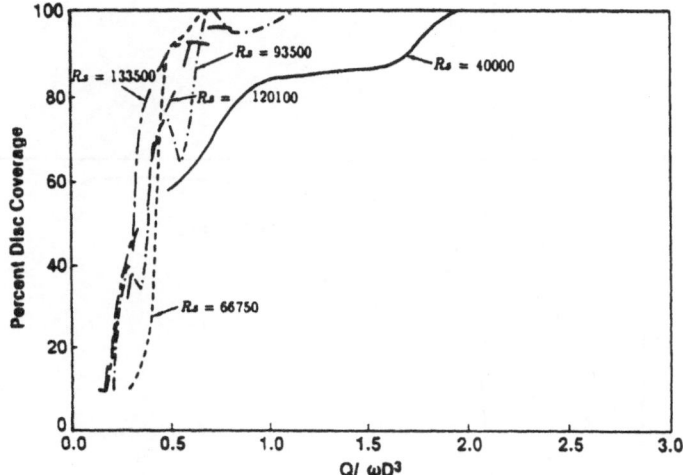

Figure 6   Percentage of water-film coverage as a function of the coverage parameter CP at the 2nd location.

3.2   Phase II

Because of the carrier geometry, this phase of the project employed one jet for every four to five wafers approximately compared to one jet per wafer in Phase I (free wafer experiments). Basically, 7 jets were used to rinse 25 wafers. During the experiments, only one or two jets were activated at any time due to the lack of enough water pressure to use the seven nozzles. A repeated pattern of the wafer arrangement with respect to the jets was identified and used.

The results from this phase are presented in terms of the following parameters.

$$Rs = v_t D / \nu$$
$$Rq = V D_j / \nu$$

where v_t is the azimuthal velocity at the edge of the rotating wafer, V is the water jet velocity, D is the wafer diameter, D_j is the nozzle diameter, and ν is the kinematic viscosity of water.

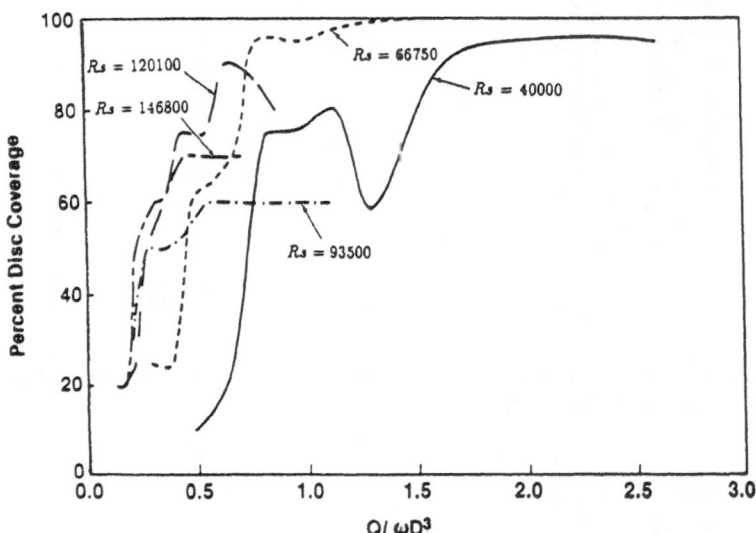

Figure 7 Percentage of water-film coverage as a function of the coverage parameter CP at the 3rd location.

Figure 8 shows the percentage coverage for 16 five inch wafers using jets #3 and 4 for different flow rates. The results show that only at an intermediate flow rate, good coverage is obtained for wafers #6-14. Figure 8 also shows that for an intermediate rotational speed a high flow rate provides a better coverage between wafers 6-14. A higher rotational speed is preferred as it provides a higher fluid velocity which increases the drag and centrifugal forces acting on the particles.

Figure 9 shows the coverage for 12 eight inch wafers using jets #2 and 3 for different flow rates and low speed. The figure shows that the coverage is different from that of Figure 8. At a higher speed (which is preferred) the film coverage gets worse (Figure 10) which indicates that the relationship between the governing parameters is by no means linear. A new dimensionless group for the coverage parameter CP (that includes the distance between the jet and wafers) is introduced. The new relationship accounts for the shielding and multiple wafers. The nondimensional coverage parameter is given as

$$CP = \frac{Q}{\omega D^3} - \frac{L}{12.7D} \qquad (3)$$

where L is the distance between the wafer and jet. Figure 11 exhibits the percentage coverage as a function of CP for wafers #2-5. The figure shows that good coverage is attainable only at $CP=0.09$ and 0.19. Figure 12 shows the coverage for wafers #6-8 and indicates that good coverage is attainable at approximately the same values for CP.

4 NUMERICAL SIMULATION

A transient three dimensional turbulent prediction procedure for the computation of swirling and recirculating flows is being used to predict the flowfield around the rotating wafers. The trajectories of the water droplets, issued from a peripheral jet, in the flowfield between two rotating disks is also being predicted to investigate the effect of different governing parameters on effectiveness of the rinsing process of the wafers. Brief description of the developed prediction procedure is presented below.

4.1 The Governing Equation

The generalized Reynolds average transport equation for incompressible flow in cylindrical coordinates can be expressed as[2]

$$\frac{\partial}{\partial t}(\rho\phi) + \frac{1}{r}\frac{\partial}{\partial r}(\rho r u\phi) + \frac{1}{r}\frac{\partial}{\partial \theta}(\rho v\phi) + \frac{\partial}{\partial z}(\rho w\phi)$$
$$= \frac{1}{r}\frac{\partial}{\partial r}\left(r\Gamma_\phi\frac{\partial\phi}{\partial r}\right) + \frac{1}{r}\frac{\partial}{\partial \theta}\left(\Gamma_\phi\frac{1}{r}\frac{\partial\phi}{\partial \theta}\right)$$

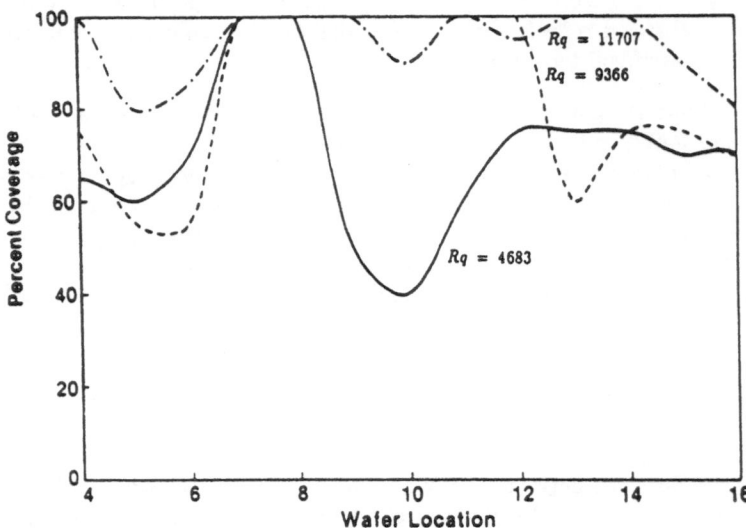

Figure 8 Percentage of water-film coverage as a function of the wafer number for an intermediate speed and three flow rates.

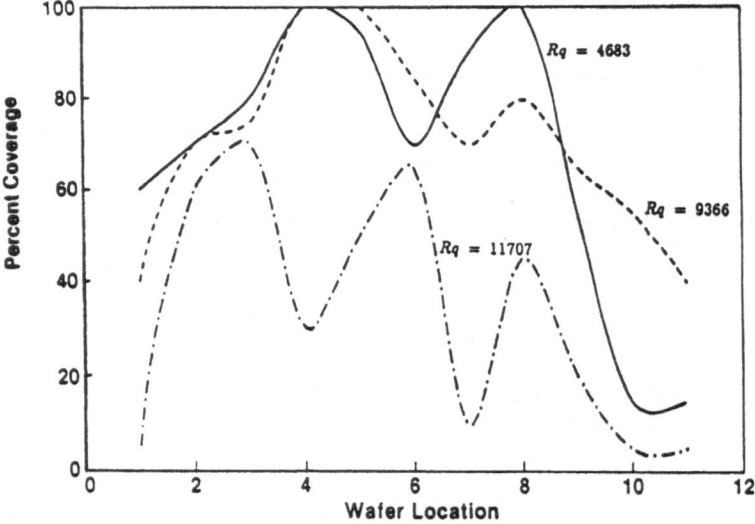

Figure 9 Percentage of water-film coverage as a function of the wafer number for a low speed and three flow rates.

$$+\frac{\partial}{\partial z}\left(\Gamma_\phi \frac{\partial \phi}{\partial z}\right) + S_\phi \tag{4}$$

where t is the time, ρ is the density and ϕ stands for any of the dependent variables u, v, w, k, and ϵ. For the continuity equation $\phi = 1$. The time-averaged velocity components u, v, and w are in the radial, azimuthal, and axial (r, θ, and z) directions, respectively. The turbulent viscosity is calculated from the k-ϵ turbulence model

$$\mu_t = \rho C_\mu k^2 / \epsilon \tag{5}$$
$$\mu_{\text{eff}} = \mu_t + \mu \tag{6}$$

where μ is the physical viscosity. The quantities k and ϵ ($= k^{3/2}/\ell$ where ℓ is a length scale) are the turbulent energy and its dissipation rate respectively, both of these being obtained from the solution of their respective transport equations. The transport equations for each of the variables differ primarily in their exchange coefficient Γ_ϕ and final source terms S_ϕ, as indicated in Table I. The constants appearing in the table are given the usual recommended values: $C_D = 1.0$, $C_\mu = 0.09$, $C_1 = 1.44$, and $C_2 = 1.92$. These values have been used in a wide variety of turbulent flow situations and have exhibited good predictive capability[3].

4.2   The Grid System

A three dimensional staggered mesh in cylindrical coordinate system is used where pressure and other scalars are placed at the center of the cell and the velocity components are placed normal to the respective surfaces of the cell. The flow domain is surrounded by a fictitious layer of cells on all sides for easy simulation of the boundary conditions. Figure 13 illustrates the total arrangement of the cylindrical coordinate mesh in $r\theta$ and rz planes.

4.3   The Boundary Conditions

At the outflow boundary, zero normal gradient condition is imposed. At the wall, rigid, no slip, conditions are imposed. At the plane of symmetry between two rotating disks, rigid free slip boundary condition is imposed. At the rotating disk, no slip condition is imposed.

4.4   The Solution Procedure

The finite difference approximations representing the partial differential equations may be written explicitly as

$$\phi = \phi^n + \Delta t[\cdots] \tag{7}$$

where $\phi = u, v, w, k$, and ϵ. Subscripts n and (blank) are used to denote values at time-level t and $t + \Delta t$, respectively. Convection, diffusion and source terms occur in the parentheses of the right hand side of Equation (7). For the turbulence quantities k and ϵ, source term is treated in the manner recommended for always positive variables[4]. The second order upwind differencing scheme is used for the convective terms in the momentum equations. In a recent study, by the authors[5], of seven finite differencing schemes for the advection terms, the second order scheme was shown to be accurate,

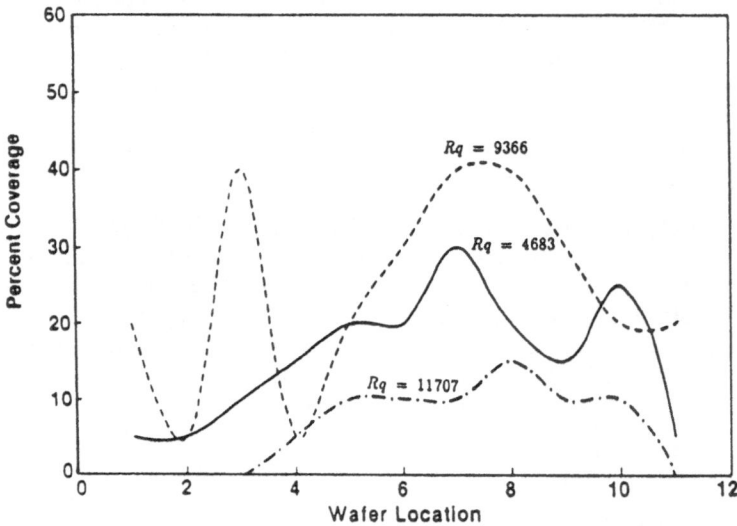

Figure 10 Percentage of water-film coverage as a function of the wafer number for a high speed and three flow rates.

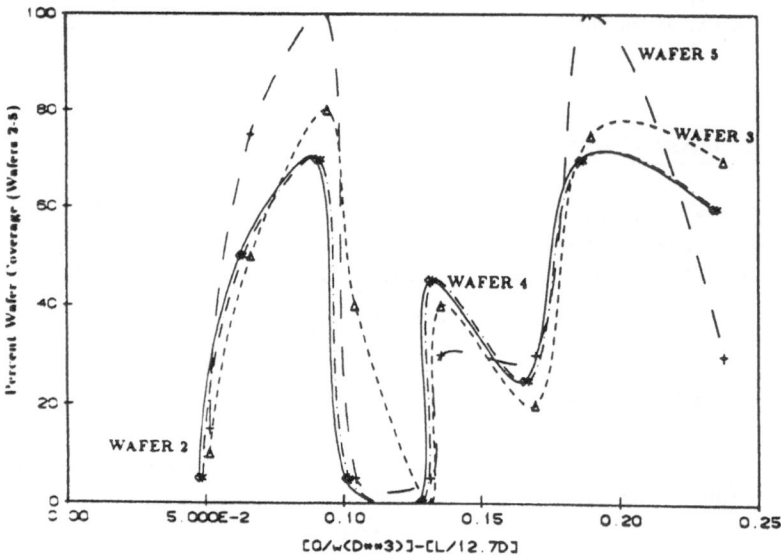

Figure 11 Percentage of water-film coverage as a function of the coverage parameter CP for multiple wafers #2-5.

Table 1. Exchange coefficients and source terms for different variables in the general transport equation.

ϕ	Γ_ϕ	S_ϕ
1	0	0
u	μ_{eff}	$\dfrac{\rho v^2}{r} - \dfrac{\partial p}{\partial r} + \rho g_r + \dfrac{1}{r}\dfrac{\partial}{\partial r}\left(r\mu_{\text{eff}}\dfrac{\partial u}{\partial r}\right) + \dfrac{1}{r}\dfrac{\partial}{\partial \theta}$ $\left[r\mu_{\text{eff}}\dfrac{\partial(v/r)}{\partial r}\right] - 2\dfrac{\mu_{\text{eff}}}{r}\left(\dfrac{1}{r}\dfrac{\partial v}{\partial \theta} + \dfrac{u}{r}\right) + \dfrac{\partial}{\partial z}\left(\mu_{\text{eff}}\dfrac{\partial w}{\partial r}\right)$
v	μ_{eff}	$-\dfrac{\rho u v}{r} - \dfrac{1}{r}\dfrac{\partial p}{\partial \theta} + \rho g_\theta + \dfrac{1}{r}\dfrac{\partial}{\partial r}\left[r\mu_{\text{eff}}\left(\dfrac{1}{r}\dfrac{\partial u}{\partial \theta} - \dfrac{v}{r}\right)\right]$ $+\dfrac{\mu_{\text{eff}}}{r}\left[r\dfrac{\partial(v/r)}{\partial r} + \dfrac{1}{r}\dfrac{\partial u}{\partial \theta}\right]$ $+\dfrac{1}{r}\dfrac{\partial}{\partial \theta}\left[\mu_{\text{eff}}\left(\dfrac{1}{r}\dfrac{\partial v}{\partial \theta} + \dfrac{2u}{r}\right)\right] + \dfrac{\partial}{\partial z}\left(\mu_{\text{eff}}\dfrac{1}{r}\dfrac{\partial w}{\partial \theta}\right)$
w	μ_{eff}	$-\dfrac{\partial p}{\partial z} + \rho g_z + \dfrac{1}{r}\dfrac{\partial}{\partial r}\left(r\mu_{\text{eff}}\dfrac{\partial u}{\partial z}\right) + \dfrac{1}{r}\dfrac{\partial}{\partial \theta}\left(\mu_{\text{eff}}\dfrac{\partial v}{\partial z}\right)$ $+\dfrac{\partial}{\partial z}\left(\mu_{\text{eff}}\dfrac{\partial w}{\partial z}\right)$
k	$\mu_{\text{eff}}/\sigma_k$	$G_k - C_D \rho \epsilon$
ϵ	$\mu_{\text{eff}}/\sigma_\epsilon$	$(C_1 G_k \epsilon - C_2 \rho \epsilon^2)/k$

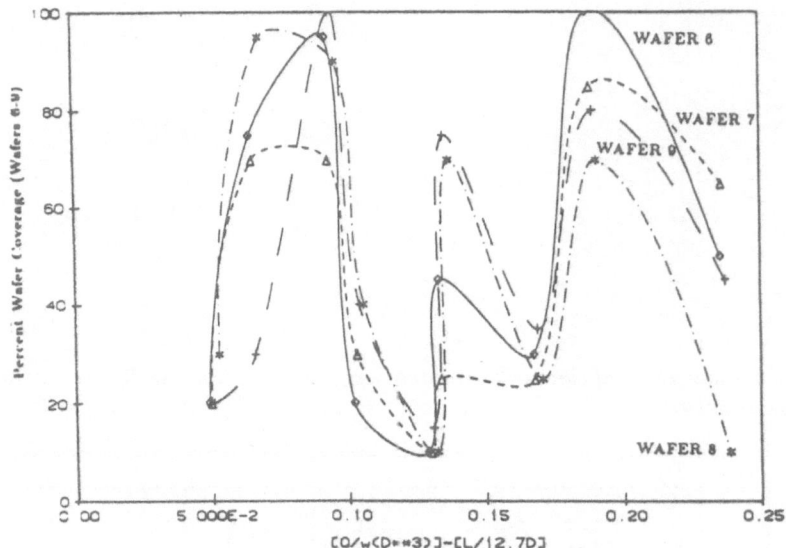

Figure 12   Percentage of water-film coverage as a function of the coverage parameter CP for multiple wafers #6-8.

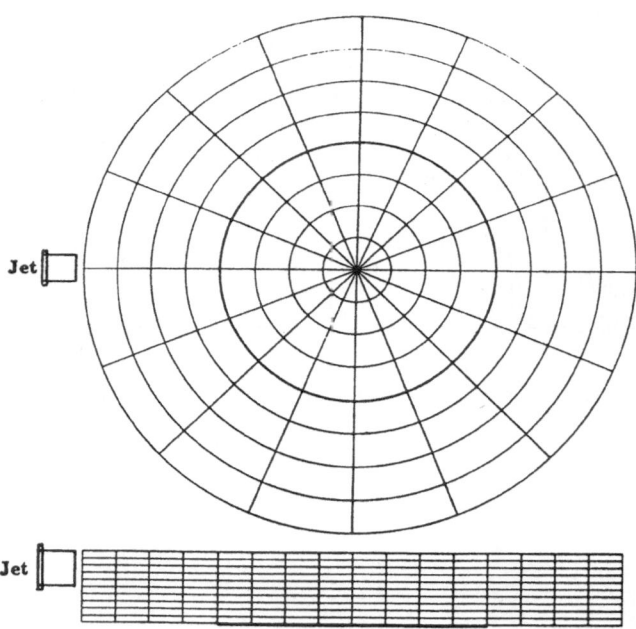

Figure 13   Grid schematic for numerical simulation of the flowfield between two rotating disks. . . . .

ρg_r, ρg_θ, and ρg_z are the components of body forces in radial, azimuthal, and axial (r, θ, and z) directions, respectively, and $C_D = 1.0$, $C_1 = 1.44$, $C_2 = 1.92$, $\sigma_k = 1.0$, $\sigma_\epsilon = 1.3$.

$$G_k = \mu_{\text{eff}} \left[2\left(\frac{\partial u}{\partial r}\right)^2 + 2\left(\frac{1}{r}\frac{\partial v}{\partial \theta} + \frac{u}{r}\right)^2 + 2\left(\frac{\partial w}{\partial z}\right)^2 \right.$$
$$\left. + \left(\frac{1}{r}\frac{\partial u}{\partial \theta} + \frac{\partial v}{\partial r} - \frac{v}{r}\right)^2 + \left(\frac{\partial v}{\partial z} + \frac{1}{r}\frac{\partial w}{\partial \theta}\right)^2 + \left(\frac{\partial w}{\partial r} + \frac{\partial u}{\partial z}\right)^2 \right]$$

produces less over- or under-shoot in the solution, uses less computer time and is easier to implement.

For each calculation cycle (time-step), the three components of velocity (initial guesses), k and ϵ at all internal points are computed from their respective transport equations. The cell pressure and velocities are then adjusted iteratively to satisfy the continuity equation. When convergence is achieved, the pressure and velocity values will be at the advanced time-level and can be used to start the calculation of the next time-step.

4.5 Prediction of the Jet Trajectories

The velocity components of a particle or droplet in a moving fluid can be computed from the three ordinary differential equations in cylindrical coordinates given below which take account of the drag, viscous, and gravitational forces.

$$\frac{du_p}{dt} = \frac{(3\overline{\rho}/4d)C_d q_r(u_f - u_p) - (1 - \overline{\rho})(g \sin \alpha)\sin \theta}{(1 + 0.5\overline{\rho})} + \frac{v_p^2}{r} \tag{8}$$

$$\frac{dv_p}{dt} = \frac{(3\overline{\rho}/4d)C_d q_r(v_f - v_p) - (1 - \overline{\rho})(g \sin \alpha)\cos \theta}{(1 + 0.5\overline{\rho})} - \frac{u_p v_p}{r} \tag{9}$$

$$\frac{dw_p}{dt} = \frac{(3\overline{\rho}/4d)C_d q_r(w_f - w_p) - (1 - \overline{\rho})(g \cos \alpha)}{(1 + 0.5\overline{\rho})} \tag{10}$$

where,

t = time

u_p = radial component of the droplet velocity

v_p = azimuthal component of the droplet velocity

w_p = axial component of the droplet velocity

u_f = radial component of the fluid velocity

v_f = azimuthal component of the fluid velocity

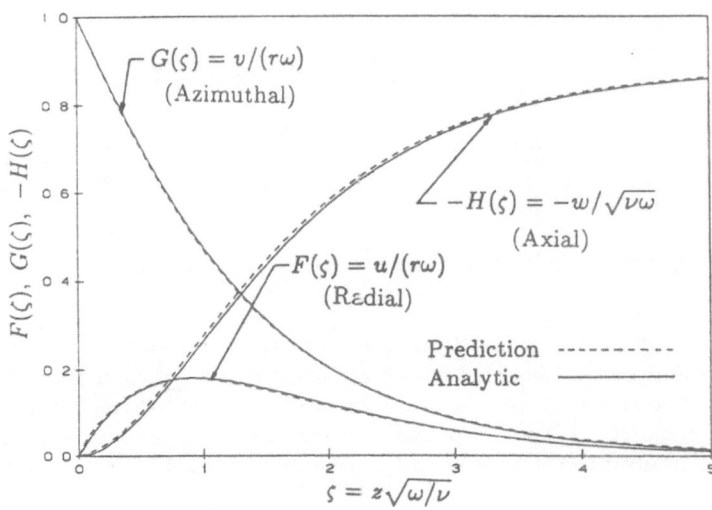

Figure 14 Predicted and analytical velocity profiles for the rotating disk problem.

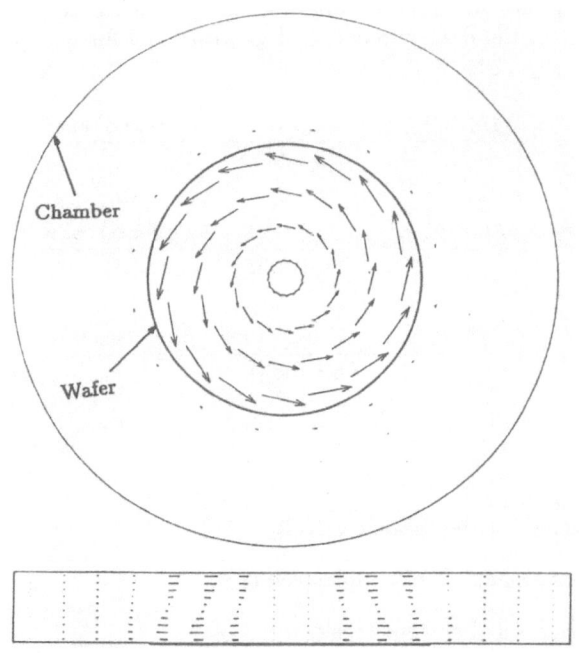

Figure 15   Velocity vector plots in the $r\theta$ and rz planes.

w_f = axial component of the fluid velocity

$\bar{\rho}$ = ratio of the fluid density to the droplet density

d = diameter of the droplet

C_d = drag coefficient

q_r = relative velocity of the droplet with respect to the fluid

α = angle between the vertical and the axial direction

A fourth-order Runge-Kutta technique is used to solve the above equations for the velocity and position of the particle after every time step. The water droplets, issued from a peripheral jet, are assigned a certain distribution of size and velocity consistent with the nozzle type and size and introduced into the flowfield between the two rotating disks. The trajectories of the droplets are then computed assuming each droplet as an individual particle.

4.6   Computational Results and Discussion

Initially the code is used to predict the laminar flowfield near an infinite rotating disk and the predicted velocity profiles are compared with the analytic profiles in Figure 14 to verify the accuracy of the scheme[6]. The figure shows that the agreement is excellent. The prediction procedure is then applied to predict the flowfield between two rotating

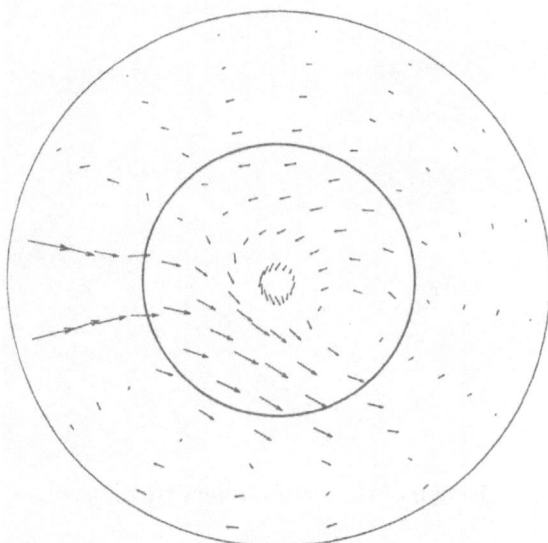

Figure 16   Velocity vector plot in the $r\theta$ plane with the lateral jet, of same fluid, activated.

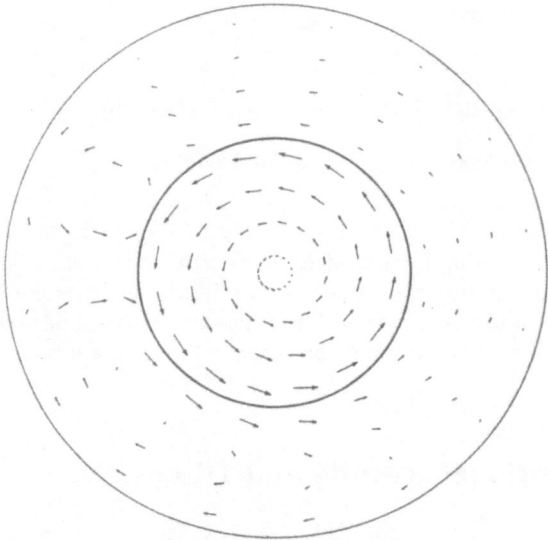

Figure 17   Velocity vector plots in the $r\theta$ plane below the jet (nearer to the disk).

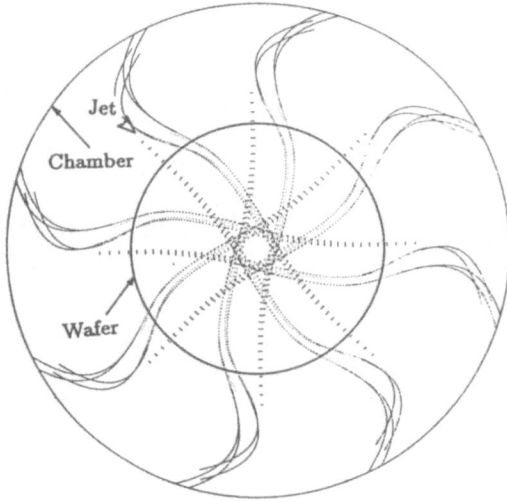

Figure 18   Projection of the water droplet trajectories on the $r\theta$ plane.

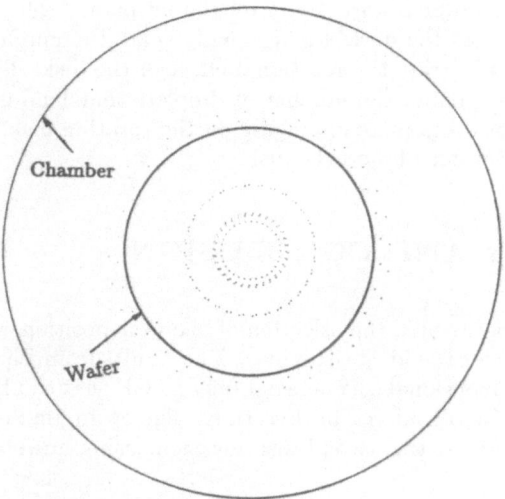

Figure 19   Plot of water droplets on the disk.

disks inside a cylindrical chamber. The radii of the disks is 0.06 m and of the chamber is 0.12 m. The axial distance of separation between the two disks is 0.004 m. The rotational speed of the disks is 500 rpm. The flow Reynolds number based on the tip radius of the disks and the azimuthal velocity at that radius is 1.2×10^4. The critical Reynolds number for transition to turbulence for this type of flow[7] is 3×10^5. The considered flowfield is therefore, laminar. The computation is done by deactivating the turbulence model in the code. Figure 15 exhibits the velocity vectors in $r\theta$ and rz planes. Prediction is also done for the case when a peripheral jet of the same fluid is introduced radially into the flow field which simulates the drying process. Figure 16 illustrates the flow field in the $r\theta$ plane containing the jet. It shows that the jet does not penetrate the core region in the center of the cylindrical chamber. Figure 17 shows a similar $r\theta$ plot below the plane of the jet. The two figures show the recirculation zones around and under the jet. The trajectories of the water droplets issued from a peripheral jet into the flowfield between the two rotating disks are then predicted. Uniform size (200 micron diameter) and velocity (5 m/s) of the droplets at the jet exit are assumed. Only 10 droplets, distributed around the circumference of the jet, are used. Figure 18 shows the projections of the droplet trajectories on the $r\theta$ plane. The figure demonstrates that the jet travels across the disk. Eight equally spaced circumferential locations for the jet are plotted to illustrate the rotational effect of the disk. The main objective of this computation is to predict the number of droplets that land on the wafer surface. Figure 19 shows a plot of droplets impinging on the rotating disk. These droplets will form the beginning of water-film on the disk.

5   SUMMARY AND CONCLUSION

Based on Phase I experiments, the selection of the best position, rotational speed and flow rate can be made for the ideal situations. The results from Phase I provide a correlation using the nondimensional coverage parameters CP that can be used to extend the data to other values not considered in this study. The optimum disk coverage in Phase I by a water film occurs at the second disk location, using an intermediate rotational speed and flow rate.

Phase II considers the actual geometry of the rinser/dryer. The experiments demonstrated the effectiveness of the newly introduced flow visualization scheme in investigating the flow pattern in extremely difficult geometries and applications. Results indicate that an intermediate rotational speed and a high flow rate will provide the best coverage using the second nozzle. The introduced coverage parameter for the shielded wafers, which represents the functional relationship between the film coverage and the operating parameters, correlates well with the data and can be used to predict the coverage for any wafer.

The effectiveness of numerical simulation in modeling the jet and the surrounding flowfield has been demonstrated under simplified assumptions. The model can be used to investigate the film coverage using the actual geometry of the application. However, experimental data for the size and velocity distribution of the water droplets at the jet exit for a particular type of nozzle, are essential for the numerical investigation. Data for the utilized nozzles in the experiments are not available.

ACKNOWLEDGEMENT

This Work was supported by IBM Corporation (Essex Junction, Vermont).

REFERENCES

[1] A. A. Busnaina, J. Edler, G. Gale, and F. W. Kern Jr., A fluid dynamic study of microcontaminant particle removal from silicon wafers, in *"Proceedings, IES 33rd Annual Technical Meeting,"* San Jose, 319–327, 1987.

[2] A. K. Gupta and D. G. Lilley, *"Flowfield Modeling and Diagnostics,"* Abacus Press, Tunbridge, Wells, U.K., 1984.

[3] B. E. Launder and D. B. Spalding, Numerical computation of turbulent flows, *Computer Methods in Applied Mechanics and Engineering,* 3, 269–289, (1974).

[4] S. V. Patankar, *"Numerical Heat Transfer and Fluid Flow,"* Hemisphere McGraw-Hill, New York, 1980.

[5] M. A. R. Sharif and A. A. Busnaina, Assessment of finite difference approximation for the advection terms in the simulation of practical flow problems, *J. Computational Physics,* 74, 143–176, (1988).

[6] F. White, *"Viscous Fluid Flow,"* 164–170, McGraw-Hill, New York, 1974.

[7] H. Schilichting, *"Boundary Layer Theory,"* 6th Edition, 174–176, McGraw-Hill, New York, 1968.

PARTICLE REMOVAL FROM SEMICONDUCTOR WAFERS USING CLEANING SOLVENTS

V. B. Menon, L. D. Michaels, R. P. Donovan, and V. L. Debler

Research Triangle Institute
P.O. Box 12194
Research Triangle Park, NC 27709

M.B. Ranade

Particle Technology, Inc.
P.O. Box 13851
Research Triangle Park, NC 27709

The overall objectives of this study were to evaluate the performance of solvents and cleaning devices for removing particles from silicon wafers, determine the limitations of wet cleaning practices used in today's semiconductor industry and suggest improvements based upon fundamental measurements of particle adhesion to surfaces. Specifically, the cleaning efficiencies of selected solvents were measured in an ultrasonic force field for particulate contaminants of well-defined sizes and properties. The influence of wafer surface properties on particle removal and the merits of the ultrasonic cleaning system were also investigated. Polar solvents such as water and ethanol-acetone (1:1 mixture) were found to be far more superior than the Freons in removing particles of widely differing compositions. Also, the force required to clean wafers with surface oxide layers was observed to be lower than that for bare silicon wafers.

INTRODUCTION

There are numerous cleaning procedures that are currently being employed for the removal of contaminants from wafers. Most of these cleaning sequences are designed for removing soluble contaminants, and not particles. In fact, some chemical cleans such as the RCA method may occasionally increase the level of particulate contamination because they are designed to remove nonparticulate contaminants instead.[1]

Contamination sources can include chemicals (e.g., acids, bases, organic solvents, photoresist materials, and water), tools and equipment (e.g., ovens, jigs, work surfaces, vacuum systems, cleaning stations, and moisture-laden compressed air lines), and residues (e.g., residual gases on the components, residual lubricants on the parts, lapping residues, mold-releasing agents, and metal and nonmetallic residues). In addition, the high-purity water used in manufacturing can become a source of contamination if it is not carefully monitored for resistivity, trace metals, ion impurities, bacteria, particles, organics, dissolved solids etc. People are a major source of contaminants, including skin flakes, hair, cosmetics, spittle, and clothing.[2]

Forces between solids are predominantly of an attractive nature and cause adhesion of particles to each other and to surfaces. These forces become increasingly significant for fine particles since Van der Waals attraction depends on the first power of particle diameter while mechanical removal forces vary with the third power of particle diameter.[3,4] The force of adhesion, which is governed by the interatomic or intermolecular forces at the interface, is dependent on surface roughness, number of polar sites on the surface, particle size and polarity, chemical nature of the particle, and nature of the medium. Table I shows the relative adhesional forces as a function of particle size for glass spheres on bare polished and oxidized silicon wafers. The procedure for estimating these values is discussed in Results and Discussion.

The forces of adhesion between particle contaminants and surfaces are generally two to three times lower in liquids than they are in air[5]. Hence, it is advantageous to use a wet cleaning process for particle removal. The factors affecting the selection of a cleaning agent include effectiveness, compatibility, flammability, toxicity, and disposability[2]. In choosing a cleaning technique it is important to have a thorough knowledge of the substrate-solvent interactions, and the nature and size of the particulate contaminant. It is also important to ensure that the wet cleaning procedure has high cleaning efficiencies, otherwise residual liquid in the region of particle-surface contact will form liquid bridges and enhance the force of adhesion upon removal of the wafer from the cleaning solvent. The effectiveness of a cleaning process thus varies depending on the technique used, the particle size, and the nature of the surface.[6]

Some of the cleaning techniques commonly used are pressure spraying, mechanical agitation with a solvent or detergent solution, vapor degreasing, ultraviolet/ozone cleaning,[7] shear stress cleaning,[8] megasonics and ultrasonics.[9,10] Stowers[10] has reported typical particle removal efficiencies for optical surfaces. Such numbers, although useful, convey little basic information since the forces holding particles at a surface are specific to each system. For example, a given solvent/cleaning

Table I. Relative Adhesional Forces as a Function of Particle Size ($Z_0 = 7$ Å)

Particle Diameter (μm) Glass Microspheres	Substrate	Medium	F_{AD} (mdynes)
1	Si	Water	0.4
	SiO$_2$	Water	0.1
5	Si	Water	2.1
	SiO$_2$	Water	0.5
10	Si	Water	4.0
	SiO$_2$	Water	1.0

method may remove silica particles of 5 μm size from a bare silicon wafer while it does not dislodge a polymeric particle of the same size from the same surface. Or, particles of one size may be cleaned from a bare silicon wafer surface by a solvent which is ineffective for removing those same particles from an oxidized wafer surface. While a large number of equipment options and solvents are becoming available for cleaning wafers, standard methods to evaluate their performance have not yet been developed.[11] Hence, the prediction of the cleaning efficiency of any given solvent/particle/wafer system cannot be made in advance of actual measurements. A systematic evaluation of commercial solvents and decontamination devices using well-defined particle/wafer systems is needed. Such an evaluation will enable a semiconductor manufacturer to select a cleaning process for a specific application with minimum of in-house experimentation.

Therefore, the objectives of the present study were to determine the cleaning efficiencies of selected solvents, for bare polished wafers and oxidized wafers using an ultrasonic cleaning technique. This paper discusses the results of our studies using the ultrasonic cleaning technique for the system of glass microspheres on bare polished and oxidized silicon wafers in four solvents, viz., DI water, a 1:1 volume mixture of ethanol and acetone, Freon-TMS, and Freon-TF.

APPARATUS AND PROCEDURE

The ultrasonic device used for the experiments is shown schematically in Figure 1. The unit consists of a flow cell made of glass which is placed within a cup horn. The horn consists of a polycarbonate cup at the bottom of which is the vibrating surface made of titanium. The cup horn has inlet and outlet ports through which cooling water can be circulated around the inner flow cell. The contaminated wafer was placed at the base of the flow cell. The power to the ultrasonic horn is provided by an amplifier with multiple settings. The amplifier can be adjusted to control and vary the energy imparted to the surface of the horn. The cleaning solvent was circulated through the flow cell in which the contaminated wafer was placed. The solvent circulation pump (CHEM-FEED, Cole Parmer) had a flowrate of 1.4 cm^3/sec which was adjusted such that it was insufficient for removal of particles from the wafer surface, but sufficient for carrying the particles away from the wafer once they had been removed by the ultrasonic field. The solvent passed through a Teflon membrane filter (Acro50 Hydrophobic; 0.2 μm) as it entered and left the cell. The cooling water for the cup horn was provided via a Millipore DI system and a Gelman 0.2 μm filter.

The glass microspheres were obtained in the form of dry powder, from Duke Scientific, Palo Alto, California, in two sizes, 1 μm and 5 μm. The particles, as supplied, had a large standard deviation. Hence, they were further classified in settling columns to get narrower cuts. The powders were dispersed dry, in air, on the surface of the wafer in such a fashion as to minimize the aggregation of particles. To achieve this, the method and apparatus suggested by Zimon[12] was employed. This device is shown schematically in Figure 2. A 2-inch wafer was placed on a holder contained within a "clean" dispersion box. The sample particle system was then weighed and placed on an elastic membrane beneath the sample holder. The amount of powder required had to be optimized for each system. The box was then closed and the membrane vibrated to disperse the powder. The powder cloud was allowed to settle onto the wafer. It was possible to deposit single particles on wafers using this simple apparatus. The deposit was examined under an optical microscope (Nikon Alphaphot) and an initial count of the number of particles was made.

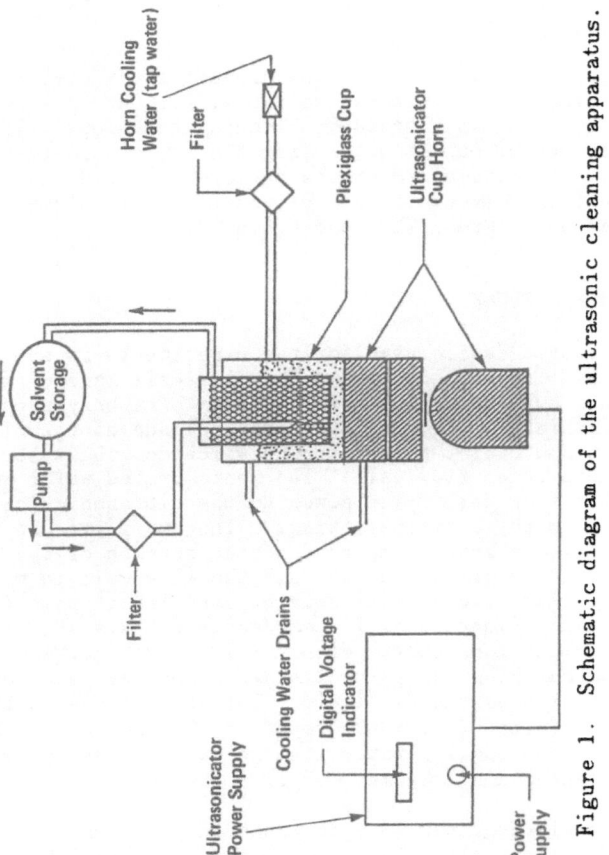

Figure 1. Schematic diagram of the ultrasonic cleaning apparatus.

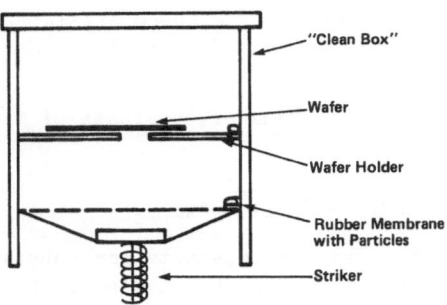

Figure 2. Schematic diagram of particle deposition apparatus.

The National Institute of Occupational Safety and Health (NIOSH) method 7400[13] developed for the systematic optical counting and sizing of asbestos fibers was modified to this purpose. Employing the counting rules adapted from this method, a counting precision of 10 to 12 percent is expected. Periodic duplicate measurements showed precision to be within this range for the current work. All sample handling and preparation was conducted in a Class 100 clean room. Recently, the entire experimental setup was moved to a Class 10 clean room at the Microelectronics Center of North Carolina and the particle counting procedure was speeded up by integrating the experiment with a wafer surface scanner (Aeronca WIS-150). Four solvents, viz., DI water, Freon-TF®, Freon-TMS® (Freon® with 6 percent methanol) and ethanol/acetone (50 percent acetone in ethanol), were used during this investigation.

The experimental procedure involved the application of a known amount of ultrasonic energy (by adjusting the power setting on the sonifier) to a wafer in the flow cell where it was immersed in the chosen solvent. From preliminary experiments it was determined that the application of the ultrasonic field for more than 2 minutes at any given power setting did not result in increased efficiencies. Hence, for all experiments, the ultrasonic field was applied for two minutes. After each power increment, the number of particles remaining on the wafer was counted. From the count before and after the cleaning process the percentage of particles removed from the wafer at any power setting was calculated. The wafer was then returned to the flow cell and vibrated at a higher power setting. This procedure was repeated until removal efficiencies in excess of 75 percent were obtained.

In order to convert the value of a power setting to acceleration units it is necessary to calibrate the ultrasonic device. Calibration of this and similar instruments has been conducted by Berliner.[14] The calibration values reported by Berliner were used in earlier work performed at Research Triangle Institute[6] and were found to give a precision in the measurement of particle accelerations of 5 percent. Consequently, these values were used in the present study. Correction calibrations were performed to account for attenuation of the ultrasonic vibration through the cooling liquid in the cup horn and the flow cell.

RESULTS AND DISCUSSION

The theories for molecular interactions are based on van der Waals dispersion interactions. Atoms in the bodies are instantaneous dipoles and the dispersion interaction between these dipoles and the induced dipoles in neighboring atoms is summed over all atoms. It is conveniently represented by a Hamaker constant, A.

The relations between the Hamaker constants of two dissimilar materials may be represented by:

$$A_{12} = \sqrt{A_{11} \cdot A_{22}} \; , \tag{1}$$

where A_{11} and A_{22} are the Hamaker constants for substances '1' and '2' respectively. In the presence of a medium denoted by '3', the net interaction is given by:

$$A_{132} = A_{12} + A_{33} - A_{13} - A_{23} \; . \tag{2}$$

The force of adhesion may be calculated from the Hamaker constant and the geometry of the materials involved in the interactions. The force of adhesion between two spheres is given by:

$$F_{AD} = \frac{A_{132}d_p}{24 \, Z_o^2} \; . \tag{3}$$

where F_{AD} is the force of adhesion, d_p is the diameter of the particle and Z_o is the distance of separation between the two spheres.

Similarly, F_{AD} for the interaction between a sphere and a plate this relation is given by Hamaker[15] as:

$$F_{AD} = \frac{A_{132}d_p}{12 \, Z_o^2} \; . \tag{4}$$

It should be noted that these theoretical computations are subject to several assumptions. Even experimental determination is not conclusive since the distance of separation Z_o is not measurable and current estimates range from 4 to 125 Å units.[5,12] Table II shows the calculation of F_{AD} for the four liquids, two wafer surfaces, and two glass microsphere diameters addressed in this report using a separation of 7.0 Å. This value of Z_o is within the range reported by Van den Tempel[16] for silicon substrates with and without surface oxide layers. Effects of microsurface roughness and deformation at the point of contact also make the calculation of 'A' from experimental measurements ambiguous. A thin surface layer of oxide can affect the adhesion forces acting between the adhering particle and that surface. The contribution of the surface layer to the adhesional force has been shown to be inconsequential for thicknesses in excess of Z_o[5]. Hence, it is appropriate to consider the case of the oxidized wafer (0.5 μm oxide layer) as equivalent to the case of a solid plate of silicon dioxide.

Deryagin[17] first used ultrasonic vibrations to improve and extend the range of adhesion forces that could be measured. By varying the power input to an ultrasonic generator (in this case, operating at 20 kHz), a wide range of surface accelerations, up to 10^6 m/sec^2, may be obtained. When driven at a known amplitude, the magnitude of surface acceleration required to remove a given fraction of the adhering particles may be measured. In turn, the adhesion strength may be calculated according to Zimon[12] as:

$$a = 4 \, \pi^2 \times 1^2 \times W \tag{5}$$

Table II. Force of Adhesion for Test Systems

Liquid	Hamaker Constant (A_{132})* (ergs x 10^{13})		F_{AD}** (mdynes)	
	Bare Surface	Oxidized Surface (0.5 μm)	Bare Surface	Oxidized Surface (0.5 μm)
5 μm Glass† Microspheres				
Water	2.42	0.63	2.1	0.5
Ethanol-Acetone	3.58	1.21	3.0	1.0
Freon-TMS	5.89	2.49	5.0	2.1
Freon-TF	5.99	2.50	5.1	2.1
1 μm Glass Microspheres				
Water	2.42	0.63	0.4	0.1
Ethanol-Acetone	3.58	1.21	0.6	0.2
Freon-TMS	5.89	2.49	1.0	0.4
Freon-TF	5.99	2.50	1.0	0.4

* Hamaker constant for the system calculated using equation 2

** F_{AD} calculated using equation 4 and Z_o = 7.0 Å

†Density of glass, 2.65 g/cm^3

where 'a' is the acceleration, l is the frequency of vibration, W is the amplitude of vibration. The force required to remove contaminant particles can then be calculated according to the relationship:

Force = mass x acceleration

The experimental results for the systems involving glass microspheres on bare polished wafers and oxidized wafers in deionized water, ethanol-acetone, Freon-TMS and Freon-TF are shown in Figures 3 through 6. In these figures, the force introduced by the ultrasonic horn is expressed in relative units. This value of the force was obtained by taking the logarithm of 1,000 times the voltage reading on the power amplifier. Removal efficiencies for oxidized wafers in water were over 80 percent without the addition of ultrasonic energy for both 1 and 5 μm microspheres. Also, for 5 μm particles in Freon-TMS high removal efficiencies were achieved without sonication on the oxidized wafer. For these systems the lowest measurable value of power input (0.002 volts) has been assumed to be the force corresponding to 50 percent cleaning efficiency. In all cases except that of Freon-TF, the oxidized substrate was easier to clean than the bare polished wafer surface. This general trend is in agreement with that predicted based on van der Waals adhesional forces.

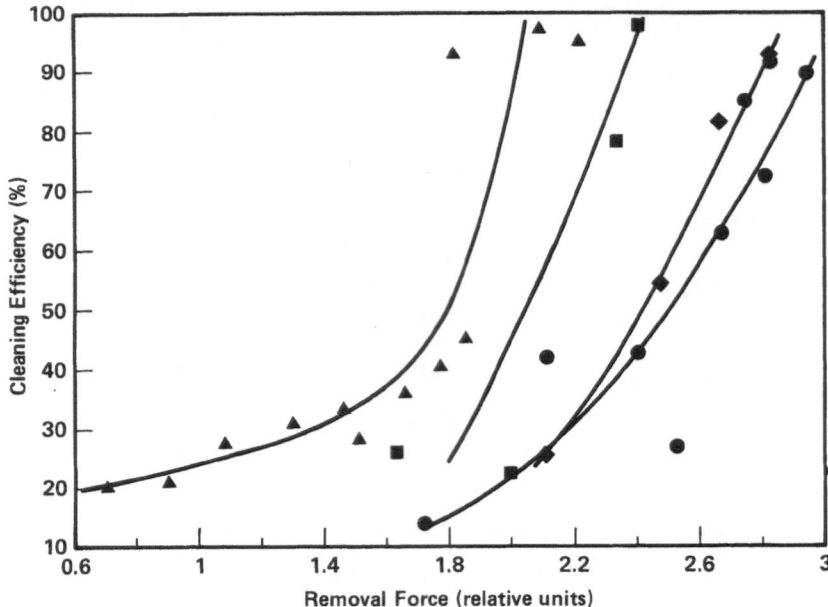

Figure 3. Comparison of the adhesion curves for 5 μm glass particles on a bare silicon wafer in different liquids.

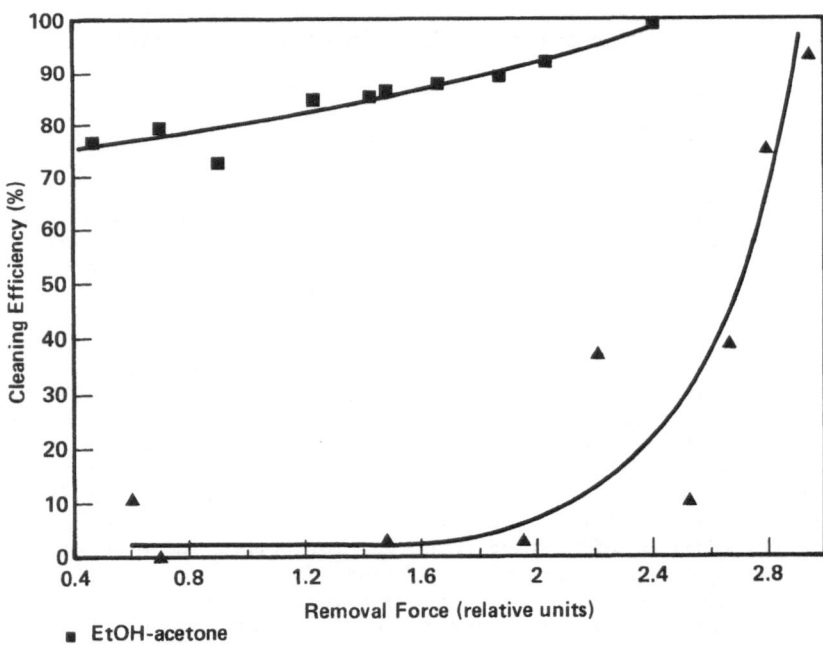

Figure 4. Force required to remove 5 μm glass particles from a silicon wafer with a 5000 Å oxide layer.

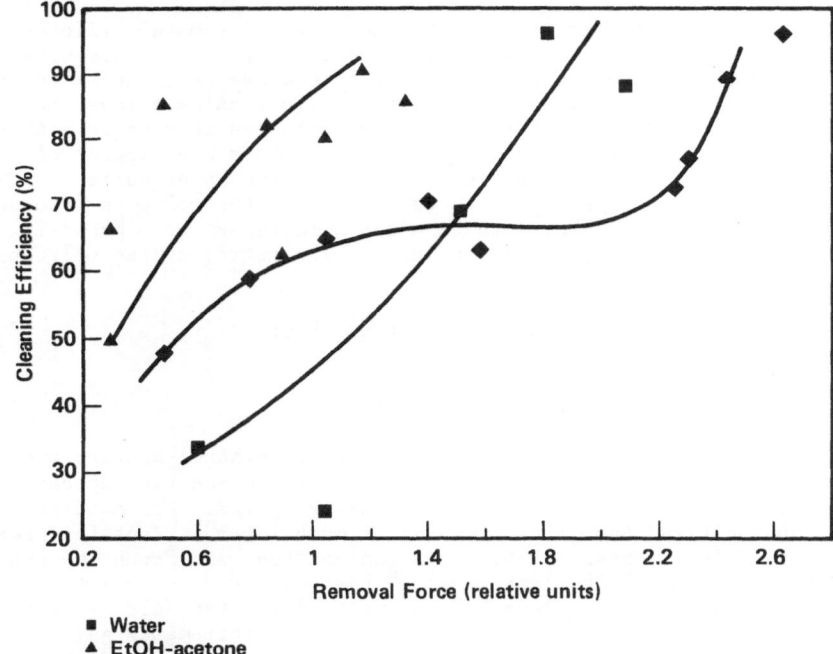

Water

▲ EtOH-acetone

◆ Freon-TMS

Figure 5. Force required to remove 1 μm glass particles from a bare silicon wafer.

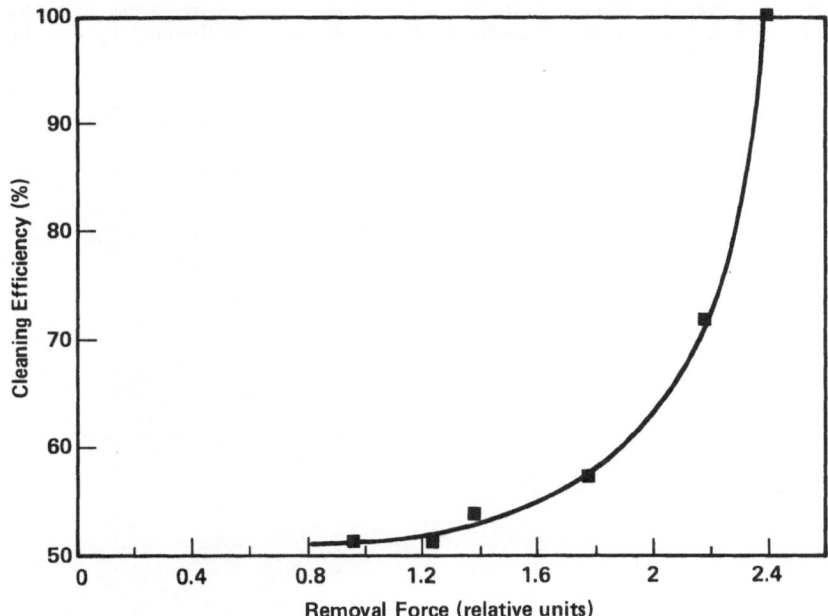

Figure 6. Force required to clean 1 μm glass microspheres from a wafer with a 5000 Å oxide layer using ethanol-acetone.

No single characteristic shape to the curve of removal efficiency versus force was noted. Consequently, direct comparison of cleaning efficiency becomes complicated. In order to compare experimental measurements of the force of adhesion with theoretical calculations, the force required to remove 50 percent of the particles is equated with the force of adhesion for that system.[12] Figure 7 shows a histogram of the force at 50 percent cleaning efficiency for the two wafer surfaces and four solvents. This figure permits the rating of the solvents in order of cleaning efficiency. For both bare and oxidized wafers, Freon-TF appears to be the least effective solvent. The rating of the solvents for bare wafers is as follows:

1. Ethanol-Acetone (best)
2. Water
3. Freon-TMS
4. Freon-TF (worst)

The differences in efficiency for water and ethanol-acetone are slight for the 5 μm/bare wafer system, while ethanol-acetone appears to be distinctly superior for the 1 μm/bare wafer system. For oxidized wafers, not only was the force of adhesion much lower (except for Freon-TF) but the effectiveness of water, ethanol-acetone and Freon-TMS appears to be about the same. Phillips et al.[18] have compared the relative merits of using Freon-TF, Freon-TMS and water for ultrasonic cleaning of stainless steel and aluminum parts. They observed that water gave the best and Freon-TF the worst cleaning efficiencies. This trend is in agreement with the results reported here.

In order to compute the theoretical force of adhesion, it is necessary to assume a value of particle-surface separation distance. Among the four solvents used in this study, the likelihood of having cavitation and electrostatic forces playing a role in particle detachment is least for Freon-TF. The relative cavitation intensities of various solvents has been reported by Niemczewski[19] and Freon-TF has a maximum cavitation intensity which less than 15 percent that of water. Hence, it can be

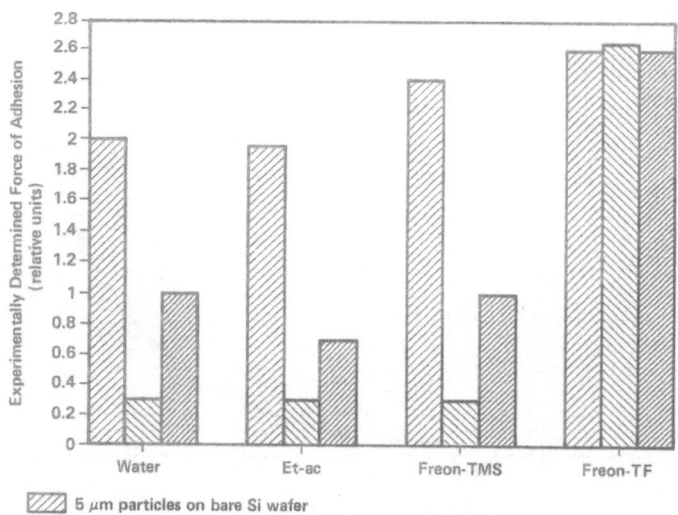

Figure 7. A comparison of cleaning efficiencies for the systems studied.

assumed that the ultrasonic force is directly acting against the van der
Waals force of adhesion. The force corresponding to 50 percent removal
of 5 μm glass microspheres from a bare wafer was, therefore, equated to
the van der Waals force in order to calculate the particle-surface separation distance. A value of 17.75 Å was obtained. This separation
distance was assumed to be the same for all systems. Table III shows the
force of detachment calculated from the mass of the individual particle
and the surface acceleration required to remove 50 percent of the particles. These values of force are compared to the theoretical force of
adhesion calculated from the system Hamaker constant. In most cases, the
force required to remove particles from the wafer surface was less than
that predicted from theory. The agreement of experiment with theory was
best for the freons and least for the two polar liquids. The experimental force was about two orders of magnitude lower for water and acetone.
This discrepancy can be attributed to the interplay of electrostatic,
surface tension (hydrophobic/hydrophilic), and cavitational forces along
with van der Waals attraction.

Electrostatic forces of repulsion between particles and surfaces
result as a consequence of ionic charges on the particle and/or substrate. For example the surface potential of silica particles in water
at a pH of 7.0 has been reported to be -30 mV.[20] The electrostatic force
of repulsion for the oxidized wafer/5 μm glass microsphere system, as
estimated from the Deryagin, Landau, Verwey, Overbeek (DLVO) theory, is

Table III. Comparison of Experimental and Theoretical (van der Waals)
Force

Liquid	Particle Diameter (μm)	Surface	F_{EXPT} (mDynes) Experimental	F_{AD} (Z_o=17.75Å) (mDynes) van der Waals
Water	5	Oxidized*	1.7×10^{-5}	0.05
	5	Bare	4.6×10^{-3}	0.33
	1	Bare	3.4×10^{-4}	0.08
Ethanol-Acetone	5	Oxidized	1.7×10^{-5}	0.16
	5	Bare	3.8×10^{-3}	0.47
	1	Bare	7.7×10^{-5}	0.12
Freon-TMS	5	Oxidized	1.7×10^{-5}	0.33
	5	Bare	0.50	0.78
	1	Bare	3.5×10^{-4}	0.16
Freon-TF	5	Oxidized	0.79	0.33
	5	Bare	0.79	0.79
	1	Bare	0.89	0.16

*5000 Å surface oxide layer deposited by chemical vapor deposition

approximately 0.04 mdynes. This value of the repulsive force is of the same order of magnitude as the van der Waals force of attraction. The net force of attraction is therefore considerably reduced. In addition, hydrophilic interactions between the wafer and the particles may also contribute to a lowering of the overall force of adhesion. Glass particles, owing to their hydrophilicity, will favor dispersion into the aqueous phase rather than attachment on a surface which is less hydrophilic.[21] A combination of electrostatic and hydrophilic interactions may explain the high efficiencies of cleaning that were observed for wafers in polar solvents even in the absence of ultrasonic energy.

CONCLUSIONS

The most significant finding of this study was that the force required for cleaning oxidized wafers in most solvents is much lower than that for bare polished wafers.

The force of adhesion of 5 μm particles on bare silicon wafers was consistently more than that of the 1 μm particles. This is in agreement with adhesion theory.

Water and ethanol-acetone appear to be better solvents than the Freons for particulate decontamination. Since ethanol and acetone cost much more than water, and the differences in cleaning efficiency between them are not significant, water would seem to be the solvent of choice.

ACKNOWLEDGMENT

This research was sponsored by the Semiconductor Research Corporation through a Manufacturing Science Contract with the Microelectronics Center of North Carolina.

REFERENCES

1. D. W. Cooper, Particulate-contamination and microelectronics manufacturing: An introduction, Aerosol Sci. Technol., 5, 287 (1986).

2. P. Viswanadham, Contamination control in disk-drive manufacturing--A quality and reliability perspective, Microcontamination, 5, 40 (1987).

3. D. W. Johnson, Evaluating low-particulate chemicals, Semiconductor Intl., 8, 168 (1985).

4. M. B. Ranade, Adhesion and removal of fine particles on surfaces, Aerosol Sci. Technol., 7, 161 (1987).

5. J. Visser, Adhesion of colloidal particles, Surface and Colloid Science, 8, 3-84 (1976).

6. M. B. Ranade, V. B. Menon, M. E. Mullins, and V. L. Debler, Adhesion and removal of particles: Effect of medium, in "Particles on Surfaces:1 Detection, Adhesion and Removal," K. L. Mittal, editor, pp. 179-191, Plenum Press, New York, 1988.

7. J. R. Vig, UV/ozone cleaning of surfaces, in "Treatise on Clean Surface Technology," K. L. Mittal, editor, Vol. 1, pp. 1-22, Plenum Press, New York, NY, 1987.

8.   R. P. Musselman and T. W. Yarborough, Shear stress cleaning surface departiculation, J. Environmental Sci., 30, 51 (1987).

9.   M. E. Mullins and M. B. Ranade, 1985, Effects of particle geometry on adhesion: Spherical and plate-like particles, paper presented at the Fine Particle Society Annual Meeting, Miami, Florida, April 1985.

10. I. F. Stowers, Advances in cleaning metal and glass surfaces to micron-level cleanliness, J. Vac. Sci. Technol., 15, 751 (1985).

11. G. Sielaff and N. Harder, A classification model for liquidborne particles in semiconductor process chemicals, Microcontamination, 4, 42 (1985).

12. A. D. Zimon, "Adhesion of Dust and Powder", 2nd Edition, Consultants Bureau, New York (1982).

13. National Institute of Occupational Safety and Health, Method # 7400, February 15, 1984.

14. S. Berliner, "Ultrasonic Power vs. Intensity in Sonication," Heat Systems Catalog, Ultrasonic, Inc. (1982).

15. H. C. Hamaker, London - van der Waals attraction between spherical particles, Physica, 4, 1088 (1937).

16. M. Van den Tempel, Interaction forces between condensed bodies in contact, Adv. Colloid Interface Sci., 3, 137 (1972).

17. B. V. Deryagin and A. D. Zimon, Adhesion of particles to planar surfaces, Kolloidny Zhurnal, 23, 544 (1961).

18. Q. T. Philips, G. J. Stone, and J. M. Baldwin, Parts cleaning: An evaluation of ultrasonic systems, in "Proceedings of the 30th Annual Technical Meeting of the Institute of Environmental Sciences," pp. 1-13, 1984.

19. B. Niemczewski, A comparison of ultrasonic cavitation intensity in liquids, Ultrasonics, 107-111 (May 1980).

20. A. C. Hall, S. H. Collins, and J. C. Melrose, The stability of aqueous wetting films in Athabasca tar sands, Soc. Pet. Eng. J., 23, 249 (1983).

21. J. N. Israelachvili, "Intermolecular and Surface Forces with Applications to Colloidal and Biological Systems," Academic Press, New York, 1985.

PARTICLE CONTRIBUTIONS OF THREE TYPES OF CLEANROOM JUMPSUITS

R. C. White and J. R. Weaver
Delco Electronics Corporation
P.O. Box 9005   M.S. Fab A
2150 E. Lincoln Road
Kokomo, IN   46904-9005

A study was performed to evaluate the relative performance of three "advanced technology" cleanroom jumpsuits for use in a Class 10 wafer fabrication facility. Simulations of actual operator activity were used in the evaluation, and performance was rated through airborne particle counting as well as witness wafer surface measurements. A dramatic reduction in particles was observed when operators wore garments laminated with an expanded PTFE membrane, without a significant loss of operator comfort.

INTRODUCTION

As the geometries of integrated circuits get smaller, more effort is needed to keep the environment around in-process wafers clean. This requires improved methods of protecting the product and environment from human contamination. One method of keeping the environment clean is to use cleanroom garments which offer an effective barrier to wearer-generated contamination and are inherently nonshedding.

The purpose of this study was twofold: 1) to determine the effect of cleanroom garments on particulate contamination at the wafer processing level; and 2) to evaluate the barrier/shedding properties of cleanroom garment fabrics.

GARMENT MATERIALS

Three advanced-technology jumpsuits were chosen for the test: 1) custom-designed 100% polyester herringbone; 2) expanded PTFE membrane laminated on polyester; and 3) coated spunbonded polyolefin. The first two garment types were laundered per manufacturers' specifications prior to use; the third type was received sealed from the manufacturer. The same type of headgear and boots were used with all three types of garments. The headgear used was a full-face "bubble hood" with a recirculated airflow system and a HEPA-filtered exhaust. The fabric portion of the hood was replaced with a PTFE-laminated piece, and the nonfabric portions were thoroughly cleaned with deionized water. This highly efficient hood was used to eliminate contamination from the head area during the study. The boots used for the study were a knee-high, PTFE-laminated upper with a molded sole. They were worn over the lower leg of the jumpsuit in all cases. Boots were changed with each garment change.

EXPERIMENT LOCATION AND EQUIPMENT

The area used for the study was a very clean, low-traffic area of a Class 10 cleanroom. The tables and associated areas were wiped with isopropyl alcohol immediately prior to each series of tests. Initially, there were three simultaneous activities, and a fourth activity was added later. Particles from the first, third, and fourth activities were counted with an optical aerosol particle counter with a lower detection limit of 0.3 micrometer. The readout panel of the particle counter was facing away from the participants to prevent feedback on their activities. The first and second activities each utilized six bare silicon witness wafers, which were analyzed using a laser surface scanner before and after each test. With a lower detection limit of 0.2 micrometer, all precounts were below fifty particles per wafer. The airborne particle count at the wafer processing level for the twenty-four hours preceding the experiment averaged less than one particle per cubic foot of air, 0.3 micrometer and larger. The adjacent areas of the cleanroom were occupied for eight of those twenty-four hours, indicating good isolation of the test area. Air currents within the area of the test were characterized with liquid nitrogen vapors to assure that the measurement points were in the path of the airflow from the operator.

TESTING PROTOCOL

Three wafer fab operators were the subjects for the testing. The operators wore the same type of garment for each series of three activities to minimize residual contamination effects. One activity was a two-hand reach, another was a one-hand back-and-forth motion, and the third activity was walking. The activities took place simultaneously, with enough distance between their locations to eliminate cross-contamination effects. Each garment was subjected to the same series of tests.

The particle counters were operated during the break time between each series of activities to verify cleandown of the area. The witness wafers were set in place immediately prior to the beginning of each series of activities and collected immediately after each test. They were transported in a closed wafer storage box, along with nine control wafers which had been left in the open box during the test.

After instructing the operators regarding test conditions and demonstrating the activity movements, instructions were given on how to don the garments properly. All garments were taken directly from the packages. The operators wore clean vinyl gloves while donning the garments, then changed to fresh vinyl gloves for the test. The operators were asked to briefly demonstrate each of the activities to verify that they understood the procedures. The operators were then instructed to step back from ,the test station until the cue was given to start the activities.

The first three activities took place for a period of fifteen minutes each. Each movement was cued by a metronome to keep a steady pace throughout the test. The first two activities were performed at plastic-laminate-covered tables; the third and fourth activities took place in an open area of the bay.

For the first activity (see Figure 1) there were three points marked on the table: one directly in front of the operator at arm's length (averaged among the three operators); another to the right of the operator also at arm's length; and a third equidistant on the left. The operator stood at the edge of the table, not leaning against it but just touching it. The activity, which simulated wafer carrier movement, was to reach straight out with both hands to the marked spot in front of her, back to the front of the garment, to the right and back, forward and back, then left and back. This pattern was continuously repeated. A 12-inch x 15-inch airborne-particle-collecting device, connected to one of the aerosol particle counters, was placed under the operators' movements. Six witness wafers, three on each side of the aerosol sampler, were used.

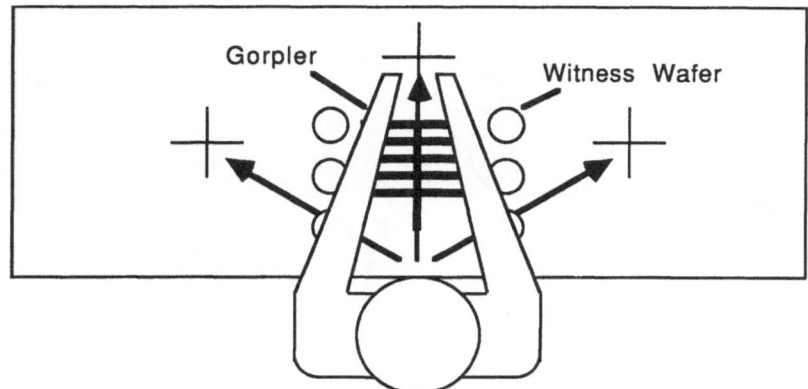

Figure 1. This diagram illustrates activity one, the "two hand reach." Particles are measured with an aerosol particle counter with a Gorpler™ attached and through the measurement of six witness wafers with an Aeronca WIS-150 surface scanner. This activity simulates the movement of cassettes within a work station.

The second activity (Figure 2) was a simulated manual wafer transfer. Two points were marked, about 12 inches from the front edge of the table, at points equidistant from a line perpendicular to the table edge in front of the operator. The marks were at arm's length from the operator. The activity was to reach with one arm to one mark and back to the front of the garment, then to the other mark and back. The operator alternated arms after every ten cycles. Ten witness wafers were arranged in two rows between the front edge of the table and the two marks.

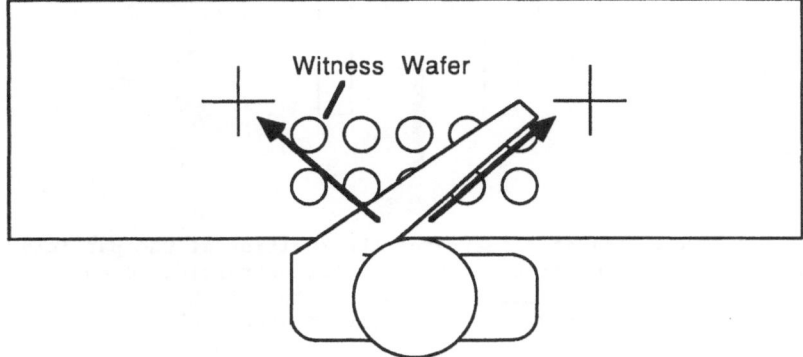

Figure 2. This diagram illustrates activity two, the "one hand reach." Particles are measured on ten witness wafers with an Aeronca WIS-150 surface scanner. This activity simulates the unloading of wafers from a cassette.

The third activity was walking (Figure 3). The operator paced five steps back and forth for an indicated distance and within path-width markers one foot apart. An aerosol particle counter was used with the probe on a tripod stand on the floor. The orifice of the probe was approximately 18 inches off the floor with the probe pointed upward at about a 45-degree angle from the floor. The location of the probe was at a point midway between the turnaround points and directly above a path-width marker. There was a slight sideways component to the airflow which directed the air toward the particle counter probe.

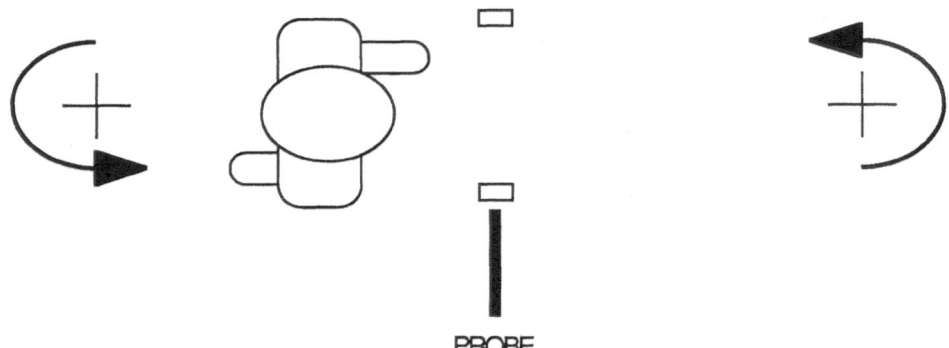

PROBE

Figure 3. This diagram illustrates the position of the particle counter probe in relation to the path designated for the operator to traverse. The particles were measured using a Climet 8040 aerosol particle counter with the probe located eighteen inches above floor level, angled upward at a forty-five degree angle.

The idea for the fourth activity (Figure 4) developed after the first series of tests. An interest developed in the measurement of the particles generated when each of the operators did a series of five deep knee bends directly in front of the particle counter probe. This activity was repeated for each garment.

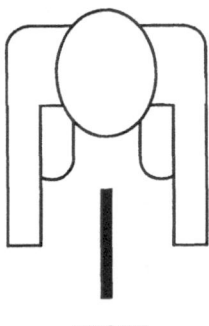

PROBE

Figure 4. This diagram illustrates the position of the particle counter probe in relation to the operator performing "deep knee bends." The probe was again located eighteen inches above floor level and angled upward at a forty-five degree angle.

TEST RESULTS

One of the first differences noted during the tests was the extreme variation in the quantity of particles collected from each of the operators, even while performing the same activity in the same garment. This variation was first noticed during the third activity. One operator shed very few particles while another shed hundreds of particles, and the third shed several thousand particles. This trend held true for all but two of the cells. Because of the measurement methodology, only activities one, three, and four distinguished among operators. These are the counts reflected in Figure 5.

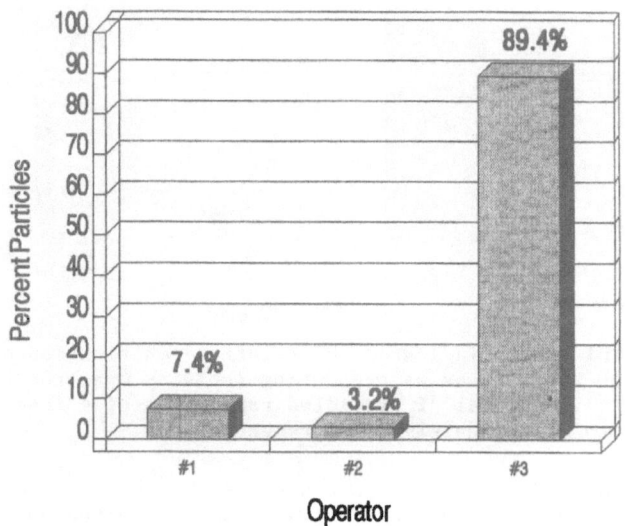

Figure 5. This chart indicated the percentage difference in total particles contributed by the three operators.

Of the 20,386 particles captured, 7.4% were from operator number one, 3.2% were from operator number two, and 89.4% were from operator number three. There was no obvious cause for this difference.

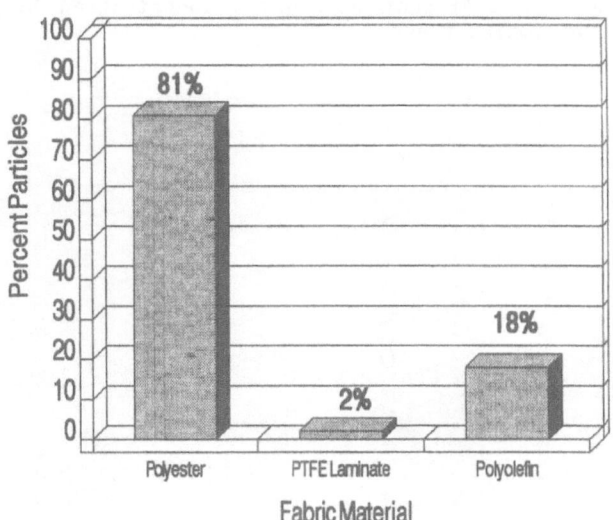

Figures 6. This chart indicated the relative contribution of aerosol particles measured by a Climet 8040 aerosol particle counter during Activity One, the two-hand reach. This is a total of particles ranging in size from 0.3 micrometer to approximately 8 micrometers.

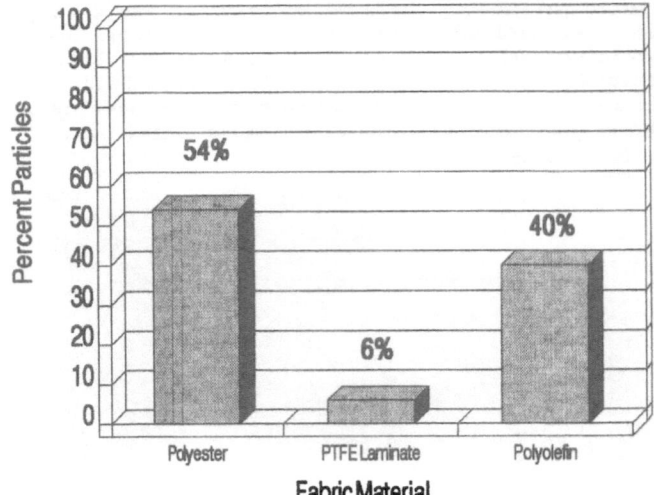

Figure 7. This chart indicates the relative contribution of particles deposited on the witness wafers during Activity One, the two-hand reach. This is a total of particles ranging in size from 0.3 micrometer to approximately 20 micrometers.

The expanded-PTFE-laminate jumpsuit was the best barrier to operator-generated contamination in all tests. The coated-spunbonded-polyolefin garment was second in all except two of the cells and was easily second overall. The polyester-herringbone garments offered the least protection in all except the two previously mentioned cells. These two exceptions appear to fit a pattern, but a cause could not be assigned.

Activity number one was monitored by both airborne and witness wafer particle counts. Although the data (Figures 6 and 7) show these methods yielded a somewhat different percentage contribution by each garment, both followed a similar trend.

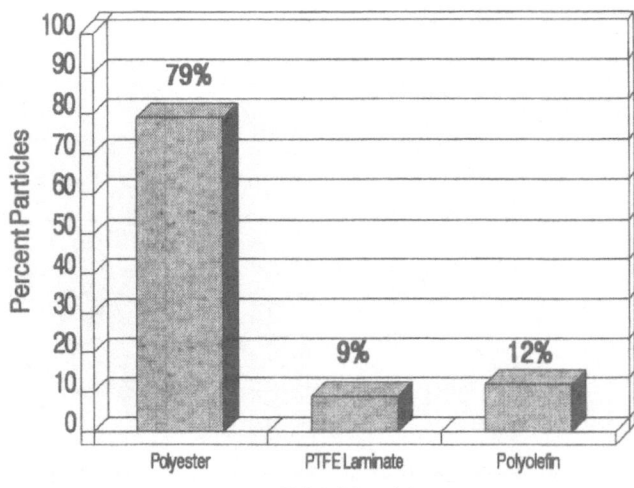

Figure 8. This chart indicates the relative contribution of particles deposited on the witness wafers during Activity Two, the one-hand reach.

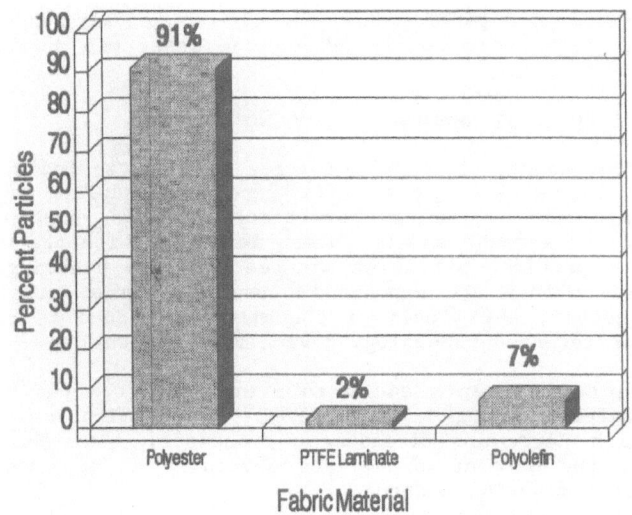

Figure 9. This chart indicates the relative contribution of aerosol
particles measured during Activity Three, walking.

Activity number two (Figure 8) indicates the same trend, but the
particle count from the coated spunbonded polyolefin was considerably
higher.

Activity number three (Figure 9) follows the general trend, with a
high percentage of the particles attributed to the polyester garment.

The deep knee bends of activity number four (Figure 10) produced the
highest percentage contribution of particles from the polyester-herringbone jumpsuit of any of the tests. The billowing effect of gross
movements was most likely a large contributing factor. Likewise, this
was the lowest percentage for the expanded-PTFE-laminate jumpsuits. The
indication is that use of the laminate and the better-fitting design
together provide a superior barrier to contamination from movements which
can cause billowing. One operator, wearing an expanded-PTFE- laminate

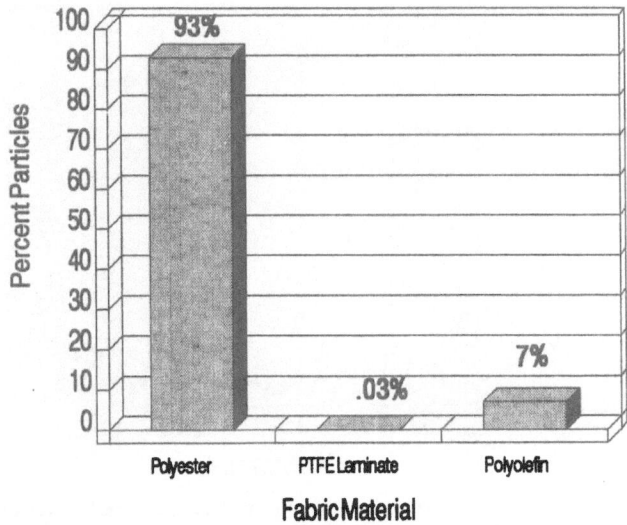

Figure 10. This chart indicates the relative contribution of aerosol
particles measured during Activity Four, deep knee bends.

jumpsuit, registered 0 parties from the five deep knee bends and a total of only 3 particles were registered on the other two operators wearing the same type of garment.

DISCUSSION AND CONCLUSIONS

The witness wafers used in activities one and two indicated that, on the average, 14 particles per wafer were added during 45 minutes of activities typical to wafer processing. This contamination was attributed to the garment and the human interface with the wafers. This is roughly one particle per three minutes of wafer handling. While the amount of time wafers are exposed to handling varies greatly from one process to another, this level of contamination added is certainly not acceptable for advanced-technology device fabrication.

The cleanest garments added an average of 1.3 particles per 45 minutes of wafer handling. This level of contamination is much more compatible with current semiconductor processes. The shedding/barrier properties of the garment chosen are obviously an important factor in wafer-level contamination control.

Another way of evaluating the garment contamination level is to analyze the impact on the cleanliness of the cleanroom. With the least-clean garments in activity one, an average of 100 particles per cubic foot of air sampled was added at table level. This increased the quantity of particles in the immediate environment by two orders of magnitude and degraded the air cleanliness from Class 10 to a doubtful Class 100 status.

The cleanest garments added less than one particle per cubic foot of the air that was sampled, thereby maintaining Class 10 conditions in the cleanroom.

Results of airborne particle counts from the walking protocol, measured at 18 inches above floor level, introduced a very large quantity of particles for the least-clean jumpsuit, but still within Class 10 specifications for the cleanest jumpsuit.

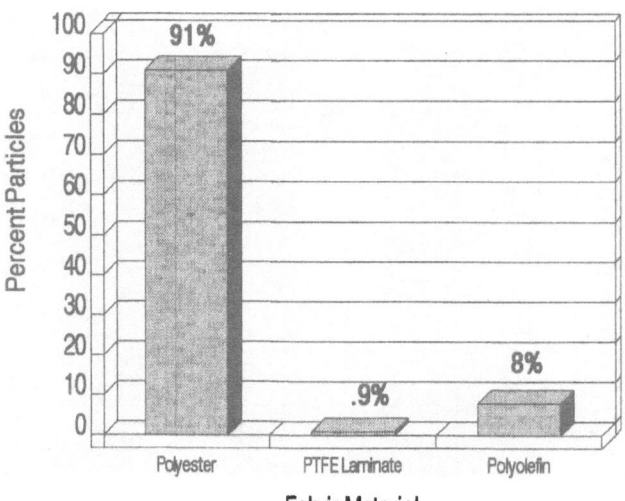

Figure 11. This chart indicates the total particle contributions from all tests, including both witness wafer and aerosol measurements.

Figure 11 shows the percentage of total particles collected from all tests, as detected by both witness-wafer and airborne methods. Of the total number of particles, an alarming 91% were attributable to the polyester garments. This result, along with the data presented above, indicated that a polyester herringbone garment, even with advanced tailoring, may not be suitable for a Class 10 cleanroom.

The coated-spunbonded-polyolefin garment was much better, having accounted for only 8% of the collected particles. Of concern were the two cells of the test in which counts from this garment were significantly higher than the other cells. While these two tests related to only one operator/garment change, further investigation into the consistency of the garments may be indicated.

The expanded-PTFE-laminate garments were exceptional performers. A total of 193 particles, less than 1% of the particles measured, were attributable to these garments.

Subjective questionnaires were given to the operators who participated in the study, asking them to rate the garments for comfort, ease of donning, ability to perform usual tasks while wearing the garment, and for garment appearance. The results among the operators were consistent, with the expanded-PTFE-laminate jumpsuits rated very high. The lowest rating was in ease of donning, where they were rated slightly above average. The polyester garments were rated second. These are the garments currently in use in the fab and the operators are quite pleased with them. The lowest rating went to the coated-spunnbonded-polyolefin garments, especially in the comfort/fit category.

Cost factors were not taken into account in this study, but there is a significant cost difference between the garments. Also not considered were wear factors.

The deterioration of the garment over time could have a substantial effect on the shedding properties and barrier efficiencies of the garment, but these effects were beyond the scope of this study.

A FINAL NOTE

The operators involved in this study were enthusiastic and dedicated. They followed instructions faithfully and carefully, and during the final series of tests when both fatigue and boredom had set in, they kept up with the beat of the metronome. Their dedication was greatly appreciated, for without them this study would not have been possible.

We sincerely thank John Bowser and Bill Hanna of W. L. Gore and Associates. Their loan of equipment and technical information without attempting to influence the study is greatly appreciated.

Finally we thank Steve Bahr of Climet Instruments, who loaned an additional 8040 Aerosol Particle Counter for the study.

PARTICULATE GENERATION IN DEVICES USED IN CLEAN MANUFACTURING

H. S. Nagaraj

IBM General Products Division
5600 Cottle Road
San Jose, CA 95193

B. L. Owens and R.J. Miller

Manufacturing Research
IBM Thomas J. Watson Research Center
Yorktown Heights, NY 10598

Particulate contamination arising from tools and product handling devices such as robots and grippers used in clean manufacturing has been receiving increased attention because of its significant effect on product yield. Although particles generated in the above mentioned devices have been recognized as major contaminants, reliable quantitative information in terms of number and size distribution of these particulates in operating systems or subsystems is seldom available.

The objective of this paper is to summarize the characterization of two devices for generation of particulate matter during their operation to illustrate the profound influence of the design on particle generation rates. The first device to be discussed is a chaindrive used in transporting a pallet of wafers in a sputtering system. The second device is a fine positioner that can be used during robotic product handling. The testing was performed in a clean environment comparable in airflow and aerosol particle concentration to a class 10 cleanroom. Airborne particulate debris were counted and sized by sampling the aerosol using an aerosol spectrometer capable of detecting particles larger than 0.2 μm (optical equivalent diameter).

While the chaindrive was found to generate a large number of particles (up to 6×10^6 particles per ft^3 of air), it was found that the fine positioner did not add any measurable particulates to the test environment; thus it is class 10 compatible.

INTRODUCTION

Particulate contamination is a major yield detractor in the ultraclean manufacturing of semiconductor wafers as well as computer information storage devices. Miniaturization in microelectronics has been driving the line widths to increasingly smaller sizes resulting in chips becoming sensitive to smaller and smaller particles. Also, the magnetic storage devices are becoming sensitive to smaller particulates because of reduction in flying heights of the read/write heads and the precision assembly required during manufacturing.

In the past several years, a great deal of work has been done to obtain a high degree of room air cleanliness and to contain contaminants generated by people working in manufacturing areas. The attention now has been focussed on particulates generated within process tools and product handling devices[1]. To bring about reduction in this form of contamination, particulate generation at the source has to be minimized. Wear of moving parts in mechanisms constitutes one of the most important sources of these particulates. Although some lubrication may have to be used at the contacting areas, this is clearly undesirable from a contamination viewpoint. Basic understanding in this area is lacking, but, a great potential exists for controlling particle generation through careful mechanical design once such information is available. As temporary measures, through critical testing and evaluation such as reported in this paper and by others[2][3], particulate sources can often be shrouded or the particles diverted away from the product using proper airflow.

This paper deals with particulate generation data in terms of number as well as size distribution from two devices whose testing has been recently reported[4][5]. The first device to be discussed is a chaindrive used to transport a pallet of wafers in a sputtering system from the loading chamber to the sputtering chamber. In the actual system, another chaindrive tranfers the pallet back to the loading chamber. The results reported here are for a chaindrive taken out of a Perkin-Elmer 4400 sputtering system. Here, any particulates present in the chamber can potentially be damaging because they may deposit on the wafer surface either during or before a thin film deposition causing defects. The second device is a robotic fine positioning device developed by Hollis[6] (Figure 1) which can be used as the terminal link of a general purpose robot with a coarse positioning capability. In the clean manufacturing application, the positioning device has to handle the product directly; thus it should be compatible with the cleanliness requirements. It is desirable for cleanroom robotic systems not to generate any particulate debris. In the absence of any rubbing contacts and with complete flexural support to permit motion, the above positioning device can be expected to operate very cleanly.

A schematic of the set-up is shown in Figure 2. The experimental set-up was designed to provide, locally, an ultraclean environment to keep the background particle counts very low. It consists of a rectangular parallelopiped shaped duct, 36 inches long, approximately 8 inch by 8 inch cross section, and built from 1/2 inch thick

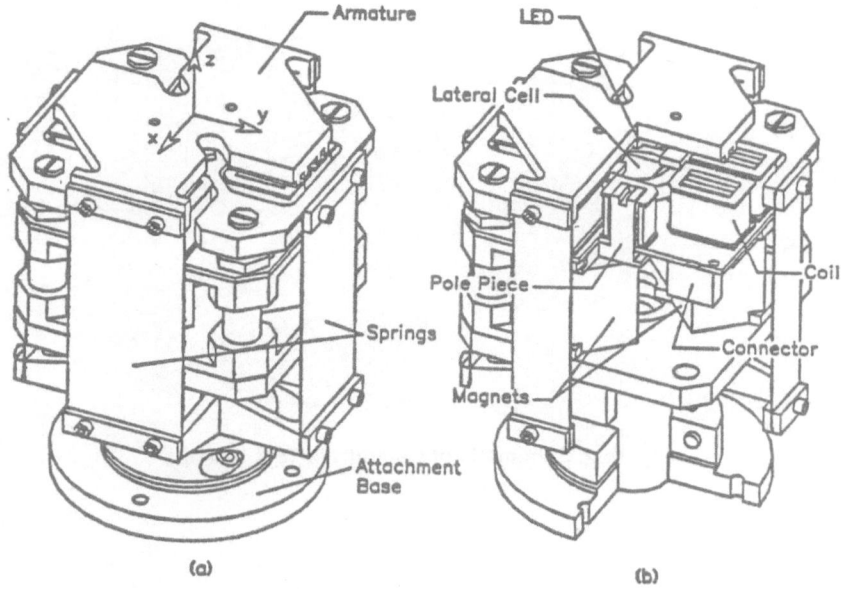

Figure 1. Two-dimensional Planar Fine Positioner: a) general view, b) cutaway view showing magnets, pole pieces, coils and sensor (from Reference 6).

Lucite (clear plastic) sheets. The duct is constructed using threaded fasteners for ease of assembly and disassembly while mounting the devices to be tested inside the duct. At one end of the duct a high efficiency particulate air (HEPA) filter and a centrifugal blower upstream of the HEPA filter are mounted. This HEPA filter is rated to be 99.97% efficient at capturing particles of 0.3 μm in diameter. The clean filtered air has time to estabish a uniform flow pattern by the time it reaches the test device which is mounted about 8 inches downstream of the filter. For the blower used in the tests, an air velocity of 130 ft/min. was measured using a hot wire anemometer. To prevent backstreaming of airborne particles in the room into the duct, the whole set-up was placed on a bench inside a vertical laminar flow clean module.

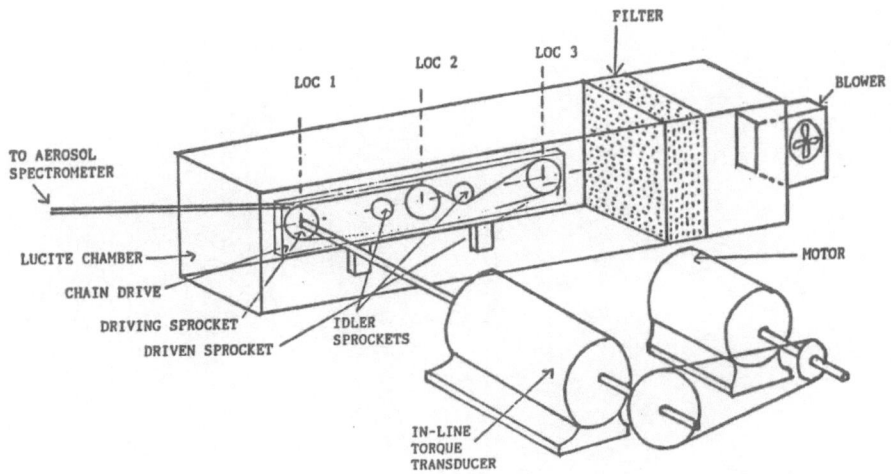

Figure 2. Schematic of Chaindrive Test Set-Up.

For the testing of the chaindrive, it was mounted firmly to the bottom plate of the lucite duct. The drive sprocket of the chaindrive was connected to a variable speed DC motor with an inline torque transducer in between. Average torque values were measured by passing the torque signal through a low pass filter with a cutoff frequency of 0.02 Hz. By depressing the tension sprockets in the chaindrive, tension and, thus, torque could be adjusted to any desired value. In actual use, the weight of the pallet increases the torque demand on the chaindrive resulting in increased tension. Speed of rotation of the drive sprocket was measured by a device which consisted of a toothed wheel mounted on the drive shaft and optoelectronic circuitry that was specially designed and built to monitor very low rotational speeds (10 rev./min) with a good resolution. To facilitate testing, the chain drive was run continuously; while in the sputtering system the drive is used only during the transport of the pallet.

Table I. Range of particle sizes in the PMS Aerosol Particle Counter.

Channel Number	Cumulative mode, μm	Differential mode, μm
1	> 0.2	0.2 to 0.3
2	> 0.3	0.3 to 0.4
3	> 0.4	0.4 to 0.5
4	> 0.5	0.5 to 0.6
5	> 0.6	0.6 to 0.8
6	> 0.8	0.8 to 1.0
7	> 1.0	1.0 to 1.2
8	> 1.2	1.2 to 1.5
9	> 1.5	1.5 to 2.0
10	> 2.0	2.0 to 3.0
11	> 3.0	3.0 to 4.0
12	> 4.0	4.0 to 5.0
13	> 5.0	5.0 to 7.0
14	> 7.0	7.0 to 9.0
15	> 9.0	9.0 to 12.0
16	>12.0	> 12.0

In the case of the fine positioner, it was firmly mounted to the top plate of the Lucite duct by means of bolts. The X and Y motions of the fine positioner were oriented parallel and perpendicular to the air flow, respectively. The wires that supply the power to the electromagnetic drive motors were snugly attached to the top plate, but, downstream of the fine positioner to minimize interference of the wires with the air flow.

Concentration of particulates generated by the device was measured using aerosol particle counting techniques. The air was sampled from the vicinity of the device (5 mm downstream) and transported via a straight stainless steel tube 0.207 inch in diameter. The tube was electrically grounded to avoid any charge build-up. A laser based aerosol spectrometer (PMS model LAS-250X) was used in all the measurements reported here. This spectrometer draws air at a constant flow rate of 0.1 ft^3/min, and counts and sizes particles greater than 0.2 μm in size by allocating them into 16 channels as shown in Table I. In the tests described in this report, the data were taken in the differential mode. The data acquisition was performed on an IBM PC via an RS 232 interface.

Table II. Net percentage of particles not lost during sampling for the
20 inches long sampling tube.

Part. Dia. μm	Particle Density, gm/cm³									
	1	2	3	4	5	6	7	8	9	10
1	99.70	99.42	99.12	98.82	98.53	98.24	97.96	97.76	97.38	97.10
2	98.90	97.83	96.75	95.69	94.65	93.63	92.59	91.57	91.57	89.58
3	97.62	95.23	93.03	90.81	88.63	86.50	84.42	82.38	83.98	78.44
4	95.86	91.88	88.06	85.72	80.83	77.41	74.10	70.89	67.80	64.79
5	93.66	87.68	82.05	76.73	71.68	66.87	62.29	57.92	53.74	49.71
6	91.02	82.79	75.19	68.14	61.57	55.41	49.61	44.12	38.93	34.00
7	88.01	77.33	67.69	58.90	56.83	43.40	36.47	29.99	23.88	18.11
8	84.65	71.38	59.69	49.24	39.79	31.16	23.19	15.78	08.84	00.34
9	80.99	65.06	51.36	39.33	28.56	18.85	09.92	01.64	00.00	00.00
10	77.06	58.45	49.67	29.30	17.35	06.58	00.00	00.00	00.00	00.00

In our earlier reports[4] [5], detailed discussions of sampling losses have been offered. Therefore, only a table (Table II) of calculated sampling efficiencies which includes both inertial loss at the tube entrance and gravitational settling loss is presented here. This table is for the 20 inch long sampling tube for particles ranging from 1 to 10 μm in diameter and densities ranging from 1 to 10 gm/cm^3. Since the gravitational loss is directly proportional to the length, for the additionally used 24 inch and the 29 inch tubes in the case of the chaindrive, the total sampling losses are 20% and 45% higher than the corresponding losses for the 20 inch tube. As can be seen from the table, larger particles do not remain suspended in the airstream and, therefore, are lost from the count. Loss due to Brownian motion[7] where particles may be transported to the walls of the tube and be lost from the aerosol was found negligible for both the chaindrive and the fine positioner.

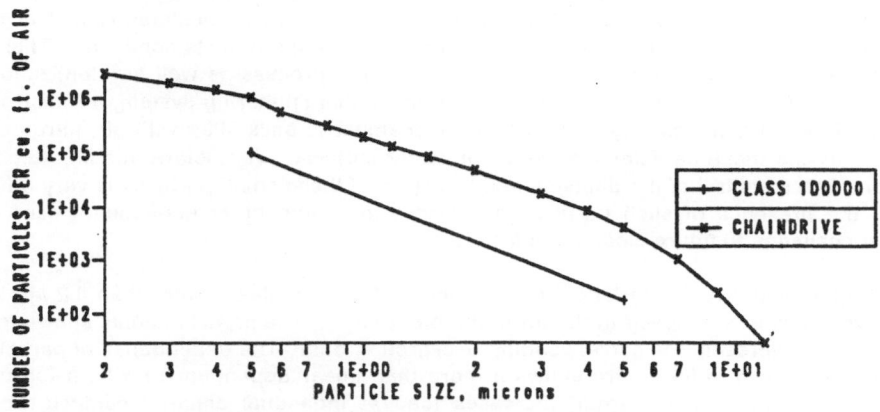

Figure 3. Number of Particles Greater than Stated Size versus Size.

Additionally, in the case of the chaindrive, larger particles do not remain airborne for as long as the smaller particles, and, thus, will not be captured by the sampling tube. Based on calculations of distances travelled before the particles settle out, it appears that for particles above 5 μm in size, some bias is introduced into aerosol sampling. In the results that will be discussed, no effort has been made to correct the data for these sampling losses. For the fine positioner, however, since no particles greater than 0.2 μm were detected the sampling losses are negligible.

DISCUSSION OF RESULTS FOR THE CHAINDRIVE

The test parameters included six speeds, three levels of torque, and three locations. Each test consisted of 60 min of running time with particle count sampling being done at 30 s intervals. Background particle counts taken under same conditions, but with the chaindrive not being operated, were found to be essentially zero. The particle count being available in 16 different size channels, distribution of cumulative particle counts versus size could be plotted as shown in Figure 3. The logarithmic scale used allows direct comparison with the class designations of federal standard FS 209B[R]. The rate of particle generation in the chain drive in this instance has produced a local concentration of particles that is an order of magnitude higher than class 100,000. While in the 0.5 to 5.0 μm range, the slope of the particle production line from the chain drive is about the same as that of class 100,000, and it appears to reduce somewhat for the 0.2 to 0.5 μm range. The size distribution depicted here which shows a preponderence (in number) of smallest size particulates is typical of particle generation in the chain drive over the entire range of tests conducted. These particles are produced during the surface interaction process as well as atomization of any lubricants present on the surfaces. In the actual sputtering system, a redistribution of the particulates may occur when the chamber is back filled with air, nitrogen, or argon and these particles may settle on wafer surfaces even before thin film deposition. Since control of the diffusion and transport of these small particles is very difficult, the presence of such particles in process tools and other equipment poses a great challenge to future clean manufacturing.

Figure 4 shows the variation in the number of all particles greater than 0.2 μm in size with torque and speed in the form of a block chart. The particle counts shown for speed = 0 represent the corresponding background data. The dependence of particle counts as shown in the figure shows a more than linear dependence on both torque and speed. Increase in torque increases load at individual asperity contacts and thereby enhances the severity of asperity interactions. Increase in speed increases the number of asperity interactions as well as the severity through increased energy dissipation locally and may also alter the mode of heat conduction. The resulting changes in local surface temperature may then alter particle generation rate through their influence upon local fracture of the material.

Figure 5 shows very clearly the variation in particle counts from one location to another. At location 1, the rate of generation of particles of all sizes is considerably higher than other locations. This is mainly a manifestation of the effect of load on particle generation because maximum torque transmission occurs at the driven sprocket and the individual links experience maximum load while articulating their way around the driven sprocket.

DISCUSSION OF RESULTS FOR THE FINE POSITIONER

Because very low rates of particle generation were expected for the fine positioner, particle count sampling was done at 10 min intervals and data were collected for 120 min for each of the two locations and five frequency combinations

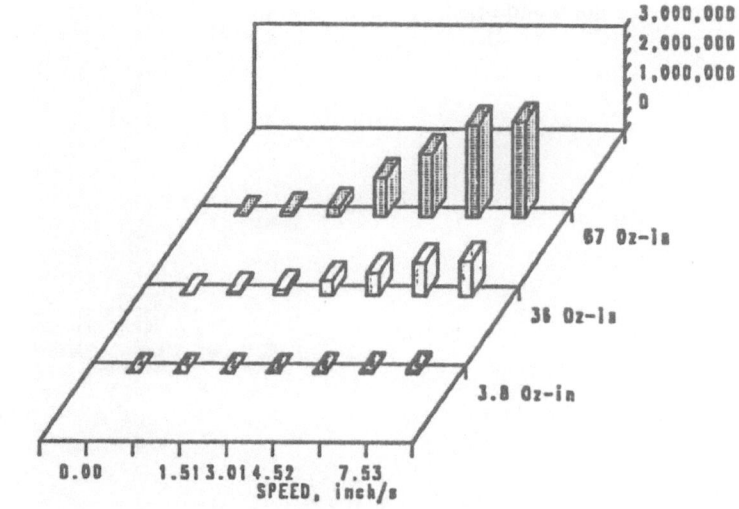

Figure 4. Number of Particles Greater Than 0.2 μm versus
Speed and Torque at Location 3.

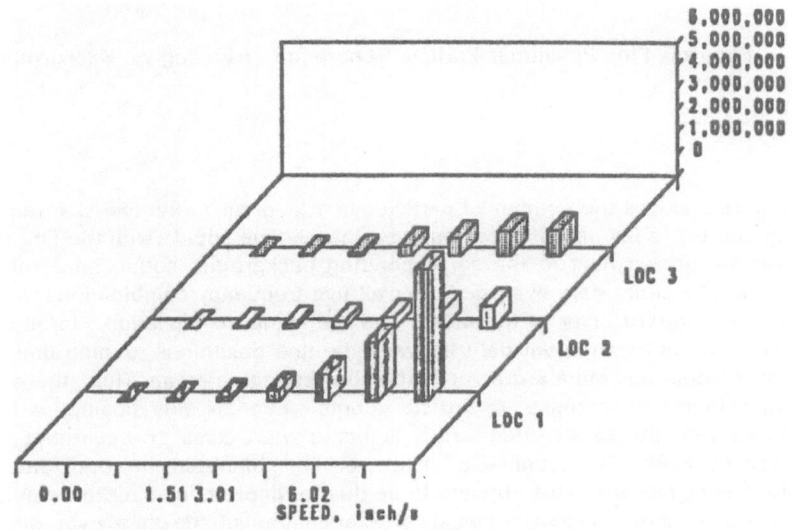

Figure 5. Number of Particles Greater Than 0.2 μm versus
Speed and Location at Torque = 36 Oz-in.

tested. With the fine positioner not running, background particle counts were measured at each probe location over an extended period of time to assess the statistical
significance of particle counts. In all the tests with the fine positioner, particles larger
than 0.4 μm were not found at all.

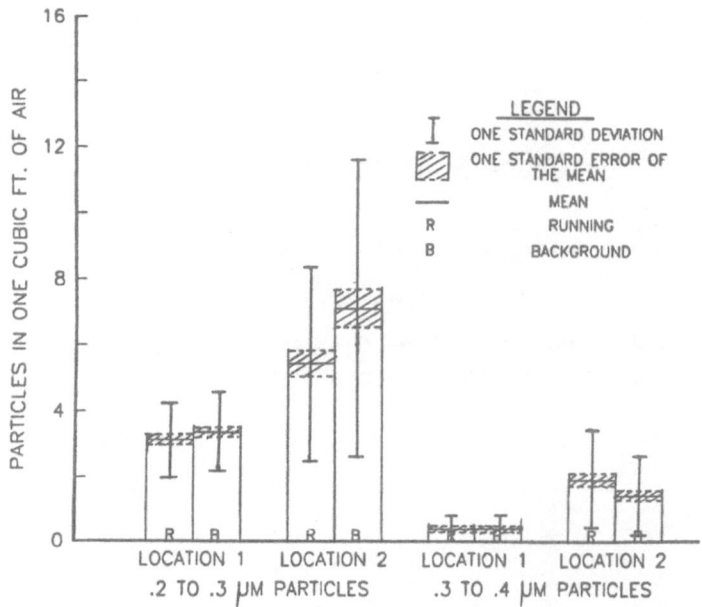

Figure 6. Fine Positioner Particle Generation - Running vs. Background.

Figure 6 shows the number of particles in 1 ft^3 of air in the two size ranges 0.2 to
0.3 μm and 0.3 to 0.4 μm. For comparison, the particle counts with the fine positioner
running are shown next to the corresponding background count data. The ordinate
represents the count data averaged over all five frequency combinations. Also shown
are mean, standard error of the mean, and the standard deviation. In all cases, the
difference in the mean count data between the fine positioner running and the background is within one standard error of the corresponding mean. Thus, there is no statistically significant increase in particle counts when the fine positioner is running
compared with the background which is better than class 10 according to federal
standard FS 209B[8] for cleanliness. However, the difference in count data between
locations is significant. This appears to be due to different local airflow configurations
at the two locations. Detailed statistical examination of the data over different frequency combinations[5] revealed no dependence of particle generation on frequency.

CONCLUSIONS

Reliable quantitative information concerning rate of particle generation and their size distribution in two devices, one that emits extremely high and the other extremely low particulates, has been successfully obtained using aerosol sampling and counting techniques. The following conclusions may be drawn from the data.

1. Mechanical design of devices can profoundly affect particle generation rates as evidenced by the results reported in this paper. However, very often kinematic and other requirements will not permit use of mechanisms such as used in the fine positioner.

2. For the chaindrive, which is perhaps a representive mechanism, the size distribution of particles produced is such that the majority is in the less than 1 μm range, the highest count being in the 0.2 to 0.3 μm. range. It is possible that particles smaller than 0.2 μm are produced, but these are too small to be detected by the aerosol spectrometer used.

3. Although particles generated in devices can somewhat be contained in enclosures, the very high rate of particle generation as seen in the chain drive (up to 6 x 10^6 particles per ft^3 of air) poses a challenge to future clean manufacturing because of diffusion and related transport of these particles to other areas.

4. Since operating variables such as load and speed have been shown to have significant influence on particle generation it appears that some optimization of particle emission is possible through better understanding of the effects of such variables.

REFERENCES

1. Richard S. Rosler, The challenge to reduce contamination in deposition systems, Microcontamination, 4 , No.4, 14(1986).

2. Louis G. Bailey and Gilbert G. Rogers, Reducing contamination in automated systems by design, Microcontamination, 3 , No.11, 80(1985).

3. Gilbert G. Rogers and Louis G. Bailey, A closed chamber method of measuring particle emissions from process equipment, Microcontamination, 5 , No.2, 42(1987).

4. H. S. Nagaraj, B. L. Owens and R. J. Miller, Characterization of particle generation in a chain drive, submitted for publication.

5. H. S. Nagaraj, R. J. Miller and B. L. Owens, Fine positioner evaluation for particle generation", submitted for publication.

6. Ralph L. Hollis, A planar XY robotic fine positioning device, in Proc. IEEE conf. on Robotics and Automation, St. Louis, Missouri, pp.329-336, March 1985.

7. William C. Hinds, "Aerosol Technology - Properties, Behavior, and Measurement of Airborne Particles", John Wiley and Sons, New York, 1982.

8. U.S.Federal Standard No. 209B, "Federal Standard Cleanroom and Work Station Requirements, Controlled Environment", Washington, D.C., 1973.

ABOUT THE CONTRIBUTORS

MAURO A. ACCOMAZZO is currently Director of the Applications Engineering Section and Technical Director of Millipore's Electronics business unit. He joined the R&D Department of Millipore Corp. in 1976 and has been responsible for the development of membrane filter devices for the microelectronics and pharmaceutical industries. He received his Ph.D. degree in Engineering from UCLA. He is Cochairman of the Particulates in Gases Subcommittee of SEMI's Gases Committee.

MARJORIE K. BALAZS is President and Chairman of the Board of Balazs Analytical Laboratory, Inc., Mountain View, CA which she founded in 1975. Before that she was with Stanford Research Institute and prior to that she had a number of appointments as Chemistry teacher and Analytical Physical Chemist. She has MS degrees in Physical Science (1961), in Chemical Education (1963) and in Chemistry (1968). She has been involved in a number of professional organizations and among the honors she has received include: First Woman Chair of SRI International Institute Staff Advisory Group; AFTA Best Paper of the Year 1977; Savvy Business Woman of the Year 1984; Bay Area Business Woman of the Year 1986; Entrepreneur Award from President Reagan, 1986; and Arthur Young/Venture Magazine: Entrepreneur of the Year, 1987.

DONALD W. BARTELSON is with Lockheed Space Operations Company at the Kennedy Space Center in Florida.

PETER G. BORDEN is with High Yield Technology in Mountain View, CA.

AHMED A. BUSNAINA is an associate professor of Mechanical Engineering and director of the Microcontamination Laboratory at Clarkson University, Potsdam, New York. He is the author of over 50 publications and presentations on computational fluid mechanics, environmental flows and combustion. He is also the author of three software packages for fluid flow simulation on personal computers and main frames.

STEVEN D. CHEUNG is a Project Engineer at Linde Division, Union Carbide Corp. Tarrytown, N.Y. (which he joined in 1985) where he is involved in contamination control research and development for the semiconductor processing industry. He received M.S. in Chemical Engineering in 1985 from State University of New York at Buffalo. He is the author of several papers on particulate contamination in semiconductor processing.

WILLIAM CHIANG has been with California Measurements since 1976 and is responsible for product development and marketing of cascade impactors for aerosol research. His earlier experience includes technical and management positions at Analog Technology Corp., Aerojet, and Jet Propulsion Laboratory covering fields of analytical instrument development, microelectronics, and space systems engineering. He received

Master of Engineering from UCLA (1966) and is a graduate of its Engineering Executive Program.

RAYMOND A. CHUAN is a Corporate Scientist at Brunswick's Defense Division. He received Ph.D. (magna cum lande) degree from the California Institute of Technology. He was Prof. of Engineering at the University of Southern California in the early 1960s and was the first Director of its Engineering Center. He has also lectured at the Von Karman Institute at Brussels and was consultant to NATO's aeronautics research group. He holds many patents in the fields of instrumentation for aerosol and other areas of scientific research and has published widely in these areas.

A.C. CLAYTON is a laboratory technician in the Center for Aerosol Technology at Research Triangle Institute in Research Triangle Park, North Carolina. He received an A.A.S. (1986) in Applied Science from Durhan Technical Institute. He now conducts experiments in particle contamination as part of a broad manufacturing science program.

DOUGLAS W. COOPER is currently with the IBM T.J. Watson Research Center in Yorktown Hts., NY which he joined in September 1983. to conduct contamination control research. Before joining IBM, he was on the faculty of Harvard for seven years. He was awarded a Ph.D. in Applied Physics (aerosol science) from Harvard in 1974. He has been involved in aerosol science research and mathematical modeling and analysis since the mid-1960's when he served at the U.S. Army's Biological Laboratories (Ft. Detrick, MD). He is the author or coauthor of over fifty articles published in scientific journals and has been an invited speaker at many meetings.

R.P. DONOVAN is a research physicist at Research Triangle Institute in Research Triangle Park, North Carolina. He received an M.S. (1959) in physics from the University of Pennsylvania. He is currently project leader on a program sponsored by the Semiconductor Research Corporation, whose purpose is to evaluate the deleterious effects of particles on silicon device manufacturing.

VICKI L. DEBLER is an engineering technician at Research Triangle Institute in Research Triangle Park, North Carolina.

DAVID S. ENSOR is director of the Center for Aerosol Technology at Research Triangle Institute in Research Triangle Park, North Carolina. He earned his Ph.D. in engineering at the University of Washington in 1972. His present responsibilities include organizing and supervising research in microcontamination, particulate control devices, air pollution control, and chemical defense.

GLENN GALE is a graduate student in Mechanical Engineering at Clarkson University. He received his BSME from Clarkson University in 1986. He is a recipient of the Ron Ostrander Award for microcontamination research from the IES.

DONALD C. GRANT is a senior consultant at Millipore Corporation. He has 12 years of experience in analysis and purification of fluids, eight of which have been with Millipore. Among his present activities is management of Millipore's particle detection and analysis laboratory. He received a B.S. degree in Chemical Egnineering from Case Western Reserve University, and is the author of more than 20 publications.

MANFRED HANGL has since 1984 been working at the Abteilung fur Apparatebau und Mechanische Verfahrenstechnik at the Technical University of Graz, Austria as an assistant and is pursuing his doctoral studies. He earned his Dipl. Ing. degree in Chemical Engineering at the Technical University of Graz in 1984. He has experience in coal gasification, while his main field is particle size analysis and he has a patent dealing with particle size analysis.

FREDERICK W. KERN, Jr. is a Staff Engineer at IBM in Burlington, Vermont. He received his BSChE from the University of Rochester. He has been responsible for clean room equipment and engineering since 1979.

MALAY K. MAZUMDER is Head of the Department of Electronics and Instrumentation and the Interim Director of the Graduate Institute of Technology at the University of Arkansas at Little Rock. Prior to his appointment at the University of Arkansas, he served with the Atomic Energy Commission as an Instrument Engineer, from 1961 to 1967. He is the inventor of the E-SPART Analyzer--a laser-based instrument for simultaneous measurement of size and charge distribution of powders. Among his other achievements is the development of a dual-beam laser Doppler velocimeter. He received his Ph.D. in instrumental sciences from the University of Arkansas, and has 50 publications and presentations in the area of laser applications and particle technology.

MARK L. MALCZEWSKI has since joining Linde in 1984 as a staff chemist been in charge of particulate measurement development programs at Linde's Particle Laboratory in Tonawanda, N.Y. He received his Ph.D. in Chemistry in 1985 from State University of New York at Buffalo. His current research interests include sampling low concentrations of particulates from pressurized gas streams, particulate generation profiles of components in gas delivery systems, and the extension of particulate measurement to gases other than air or nitrogen.

W.T. McDERMOTT is with Air Products and Chemicals, Inc., in Allentown, PA.

VENUGOPAL B. MENON is a research engineer in the Center for Aerosol Technology, Research Triangle Institute (RTI), Research Triangle Park, NC. He received his Ph.D. (1986) degree in chemical engineering from Illinois Institute of Technology, Chicago. He is currently the project leader on a program, sponsored by the Semiconductor Research Corporation, that is concerned with the cleaning of silicon wafers and other semiconductor surfaces. He is also involved in projects related to advanced materials processing and hazardous waste management.

LINDA D. MICHAELS is an environmental scientist in the Center for Aerosol Technology at Research Triangle Institute. She received an M.S. in environmental science from the University of North Carolina in 1984 and is currently involved in contamination control studies involving the measurement and removal of particles from semiconductor processing liquids.

R.J. MILLER is with Manufacturing Research, IBM Thomas J. Watson Research Center in Yorktown Hts., NY.

*KASHMIRI LAL MITTAL** is presently employed at the IBM US Technical Education in Thornwood, N.Y. He received his M.Sc. (First Class First) in 1966 from Indian Institute of Technology, New Delhi, and Ph.D. in Colloid Chemistry in 1970 from the University of Southern California. In the last 15 years, he organized and chaired a number of very successful

---

* As the editor of this volume.

international symposia and in addition to this volume, he has edited 25 more books as follows: Adsorption at Interfaces, and Colloidal Dispersions and Micellar Behavior (1975); Micellization, Solubilization, and Microemulsions, Volumes 1 & 2 (1977); Adhesion Measurement of Thin Films, Thick Films and Bulk Coatings (1978); Surface Contamination: Genesis, Detection, and Control, Volumes 1 & 2 (1979); Solution Chemistry of Surfactants, Volumes 1 & 2 (1979); Solution Behavior of Surfactants -- Theoretical and Applied Aspects, Volumes 1 & 2 (1982); Physicochemical Aspects of Polymer Surface, Volumes 1 & 2 (1983); Adhesion Aspects of Polymeric Coatings, (1983); Surfactants in Solution, Volumes 1, 2 & 3 (1984), Adhesive Joints: Formation, Characteristics, and Testing (1984), Polyimides: Synthesis, Characterization and Applications, Volumes 1 & 2 (1984); Surfactants in Solution, Volumes 4, 5 & 6 (1986); Surface and Colloid Science in Computer Technology (1987); and Particles on Surfaces 1: Detection, Adhesion and Removal, (1988). Also he is Editor of the Series, Treatise on Clean Surface Technology, the premier volume appeared in 1987. In addition to these books he has published about 60 papers in the areas of surface and colloid chemistry, adhesion, polymers, etc. He has given many invited talks on the multifarious facets of surface science, particularly adhesion, on the invitation of various societies and organizations in many countries all over the world, and is always a sought-after speaker. He is a Fellow of the American Institute of Chemists and Indian Chemical Society, is listed in American Men and Women of Science, Who's Who in the East, Men of Achievement and many other reference works.. He is or has been a member of the Editorial Boards of a number of scientific and technical journals, and is the Editor of the Journal of Adhesion Science and Technology, which made its debut in 1987.

JOHN MUNSON is with High Yield Technology in Mountain View, CA.

H.S. NAGARAJ is currently with the General Products Division of IBM in San Jose, CA working on various tribology related problems in storage devices. Before coming to San Jose, he was with the Research Division of IBM in Yorktown Hts., NY and was involved in the fundamental aspects of particulate contaminants generated by mechanical wear. He received his Ph.D. from the Georgia Institute of Technology in 1976 specializing in tribology and Mechanical Engineering. Prior to coming to IBM, he worked for Mechanical Technology, Inc. and General Motors Research Laboratories. His research interests are in the fields of tribology, solid and fluid mechanics and applied physics, and he has numerous technical publications to his credit.

B.L. OWENS is with Manufacturing Research, IBM Thomas J. Watson Research Center in Yorktown Hts.., NY.

M.B. (ARUN) RANADE is President of Particle Technology, Inc., Washington D.C. and consults in the area of aerosol and particulate science. Prior to his current position, he was Director, Center for Separation Processes Research at the Research Triangle Institute (RTI) in Research Triangle Park, NC. He was educated in India and in the USA in chemical engineering and researched at IITRI, Chicago for a number of years before joining RTI. He is Adjunct Professor at the University of North Carolina as well as the University of Maryland. He has published, lectured and consulted extensively on many aspects of particle technology and is internationally acknowledged in his field.

A. SCHWARZ is with Air Products and Chemicals, Inc. in Allentown, PA.

G.S. SETTLES is currently Associate Professor of Mechanical Engineering at the Pennsylvania State University. He received his Ph.D. from Princeton University in 1975 in Mechanical and Aerospace Engineering. He has held the positions of Research Scientist with Flow Research, Inc. (1975-1977) and Research Engineer and Lecturer at Princeton University (1977-1983). He is the author/coauthor of over 50 publications on experimental fluid dynamics and flow visualization. His current research concerns swept shock wave/boundary layer interactions and the application of flow visualization to industrial clean-room contamination problems.

M.A.R. SHARIF is a Ph.D. candiate in Mechanical Engineering at Clarkson University. He received M. Sc. (Mech. Engg.) from the University of Calgary, Canada in 1984.

A.M. SIAG is with the Graduate Institute of Technology, University of Arkansas at Little Rock, Little Rock, Arkansas.

GERNOT STAUDINGER has since 1978 headed the Abteilung fur Apparatebau und Mechanische Verfahrenstechnik at the Technical University of Graz, Graz, Austria. He earned his Dipl. Ing. and doctors degrees at the Technical University of Graz. He has experience in combustion processes, flue gas desulfurization, coal gasification and particle size analysis, and has industrial experience in both research and planning of chemical plants at Shell Research in Amsterdam and Chemie Linz. His research is focused on coal gasification, flue gas desulfurization and particle size analysis.

R.M. THOROGOOD is with Air Products and Chemicals, Inc. in Allentown, PA.

RICHARD A. VAN SLOOTEN is a senior engineering associate in the Linde Division of Union Carbide Corp., Tonawanda, NY. Before coming to Union Carbide in 1976, he was with Calspan. He received his Ph.D. in Engineering Science in 1970 from the State University of New York at Buffalo. His current responsibilities as head of the analytical methods group include the development of mathematical models of chemically reacting flows and statistical analysis of impurities in high-purity atmospheric gases. A member of the New York Society of Professional Engineers, he holds one patent and has authored numerous technical papers on the application of random process to mechanical systems.

G.G. VIA is currently a Senior Engineer at the IBM Federal Systems Division, Manassas, Virginia, where he is in charge of advanced lithographic systems and contamination control in clean-room environments. He has held major responsibility for the design of the new Submicrometer Lithographic Clean-Room Facility at Manassas. He is a Doctor of Physics from Pisa University, Italy (1957).

JOHN WEAVER is the Supervisor of Microcontamination Control and Facility Development Engineering at Delco Electronics Corporation in Kokomo, Indiana. He received a B.S. in Chemistry from Adrian College, Adrian, Michigan in 1972. He was involved in semiconductor process engineering and process development with RCA Solid State and Hughes Aircraft Company prior to coming to Delco in 1977. He holds patents in the semiconductor processing field and has authored technical papers in that area and in the microcontamination control field.

RICHARD WHITE is a Contamination Control Specialist at Delco Electronics Corporation in Kokomo, IN. He has background in training, videography and more recently in contamination control technology. He has Associates degrees in both education and business administration from Owens Technical College., and he is currently pursuing further degrees at the University of Toledo. He was involved in setting up the overall training program for occupation of Delco's 60,000 square foot Class 10 wafer fabrication cleanroom.

CARL WILLIS has since 1983 worked for Texas Instruments in Dallas on the reduction of particles in wet chemicals. In 1988 he was elected to the technical staff of TI's Semiconductor Group. He received his M.S. in Chemical Engineering in 1983.

T. YAMAMOTO is a research engineer with Research Triangle Institute in Research Triangle Park, North Carolina. Prior to joining RTI, he was an adjunct professor in the Mechanical Engineering Department and research engineer at Denver Research Institute of the University of Denver, Colorado. He earned his Ph.D. degree in Mechanical Engineering from Ohio State University (1979). He is involved in various aspects of the electrostatic precipitator, electrostatic augmentation of fabric filtration, and indoor air pollution.

WALTER A. ZORN is currently a senior associate chemist in the site assurance surface science laboratory at IBM's General Technology Division in Essex Jct., VT. He joined IBM in 1977 and has worked in many areas including solvent and gas analysis. he received his B.S. in chemistry from the State University of New York at Plattsburgh in 1982.